Comparative Analyses of Ecosystems

Jonathan Cole   Gary Lovett   Stuart Findlay
Editors

Comparative Analyses of Ecosystems

Patterns, Mechanisms, and Theories

With 75 Illustrations, 1 in Full Color

Springer-Verlag
New York Berlin Heidelberg London
Paris Tokyo Hong Kong Barcelona

Jonathan Cole, Gary Lovett, and Stuart Findlay
Institute of Ecosystem Studies
New York Botanical Garden
Millbrook, NY 12545-0129, USA

Text Coordinator: Julie C. Morgan

Library of Congress Cataloging-in-Publication Data
Comparative analyses of ecosystems : patterns, mechanisms, and
    theories / Jonathan J. Cole, Gary M. Lovett, and Stuart E. G.
    Findlay, editors ; Julie C. Morgan, text coordinator.
        p.   cm.
    Papers from the third Cary Conference, held in Millbrook, N.Y., in
1989.
    Includes bibliographical references and index.
    ISBN 0-387-97488-1. — ISBN 3-540-97488-1 (Berlin)
    1. Ecology—Congresses. 2. Ecology—Research—Congresses.
I. Cole, Jonathan J. II. Lovett, Gary M. III. Findlay, Stuart E. G.
IV. Cary Conference (3rd : 1989 : Millbrook, N.Y.)
QH540.C6   1991   90-23220
574.5—dc20

Printed on acid-free paper.

Typeset by Bytheway Typesetting Services, Norwich, NY.
Printed and bound by Edwards Brothers, Inc., Ann Arbor, MI.
Printed in the United States of America.

9 8 7 6 5 4 3 2 1

ISBN 0-387-97488-1 Springer-Verlag New York Berlin Heidelberg
ISBN 3-540-97488-1 Springer-Verlag Berlin Heidelberg New York

Cary Conference • 1989

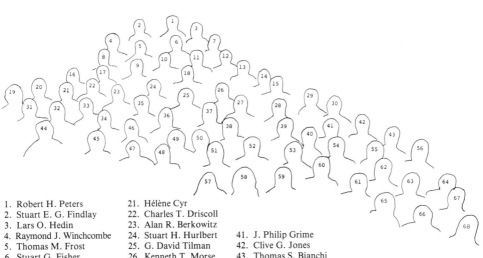

1. Robert H. Peters
2. Stuart E. G. Findlay
3. Lars O. Hedin
4. Raymond J. Winchcombe
5. Thomas M. Frost
6. Stuart G. Fisher
7. Clifford A. Ochs
8. Joel E. Cohen
9. Carlos M. Duarte
10. Michael L. Pace
11. Robert W. Howarth
12. Jacob Kalff
13. Richard H. Waring
14. S. V. Smith
15. Robin L. Welcomme
16. Nina M. Caraco
17. Jerzy Kolasa
18. Ariel E. Lugo
19. Jan Mittan
20. Stephen B. Baines

21. Hélène Cyr
22. Charles T. Driscoll
23. Alan R. Berkowitz
24. Stuart H. Hurlbert
25. G. David Tilman
26. Kenneth T. Morse
27. Randall W. Myster
28. Edward A. Ames
29. William D. Williams
30. O. W. Heal
31. John A. Downing
32. Charles D. Canham
33. John M. Melack
34. W. Gary Sprules
35. Kenneth H. Reckhow
36. S. J. McNaughton
37. Joseph S. Warner
38. Jerry M. Melillo
39. Brian Huntley
40. Martin Christ

41. J. Philip Grime
42. Clive G. Jones
43. Thomas S. Bianchi
44. Gene E. Likens
45. Pamela A. Matson
46. Carleton S. White
47. Lawrence R. Pomeroy
48. Juan J. Armesto
49. Kathleen C. Weathers
50. Stephen R. Carpenter
51. Jonathan J. Cole
52. James T. Callahan
53. Mark J. McDonnell
54. Dale W. Johnson
55. William J. Parton
56. William H. Schlesinger
57. Cathy M. Tate

58. Sybil P. Seitzinger
59. Judy L. Meyer
60. John J. Magnuson
61. Steward T. A. Pickett
62. Bertrand Boeken
63. Wilfried Wolff
64. Gary M. Lovett
65. Ellie Prepas
66. David L. Strayer
67. Robert E. Ulanowicz
68. Nancy Feldsine
Not available for photograph:
    William Robertson, IV
    Peter M. Vitousek

Preface

The Cary Conferences were conceived and initiated as a forum for comprehensive discussion of ecological issues from a more philosophical point of view. It is clear that the diverse themes in ecology can profit from such focused, interactive discussions. We believe that the environment at the Institute of Ecosystem Studies can facilitate such discussions and, thus, have made our staff and facilities available for this purpose on a biennial basis.

Comparisons across systems, the subject of this Conference, is a particularly fitting topic for a Cary Conference. Integration of insights from scientists working on diverse ecosystems can lead to a broader, if not a fuller, ecological understanding. We hope that the examples provided in this book will be a stimulus to workers in the field to try this approach in seeking to make sense of the multifaceted complexity of natural systems. Most of the papers in this volume were presented at the Conference, but several were written by the conferees as a result of the Conference. Some papers are case studies that utilize a comparative approach, others address the theory, philosophy or limitations of the approach.

The Conference produced a new awareness and sense of excitement about the value of the cross-system approach. It also provided some powerful insights regarding ecosystem function, and we hope that this book will evoke the same responses in its readers. Comparative ecosystem analysis is not the only way to study complexity in ecosystems, but clearly it is one of the important tools.

Gene E. Likens

Acknowledgments

Three elements must be combined successfully to make a Cary Conference work: careful planning of the scientific content of the meeting, meticulous attention to local arrangements, and adequate financial support. We were fortunate to have all three. The scientific content of the Third Cary Conference was planned by our steering committee of Clive G. Jones, Gene E. Likens, Jerry M. Melillo, Harold A. Mooney, Scott W. Nixon, Michael L. Pace, and David Strayer (chairman), in consultation with the editors of this book: Jonathan J. Cole, Stuart Findlay, and Gary M. Lovett. Our conference coordinators, Jan Mittan and Nancy Feldsine, did a splendid job arranging travel, meals, and entertainment for conference participants, as well as dealing with the innumerable small crises that arose during the conference. They were aided in their job by many people on the staff of the Institute of Ecosystem Studies, especially Henry Behrens, Jo Ann Bianchi, Carol Boice, David Bulkeley, Glyn Cloyd, Reb Powell, Owen Vose, Joseph Warner, and Kathie Weathers; and by a tireless crew of graduate students: Stephen Baines, Martin Christ, Hélène Cyr, Randall Myster, and Cliff Ochs. We thank all of these people for their help with the Conference. Special thanks goes to Julie Morgan for coordinating many aspects of preparing this book and for communicating with reluctant authors and reviewers.

Generous financial support for the Conference was provided by the National Science Foundation and the Mary Flagler Cary Charitable Trust.

Contents

Contributors

Juan J. Armesto Laboratorio de Sistematica y Ecología Vegetal, Facultad de Ciencias, Universidad de Chile, Santiago, Chile

Alan R. Berkowitz Institute of Ecosystem Studies, The New York Botanical Garden, Mary Flagler Cary Arboretum, Millbrook, NY 12545, USA

Thomas S. Bianchi Institute of Ecosystem Studies, the New York Botanical Garden, Mary Flagler Cary Arboretum, Millbrook, NY 12545, USA. Present Address: Department of Biology, Lamar University, Beaumont, TX 77710, USA

Bertrand Boeken Institute of Ecosystem Studies, The New York Botanical Garden, Mary Flagler Cary Arboretum, Millbrook, NY 12545, USA

Sandra Brown Department of Forestry, University of Illinois, Urbana, IL 61801, USA

James T. Callahan Ecosystem Studies, National Science Foundation, Washington, D.C. 20550, USA

Nina Caraco Institute of Ecosystem Studies, The New York Botanical Garden, Mary Flagler Cary Arboretum, Millbrook, NY 12545, USA

Stephen R. Carpenter Department of Biological Sciences, University of Notre Dame, Notre Dame, IN 46556, USA, and Center for Limnology, University of Wisconsin, Madison, WI 53706, USA

Jonathan J. Cole Institute of Ecosystem Studies, The New York Botanical Garden, Mary Flagler Cary Arboretum, Millbrook, NY 12545, USA

John A. Downing Université de Montréal, Département de Sciences Biologiques, Montréal, Québec H3C 3J7, Canada

Charles T. Driscoll Department of Civil and Environmental Engineering, Syracuse University, Syracuse, NY 13244, USA

Carlos M. Duarte Centro de Estudios Avanzados de Blanes, CSIC, Camino de Santa Bárbara s/n, Gerona, Spain

Stuart G. Fisher Department of Zoology, Arizona State University, Tempe, AZ 85287, USA

Douglas A. Frank Biological Research Laboratories, Syracuse University, Syracuse, NY 13244, USA

Thomas M. Frost Center for Limnology, University of Wisconsin, Madison, WI 53706, USA

J. Philip Grime Unit of Comparative Plant Ecology, Department of Animal and Plant Sciences, The University, Sheffield S10 2TN, England, UK

Nancy B. Grimm Department of Zoology, Arizona State University, Tempe, AZ 85287, USA

Oliver W. Heal Institute of Terrestrial Ecology, Edinburgh Research Station, Bush Estate, Penicuik, Midlothian EH26 OQB, Scotland, UK

Lars O. Hedin W.K. Kellog Biological Station, Michigan State University, Hickory Corners, MI 49060

Robert W. Howarth Section of Ecology and Systematics, Division of Biological Sciences, Corson Hall, Cornell University, Ithaca, NY 14853, USA

Clive G. Jones Institute of Ecosystem Studies, The New York Botanical Garden, Mary Flagler Cary Arboretum, Millbrook, NY 12545, USA

James F. Kitchell Center for Limnology, University of Wisconsin, Madison, WI 53706, USA

Jerzy Kolasa Department of Biology, McMaster University, Hamilton, Ontario L8S 4K1, Canada

Timothy K. Kratz Center for Limnology, University of Wisconsin, Madison, WI 53706, USA

Gene E. Likens Institute of Ecosystem Studies, The New York Botanical Garden, Mary Flagler Cary Arboretum, Millbrook, NY 12545, USA

Ariel E. Lugo Institute of Tropical Forestry, Southern Forest Experiment Station, USDA Forest Service, Río Piedras, Puerto Rico 00928

Pamela A. Matson Ecosystem Science and Technology Branch, NASA-Ames Research Center, Moffett Field, CA 94035, USA

Mark J. McDonnell Institute of Ecosystem Studies, The New York Botanical Garden, Mary Flagler Cary Arboretum, Millbrook, NY 12545, USA

Samuel J. McNaughton Biological Research Laboratories, Syracuse University, Syracuse, NY 13244, USA

Martin Oesterheld Biological Research Laboratories, Syracuse University, Syracuse, NY 13244, USA

Michael L. Pace Institute of Ecosystem Studies, The New York Botanical Garden, Mary Flagler Cary Arboretum, Millbrook, NY 12545, USA

Robert H. Peters McGill University, Département de Biologie, Montréal, Québec H3A 1B1, Canada

Steward T. A. Pickett Institute of Ecosystem Studies, The New York Botanical Garden, Mary Flagler Cary Arboretum, Millbrook, NY 12545, USA

Lawrence R. Pomeroy Department of Zoology and Institute of Ecology, University of Georgia, Athens, GA 30602, USA

Ellie Prepas Department of Zoology, Biological Sciences Center, University of Alberta, Alberta T6G ZE9, Canada

David W. Schindler Freshwater Institute, Winnipeg, Manitoba, R3T 2N6, Canada

John Shearer Freshwater Institute, Winnipeg, Manitoba, R3T 2N6, Canada

Stephen V. Smith Hawaii Institute of Marine Biology, University of Hawaii, Honolulu, HI 96822, USA

W. Gary Sprules Department of Zoology, Erindale College, University of Toronto, Mississauga, Ontario L5L 1C6, Canada

David L. Strayer Institute of Ecosystem Studies, The New York Botanical Garden, Mary Flagler Cary Arboretum, Millbrook, NY 12545, USA

Cathy M. Tate Division of Biology, Ackert Hall, Kansas State University, Manhattan, KS 66506, USA

Robert E. Ulanowicz Chesapeake Biological Laboratory, University of Maryland, Solomons, MD 20688, USA

Michael J. Vanni Center for Limnology, University of Wisconsin, Madison, WI 53706, USA; Freshwater Institute, Winnipeg, Manitoba, R3T 2N6, Canada; and Department of Zoology, Miami University, Oxford, OH 45056, USA

Peter M. Vitousek Department of Biological Sciences, Stanford University, Stanford, CA 94305, USA

Richard H. Waring College of Forestry, Oregon State University, Corvallis, OR 97331, USA

Robin Welcomme FAO, Via delle Terme di Caracalla, Rome 00100, Italy

Kevin J. Williams   Biological Research Laboratories, Syracuse University, Syracuse, NY 13244, USA

Fredrik Wulff   Institute of Marine Ecology, ASKÖ Laboratory, University of Stockholm, 10691 Stockholm, Sweden

Ann P. Zimmerman   Ramsay Wright Zoological Laboratory, University of Toronto, Toronto, Ontario M5S 1A1, Canada

Part I Overview of the Comparative Approach

1
Comparative Ecology and Undiscovered Public Knowledge

DAVID L. STRAYER

The Cary Conferences were begun in 1985 to provide a forum for critical discussion and evaluation of major issues in ecology. Many ecologists thought that conventional scientific meetings and workshops offered only limited opportunities for such discussions despite the apparent importance of confronting these large issues. In response to this need, under the leadership of Gene Likens the Institute of Ecosystem Studies initiated the Cary Conferences to bring together small groups of ecologists to confront major issues in a setting conducive to informal, productive exchange. We at the Institute plan to hold Cary Conferences about every 2 years and welcome suggestions for topics for future conferences. We hope that the Cary Conferences will help to provide direction and leadership to the field of ecology.

The first two Cary Conferences considered the status and future of ecosystem science (Likens et al. 1987) and approaches and alternatives for understanding long-term ecological phenomena (Likens 1988). There are many points of complementarity between these first two conferences and the present conference on comparative analyses of ecosystems. For instance, the conferences on long-term studies and on comparative analyses examined two very different approaches to ecological problem solving, one intensive and the other extensive, that can be especially powerful when used together. Some aspects of the relationship between intensive, long-term studies of a single site and broad, cross-system comparisons were treated at the Second Cary Conference by Steward Pickett (1988).

The inspiration for holding a conference on comparative ecology came from a recent flurry of papers on the subject and a growing interest in the comparative approach among several staff members at the Institute (e.g., Cole et al. 1988). Our interest in comparative studies was sparked in part by a series of fine papers from the empirical limnology school at McGill University [see Peters (1986) for a partial listing] and by a recent symposium comparing marine and freshwater systems (Nixon 1988). These studies have produced new insights into the workings of aquatic ecosystems and showed us that it was possible to do useful ecology in the library as well as on the lake. One point that seemed clear to us from reading these papers was that the potential for comparative analyses had not yet been exhausted, at least in aquatic sciences, and that many problems remained to be tackled using a cross-system approach.

Why is this? In some ways, we might have expected the 1980s to be the Golden Age of comparative ecology. Especially since 1960, ecologists had been busy collecting quantitative data on ecological systems from around the world. By the mid-1980s, the ecological literature contained hundreds to tens of thousands of measurements of such basic quantities as animal abundance, primary production, spatial patterning of communities, nutrient concentrations, and so on.

Paralleling this rise in ecological research was a tremendous growth in computing power, so that most of us had available, often literally at our fingertips, more than enough computing power to perform sophisticated statistical and modeling exercises. Why was this mass of ecological data not heavily used for quantitative comparative analyses?

I can think of at least three reasons why comparative ecology has not lived up to its potential. First, the information relevant to any particular comparative analysis does not exist in pure form but rather is embedded in an enormous and fragmented literature. It would be relatively straightforward for me to conduct a comparative analysis of the limits to primary production in freshwater ecosystems if all the relevant information (and nothing else) were published in a journal called "Journal of Everything Relevant to Freshwater Primary Production." I would simply take out a subscription. Unfortunately, the information on this subject is buried in a vast and growing literature. Root (1987) recently estimated that the primary literature in ecology in English alone now exceeds 40,000 pages per year and that there are about 20 scientific societies with a strong interest in ecology. Thus, the same growth in ecological research that has provided us with so many raw materials for comparative analyses has discouraged such analyses by making it very hard (if not impossible) for individual ecologists to find, use, or even be aware of potentially relevant information.

Swanson (1986a,b, 1987, 1988) has described some interesting consequences of a field, in this case medicine, being overwhelmed by a large, fragmented literature. Swanson has been searching for cases in which pieces of information are logically related but have not been connected by researchers. For example, he found 25 articles showing that an increase in dietary fish oil causes certain changes in properties of blood and 34 articles showing that these same changes in blood properties ease the symptoms of a circulatory disorder called Raynaud's syndrome. Yet no one had made the obvious suggestion that adding fish oil to the diet might ease the symptoms of Raynaud's syndrome, because there were no bibliographic connections between these two sets of articles. No one reads about both fish oil and Raynaud's syndrome. Swanson calls this phenomenon undiscovered public knowledge—knowledge that is freely available to any reasonably intelligent reader who would happen to read the right combination of papers — and has been working on searching methods to find such undiscovered public knowledge. [Incidentally, Swanson's hypothesis about Raynaud's syndrome, formulated in the library by a librarian, has since been tested in clinical trials with promising results (DiGiacomo et al. 1989)].

We believe that the vast growth of ecological literature since 1960 has brought with it an increase in undiscovered public knowledge in ecology. Furthermore, because ecological knowledge often is compartmentalized by habitat (that is, a limnologist working on a lake in New York typically knows a lot about northern temperate zone lakes, somewhat less about lakes in general, and only a little about grasslands), we believe that a deliberately comparative approach is a good way to flush out this kind of information.

Second, as anyone who has done a cross-system comparison knows, published data are not always perfectly comparable or of suitable quality for use, nor is the appropriate method for comparison always as straightforward as simple linear regression (see Chapters 5 and 3 by Carpenter et al. and Downing, respectively, this volume, for examples of some more sophisticated statistical techniques that may be helpful in comparative analyses). As a result, it may take considerable patience, perseverance, and creativity to complete a cross-system comparison even after appropriate data are located. Even when it is done entirely in the library and office, comparative ecology can be hard work.

Third, as a number of the participants at the conference suggested on their registration forms, there is a philosophical or even visceral opposition to comparative ecology in some quarters. It is apparent from some published exchanges (e.g., Lehman 1986; Peters 1986) that the comparative method is not always regarded as an adequate and respectable way to do science.

All this suggested to us that the comparative approach probably is underused by ecologists and that it would be a good time to take a critical look at the approach. For the purposes of the conference, we defined our subject as quantitative analyses of important ecosystem phenomena across systems, while recognizing that the field of comparative ecology as a whole is much broader than this limited definition (cf. Rorison et al. 1987; Pagel and Harvey 1988).

Our specific goals follow:

1. To encourage wider application of the comparative approach in ecology. To do this, we commissioned 10 of our brightest ecosystem ecologists to prepare new comparative analyses and present them at the conference. These analyses are represented by Parts II through IV of this volume. In addition, we hope that the results of this conference will inspire other ecologists to consider adding the comparative approach to their repertoires.
2. To address methodological and philosophical difficulties in using a com-

parative approach and to explore the potential and limitations of this approach. These themes ran as an undercurrent throughout the conference and are represented most notably in the chapters by Downing, Peters et al., Caraco et al., and Carpenter et al. (this volume).

3. As in previous Cary Conferences (Likens 1988; Likens et al. 1987), to expose ecologists working in diverse areas of ecology to each other to encourage cross-fertilization and, we hope, to find some regions of "undiscovered public knowledge."

REFERENCES

Cole, J.J., S. Findlay, and M.L. Pace. (1988). Bacterial production in fresh and saltwater ecosystems: a cross-system overview. *Mar. Ecol. Prog. Ser.* 43:1–10.

DiGiacomo, R.A., J.M. Kremer, and D.M. Shah. (1989). Fish-oil dietary supplementation in patients with Raynaud's phenomenon: a double-blind, controlled, prospective study. *Am. J. Med.* 86:158–164.

Lehman, J.T. (1986). The goal of understanding in limnology. *Limnol. Oceanogr.* 31:1160–1166.

Likens, G.E., ed. (1988). *Long-Term Studies in Ecology: Approaches and Alternatives.* Springer-Verlag, New York.

Likens, G.E., J.J. Cole, J. Kolasa, J.B. McAninch, M.J. McDonnell, G.G. Parker, and D.L. Strayer. (1987). *Status and Future of Ecosystem Science.* Occasional Publication of the Institute of Ecosystem Studies 3, Millbrook, NY.

Nixon, S.W., ed. (1988). Comparative ecology of freshwater and marine ecosystems. *Limnol. Oceanogr.* 33:649–1025.

Pagel, M.D., and P.H. Harvey. (1988). Recent developments in the analysis of comparative data. *Q. Rev. Biol.* 63:413–440.

Peters, R.H. (1986). The role of prediction in limnology. *Limnol. Oceanogr.* 31:1143–1159.

Pickett, S.T.A. (1988). Space-for-time substitution as an alternative to long-term studies. In: *Long-Term Studies in Ecology: Approaches and Alternatives*, G.E. Likens, ed. Springer-Verlag, New York, pp. 110–135.

Root, R.B. (1987). The challenge of increasing information and specialization. *Bull. Ecol. Soc. Am.* 68:538–543.

Rorison, I.H., J.P. Grime, R. Hunt, G.A.F. Hendrey, and D.H. Lewis, eds. (1987). *Frontiers of Comparative Plant Ecology. New Phytol.* (Suppl.) 106:1–300.

Swanson, D.R. (1986a). Undiscovered public knowledge. *Library Q.* 56:103–118.

Swanson, D.R. (1986b). Fish oil, Raynaud's syndrome, and undiscovered public knowledge. *Perspect. Biol. Med.* 30:7–18.

Swanson, D.R. (1987). Two medical literatures that are logically but not bibliographically connected. *J. Am. Soc. Inf. Sci.* 38:228–233.

Swanson, D.R. (1988). Online search for logically-related noninteractive medical literatures: a systematic trial-and-error strategy. *J. Am. Soc. Inf. Sci.* 40:356–358.

2
Comparative Analysis of Ecosystems: Past Lessons and Future Directions

OLIVER W. HEAL AND J. PHILIP GRIME

Introduction

Comparison of the characteristics of different systems is one of the oldest approaches in ecology. Its value is undisputable. The alternative of a non-comparative approach, that is, analysis of a single system while having the capacity to define particular properties and relationships, must eventually mature into comparison to determine the generality of such properties and relationships (Bradshaw 1987). The question is how, rather than why, we compare systems.

This chapter briefly examines two approaches that may be used to compare ecosystems. The first compares existing ecosystems by characterizing and quantifying major trophic components and the fluxes of energy and chemical constituents among them; this was the primary methodology of the International Biological Programme (IBP) and is used to illustrate the strengths and weaknesses of comparative analysis. The second philosophy interprets ecosystem properties as a consequence of the attributes of component organisms and depends upon current attempts to devise a functional classification of organisms with universal applicability across trophic types and across ecosystems.

The IBP Experience of Cross-System Comparisons

The International Biological Programme was designed to address the biological basis of productivity and human welfare. It is worth remembering that it was not designed as a study of ecosystems. Considerable effort was put into understanding productivity at the species and process level, and the sixteenth volume in a synthesis series is the first to consider productivity in the context of ecosystems (Goodall and Perry 1981). The so-called Biome Programs analyzed productivity within ecosystems, with more emphasis on primary than on secondary production. These programs represent the most comprehensive attempt at a coordinated cross-system comparison of ecosystem attributes, and one that depended strongly on an analytical approach involving dissection of the system and description of the components. Recognizing that the programs evolved over a decade of activity and that a systems approach was overlain after their inception (Cragg 1981), what are the lessons to be learned from these Biome studies?

Three interrelated issues raised by IBP are identified as central to the question of how we can improve cross-system comparisons: site selection, research approach, and the juxtaposition of data and theory.

SITE SELECTION

The ability to make valid comparisons, to assess variability and to determine relationships depends on rigorous site selection. The IBP selection covered a wide range of environmental conditions; data from 166 sites were compiled by the Forest Biome. Selection was based primarily on national priorities but, as in the case of Canada's choice of a high Arctic site (Devon Island) for the Tundra Biome, international needs influenced selection and allowed extension of the range of environmental conditions investigated.

The initial selection of sites involved classification based primarily on vegetation type. In contrast, examination of the final comparisons shows repeated emphasis on the environmental, particularly climatic gradients to which ecosystem processes are related. Exploration of the environmental characteristics in the Tundra Biome (Fig. 2.1) shows replication of some conditions and omission of other sets of conditions. It also shows that local environmental variations, for example, in patterned ground, were often more similar to geographically distant sites than to those immediately adjacent. Local variability provided important environmental gradients to explore rate-determining factors but created problems when summary comparisons were made between ecosystems (French 1981).

How do we express the patch in relation to the whole system, and how do we define the system boundaries? The scale of the shifting mosaic in many tundra sites allows the system dynamics to be interpreted reasonably by analysis of small areas, but this does not necessarily apply when considering forests, for example. Thus, there are problems in selecting areas for analysis that are appropriate to the scale of the dynamics and which allow comparison between systems. We still select "sites" on the assumption that they represent comparable systems.

Cross-system comparison of ecosystem components relied to a considerable degree on correlation with environmental characteristics, that is, environmental gradients. This was not fully recognized when sites were selected. It was probably only in the U.S. Grassland Biome that there was a clear design of site selection to encompass the range of environmental variation. Selection was combined with a varying degree of research into an intensive-extensive network, a design with considerable power (Van Dyne 1972). This approach has had limited application, although more recently Meentemeyer

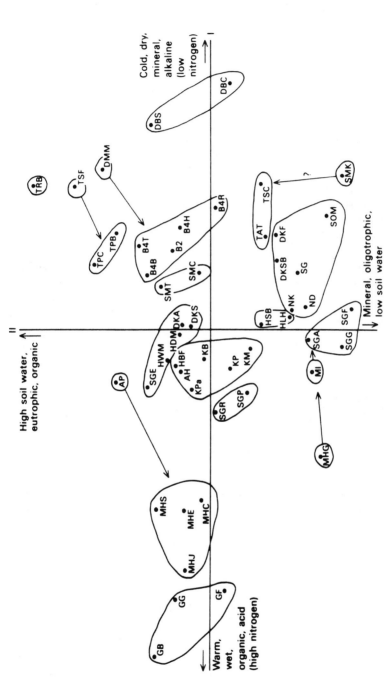

FIGURE 2.1. Distribution of IBP Tundra Biome sites in relation to environmental variation defined by principal components analysis of climate and soil variables (from French 1981). Primary clusters of closely related sites are encircled; *arrows* indicate nearest linkages of "outlier" sites. Initials identifying main sites: A, Abisko, Sweden; B, Point Barrow, Alaska; D, Devon Island, Canada; DK, Disko Island, Greenland; G, Glenamoy, Ireland; H, Hardangervidda, Norway; K, Kevo, Finland; M, Macquarie Island; MH, Moor House, U.K.; N, Niwot Ridge, U.S.A.; S, Signy Island, South Orkney Islands; SG, South Georgia; T, Tareya, U.S.S.R.

and Berg (1986) in comparing forest litter decay rates selected a suite of sites along a trans-European transect representing a defined environmental gradient (Fig. 2.2).

RESEARCH APPROACH

The Biome Programs of IBP put considerable emphasis on field measurements of undisturbed systems; major field manipulations were the exception. The rather descriptive emphasis probably reflected the early stage of ecosystem research, but also the strength of the classical European approach, for example in phytosociology and pedology. Debate on the philosophy of alternative approaches in ecosystem research are adequately covered elsewhere (e.g., Pomeroy et al. 1988), but it is certainly true that the limited use of field manipulations severely constrained cross-system comparisons of ecosystem function.

The balance between observation and experiment also varied greatly in the study of system components. Predictably, it was the combination of the two that was the most informative. For example, the laboratory-determined photosynthetic response of *Calluna* was confirmed by independent field measurements (Fig. 2.3). The combination of experiments and observations also allowed for comparison of response to climatic variations in time and space (Grace and Marks 1978). Similarly, the respiration response surface of decomposing litter allowed for prediction of field weight loss in different countries and comparison of their rates and controls (Table 2.1).

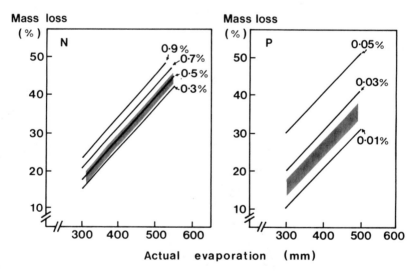

FIGURE 2.2. First-year litter-mass loss of Scots pine needles in different forests as function of actual evapotranspiration and initial nitrogen and phosphorus concentrations. Shaded areas indicate concentrations normally found in Scots pine needle litter in unfertilized stands from (Meentemeyer and Berg 1986).

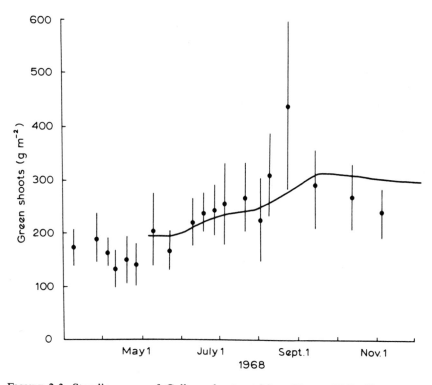

FIGURE 2.3. Standing crop of *Calluna* shoots at Moor House, U.K. (dots; mean and 95% confidence limits) and as calculated from *Calluna* growth model (−) (from Grace and Marks 1978).

TABLE 2.1. Comparison of measured annual weight loss (%) from various litters with predicted respiration and leaching losses (from Bunnel and Scoullar 1981).

	Point Barrow, Alaska	Moor House, U.K.	Moor House, U.K.	Abisko, Sweden
Litter	Total litter	*Calluna vulgaris*	*Rubus chamaemorus*	*Rubus chamaemorus*
Measured weight loss	14.6[a] 15.0[b]	15–20[c] 8[d]	36–38	32
Simulated respiration loss	13.4	7.1	30.1	23.2
Simulated leaching loss	4.6	3.7	2.9	0.5
Simulated as percent of measured	120	99	92	74

[a]*Dupontia fischeri.*
[b]*Carex aquatilis.*
[c]Shoots.
[d]Stems.

In these two examples of production and decomposition, models provided the essential interface between observations and laboratory data. This also applied when production and decomposition data were combined to predict peat accumulation at Moor House and other tundra sites (Clymo 1978). Initial integration of the data exposed anomalies (i.e., the model predicted no peat accumulation) and defined further experimental research, which then allowed more realistic predictions. These examples point out that comparative studies are strongest if they involve integration of results from different approaches or disciplines—when they combine descriptive with manipulative and mechanistic studies and test the logical consequences arising from the results, and when comparative analysis is followed by synthesis.

Juxtaposition of Data and Theory

IBP had a limited theoretical base and, with hindsight, was not strongly motivated to test hypotheses. To a considerable extent, accumulated data were used to explore possible relationships and to see how far these confirmed or refined perceived wisdom about ecosystem structure and function. For example, in the Forest Biome, the 166 site data sets provided the basis for comparisons among systems (Reichle 1981). Relationships of individual components were derived (e.g., Fig. 2.4A) and, in a few cases, data were sufficient to allow comparison of parameters that describe ecosystem function (Fig. 2.4B) (O'Neill and DeAngelis 1981). Much data redundancy occurred during synthesis; very few systems could be compared through parameters that were identifiable as characterizing the whole system (Table 2.2). The whole-system models were reasonably effective for within-system synthesis, but the magnitude of the task did not allow satisfying cross-system comparisons (Bunnell and Scoullar 1975).

In some cases, isolated data sets were accumulated to explore the extent to which a wide variety of detailed component studies could be combined within a theoretical framework to provide a quantitative description of ecosystem structure and function. For example, Heal and MacLean (1975) incorporated data on ingestion, assimilation, and production efficiencies of various species into a general trophic pyramid. The model emphasized the importance of recycling through the poorly defined soil feeding relationships. It was driven by observed values of net primary production, and the result was a comparison of observed and predicted annual secondary production. There are many reservations with this exercise, but it was one of the few options that were available to combine data and "theory" on secondary productivity.

More rigorous analyses of trophic relationships have included effects of chain length, species number, and turnover time on the stability and organization of different systems (DeAngelis 1975; Pimm and Lawton 1977; Pimm 1984; Paine 1988). Comparative ecosystem analysis will con-

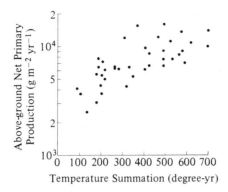

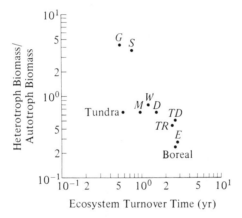

FIGURE 2.4. Cross-system comparison of (A) primary production and (B) ecosystem attributes. (A) Aboveground net primary production as function of temperature summation for IBP Woodland Data Set. Stands more than 130 years old and managed systems are omitted. (B) Ratio of heterotroph to autotroph biomass as function of total ecosystem turnover time for forest stands (from O'Neill and De Angelis 1981).

tinue to provide data for theoretical consideration, but the essential interplay between producers and consumers has not been revealed by the data, especially at the level of ecosystems. Conclusions have been sought from accumulated data and have tended to rely largely on inductive reasoning (Pomeroy et al. 1988).

The analytical approach has been successful in providing a wealth of information about ecosystem structure and function. It has been particularly effective in (i) comparison of attributes of individual components between systems, whether of distribution, numbers, or biomass of species, or the rate of a process and its control; and (ii) understanding the functioning of individual ecosystems, which has involved integration of information

TABLE 2.2. Tabular model of ecological succession: trends to be expected in development of ecosystems (after Odum 1969).

Ecosystem attributes	Developmental stages	Mature stages
Community energetics		
1. Gross production/community respiration ratio	> or < 1	Approaches unity
2. Gross production/biomass ratio	High	Low
3. Biomass supported/unit energy flow ratio	Low	High
4. Net community production	High	Low
5. Food chain	Simple, linear	Complex, weblike
Community structure		
6. Total organic matter	Small	Large
7. Inorganic nutrients	Mainly in soil minerals	Mainly in organic matter
8. Species diversity-variety component	Low	High
9. Species diversity-evenness component	Low	High
10. Biochemical diversity	Low	High
11. Stratification and pattern diversity	Poorly organized	Well organized
Life history		
12. Niche specialization	Broad	Narrow
13. Size of organism	Small	Large
14. Life cycles	Short, simple	Long, complex
Nutrient cycling		
15. Mineral cycles	Open	Closed
16. Nutrient exchange rate between organisms and environment	Rapid	Slow
17. Role of detritus in nutrient regeneration	Unimportant	Important
Selection pressure		
18. Growth form	For rapid growth ("*r* selection")	For feedback control ("*K* selection")
19. Production	Quantity	Quality
Overall homeostatis		
20. Internal symbiosis	Undeveloped	Developed
21. Nutrient conservation	Poor	Good
22. Stability (resistance to external perturbations)	Poor	Good
23. Entropy	High	Low
24. Information	Low	High

across disciplines, but largely within each system. The holistic approach has had much more limited success in making cross-system comparisons of major ecosystem attributes, in the sense of Odum (1969; Table 2.2). The concepts of succession and responses to disturbance and of environmental controls across Biomes remain, but are not satisfyingly described by such summaries as Table 2.3. The more dynamic expressions (for examples, see

Fig. 2.5) and more rigorous quantitative models have yet to be fully used in cross-system comparisons. Is this the next phase of development?

Comparative Ecosystem Synthesis?

In conventional studies of ecosystems, properties and functional insights are assembled from information about specific component processes. An alternative approach, sometimes termed reductionist (Bradshaw 1987) has its origins in comparative studies of the biology and ecology of individual populations of plants and animals. Few would dispute the assertion that knowledge of the life history, resource demands, climatic tolerance, and regenerative and dispersal characteristics of an organism can form the basis for predictions of its biogeography, local ecology, and responsiveness to management. More controversial, but with a distinguished pedigree (Tansley 1935; Elton 1966; Odum 1983), is the proposition that the same kind of knowledge of the genetic makeup of organisms may allow prediction of the role of organisms in ecosystems and can also lead to useful inferences about the gross properties of whole ecosystems.

This reductionist approach will be of little use to the comparative study of ecosystems if it requires detailed knowledge of each of the millions of organisms that occupy the planet. To permit an operational method, classical taxonomy will have to give way to functional classification. Arguments continue with regard to the validity and generality of the various functional classifications that are now under consideration. It is nevertheless quite revealing to find that the current literature contains examples of an essentially similar classification applied to organisms as taxonomically disparate as freshwater phytoplankton (Reynolds 1989), tropical lichens (Rogers 1990), and heathland plants (Clement and Touffet 1990).

Such functional classifications of component organisms have already been used (Grime 1979, 1987; Leps et al. 1982; Cooke and Rayner 1984) to understand or predict community and ecosystem properties such as the rate and trajectory of successional change or the resistance and resilience of ecosystems subjected to nutritional, climatic, or mechanical disturbance. Analysis usually involves relatively simple deductions relating attributes of constituent organisms (e.g., potential growth rate, morphology, longevity, reproductive allocation) to the duration and quality of the opportunities for resource capture, growth and regeneration afforded by particular habitats.

A more challenging task is to use this same reductionist approach to analyze and predict even more general characteristics of ecosystems such as the cycling and retention of chemical components. Here we offer one example of this approach. This arises from current efforts to explain the strong persistence of ^{137}Cs originating from the Chernobyl accident in the upland

TABLE 2.3. Primary production, organic matter standing crops, and turnover in major biomes (from Heal and Ineson 1984).

	Net primary production (t ha⁻¹ yr⁻¹)	Litter input (t ha⁻¹ yr⁻¹)	Standing dead (t ha⁻¹)	Litter (t ha⁻¹)	Soil organic matter (t ha⁻¹)	K_l^a	K_t^b
Tundra	4.0	1.7	1.8	28.0	200.0	0.06	0.017
Boreal forest	8.0	5.8	1.3	35.0	150.0	0.17	0.043
Temperate forest	13.0	8.5	7.9	30.0	120.0	0.28	0.082
Temperate grassland	15.0	7.3	–	4.0	220.0	1.78	0.065
Desert	2.0	1.3	–	1.0	80.0	1.30	0.025
Tropical forest	17.0	15.8	13.5	7.5	85.0	2.11	0.160

K_l^a, litter input ÷ accumulated litter.
K_t^b, net primary production ÷ standing dead + litter + soil organic matter.

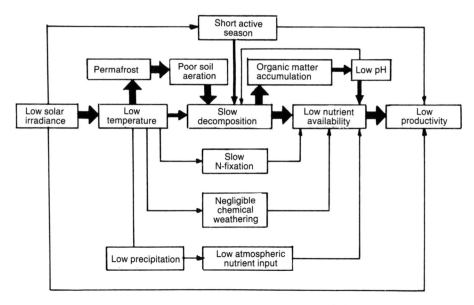

FIGURE 2.5. Functional description of ecosystem: causal relationships between low solar irradiance and low primary productivity of arctic coastal tundra. Thickness of arrows indicates magnitude of effect (from Chapin et al. 1980.)

infertile pastures of Cumberland, U.K. While the dominance of mineral or organic soils with different capacity to absorb and immobilize Cs is an important distinction between lowland and upland soils, the functional characteristics of the plant species can help to explain the persistence of Chernobyl Cs in the upland vegetation and, in particular, in species within that vegetation.

As Grime and Hodgson (1987) pointed out, in ecosystems of low productivity, confinement of primary production to long-lived, slow-growing plants (stress-tolerators sensu Grime 1979) is likely to have predictable consequences for other major ecosystem components. This is because stress-tolerant plants produce tissues that are potentially long lived and are strongly protected against herbivores (Chadwick 1960; Gimingham 1960; Grime et al. 1968; Williamson 1976; Al-Mufti et al. 1977; Sydes 1984; Profitt 1985). These defenses are usually physical in nature (Coley 1983; Coley et al. 1985) and appear to remain operational against decomposing organisms when they are assimilated into the litter component. It seems likely, therefore, that the resistant nature of the living and dead tissues of stress-tolerant plants will often dictate parallels between different trophic components of an unproductive ecosystem in life history, economics of resource capture, and population dynamics. The resulting slow dynamics will lead inevitably to long residence times for chemically stable pollutants. Table 2.4 attempts to explain in greater detail the way in which the slow

TABLE 2.4. Some consistent differences between plants of fertile and infertile soils and their relevance to the persistence of ^{137}Cs in unproductive pastures affected by the Chernobyl accident.

Plant characteristics	Species of fertile soils	Species of infertile soils	Implications for persistence of ^{137}Cs in upland areas
1. Life form	Herbs, shrubs, and trees	Lichens, bryophytes, herbs, shrubs, and trees	Lichens and bryophytes intercept and absorb ^{137}Cs directly from rainwater through their above-ground surfaces. They also sequester metals for long periods, are unpalatable, and decay slowly.
2. Life span of whole plant	Often short (<5 years)	Long (>5 years)	^{137}Cs will persist in plants of infertile soils because it is incorporated into potentially long-lived tissues.
3. Life span of individual leaves and roots	Short (<1 year)	Long (1–3 years)	
4. Leaf phenology	Well-defined peak of growth each spring and summer	Evergreens often showing no seasonal change in biomass	Presence of living foliage throughout the year in species of infertile soils will permit entrapment of ^{137}Cs regardless of the season of deposition, particularly where lichen or bryophyte thallus or higher plant leaves have hairy, rough, or grooved surfaces.
5. Maximum potential relative growth	Rapid	Slow	Slow growth of plants of infertile soils will mean lower rates of tissue turnover (see 2, 3, 4, 8) and tendency to retain ^{137}Cs for long periods.
6. Uptake of mineral nutrients and other ions	Strongly seasonal and mostly in spring and summer	Opportunistic	Presence of functional roots throughout year will allow accumulation of ^{137}Cs from mineralization pulses regardless of their season of release into soil solution.

7. Mycorrhizal infection of root system	Light	Heavy	Mycorrhizal fungi are the effective absorbing surface for root systems of plants of infertile soils and have very high affinity for metals such as ^{137}Cs. They form an effective network throughout surface soil and produce residues resistant to decay (see 9 and 10).
8. Storage of mineral nutrients	Most mineral nutrients are rapidly incorporated in growth but a proportion is stored and forms capital for expansion of growth in following growing season. Little internal recycling from old leaves to new.	Storage systems in leaves, stems, or roots. Some recycling of minerals from old leaves to new.	Because of weak coupling between mineral uptake and utilization in growth, ^{137}Cs tends to be retained in plant biomass. Some internal recycling of ^{137}Cs is likely to occur, further retaining ^{137}Cs in living tissues.
9. Palatability	High	Low	Low palatability dictates lower density of sheep in lowland pastures. Much of ^{137}Cs is likely to reside in physically repellent plant tissues that tend to be avoided by animals except in winter when supply of more palatable leaves is minimal.
10. Rate of litter decomposition	High	Low	Leaf toughness, which protects canopy of slow-growing plants of infertile pastures from heavy defoliation by sheep and invertebrates, remains operational when leaves die and fall onto the ground surface. In consequence, there is reduced rate of decay and ^{137}Cs would be expected to be recycled back into plants at slow rate. Mycorrhizal roots (especially those associated with ericaceous plants, e.g., heather) tend to be very resistant to decay and will retain ^{137}Cs on melanized cell walls of fungal residues.

dynamics of unproductive ecosystems are leading to a greater persistence of ^{137}Cs than would be predicted from ecosystem models based on productive ecosystems.

Conclusions

The initial section of this chapter focused on the strengths and weaknesses of the comparative approach to ecoystem ecology and the problems associated with making comparisons across ecosystems belonging to the same class, across ecosystems belonging to different classes, or across disciplines. In exploring those questions, the key features which we have identified are these:

1. *Comparative analysis of ecosystems* explicitly requires isolation of the components of the system. The research is done on the components (Fig. 2.6), but the understanding of how the system functions depends on integration of information about the components. The effort in integration is usually relatively small; it is difficult and requires interaction with researchers working on different components.

2. Because most effort is on components, and because these are usually equated with disciplines, *cross-system comparison of components* is well developed. They are particularly valuable when they combine different approaches, for example, laboratory physiology studies and field observation. However, comparisons between systems usually examine relationships with controlling environmental gradients rather than with the initial "classes" of ecosystem on which the study areas were selected. Thus, site selection and research approach are critical and need to be flexible.

3. *Cross-system comparison of ecosystems* depends first on integration of information within each system to define the characteristics of the functional whole. This is where the main *comparative synthesis of ecosystems* is identified as a fundamentally different approach from compara-

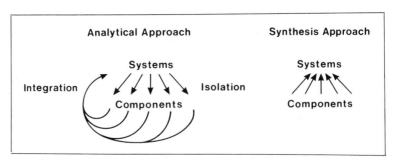

FIGURE 2.6. Comparison of analytical and synthesis approaches of ecosystem analysis.

tive analysis (Fig. 2.6). There is now real prospect of a universal functional classification of organisms that can lead to logical rules relating the characteristics of primary producers to those of the main contributors to the other trophic components. Identification of the major plant strategies within an ecosystem can be expected to allow the development of predictions of direct relevance to plant-herbivore and plant-decomposer interactions and to broaden features of ecosystem structure on function. The potential of this "bottom-up" approach to determining the emergent properties of communities and ecosystems offers a viable and complementary approach to comparative analysis.

REFERENCES

Al-Mufti, M.M., C.L. Sydes, S.B. Furness, J.P. Grime, and S.R. Band. (1977). A quantitative analysis of shoot phenology and dominance in herbaceous vegetation. *J. Ecol.* 65:759-791.

Bradshaw, A.D. (1987). Comparison — its scope and limits. In: *Frontiers of Comparative Plant Ecology*, I.H. Rorison, J.P. Grime, R. Hunt, G.A.F. Hendry, and D.H. Lewis, eds. Academic Press, London, pp. 3-21.

Bunnell, F.L. and K.A. Scoullar. (1975). ABISKO II. A Computer simulation model of carbon flux in tundra ecosystems. In: *Structure and Function of Tundra Ecosystems*, T. Rosswall and O.W. Heal, eds. *Ecol. Bull.* (Stockholm) 20:425-448.

Bunnell, F.L. and K.A. Scoullar. (1981). Between-site comparisons of carbon flux in tundra using simulation models. In: *Tundra Ecosystems: A Comparative Analysis*, L.C. Bliss, O.W. Heal, and J.J. Moore, eds. Cambridge University Press, Cambridge, pp. 685-715.

Chadwick, M.J. (1960). Biological flora of the British Isles. *Nardus stricta* L. *J. Ecol.* 48:225-267.

Chapin, F.S., P.C. Miller, W.D. Billings, and P.I. Coyne. (1980). Carbon and nutrient budgets and their control in coastal tundra. In: *An Arctic Ecosystem, the Coastal Tundra at Barrow, Alaska*, J. Brown, P.C. Miller, L.L. Tieszen, and F.L. Bunnel, eds. Dowden, Hutchinson and Ross, Stroudsburg, pp. 458-482.

Clement, B. and J. Touffet. (1990). Plant strategies and secondary succession on Brittany heathlands after severe fire. *J. Vegetat. Sci.* 1:195-202.

Clymo, R.S. (1978). A model of peat bog growth. In: *Production Ecology of British Moors and Montane Grassland*, O.W. Heal and D.F. Perkins, eds. Springer-Verlag, London, pp. 187-223.

Coley, P.D. (1983). Herbivore and defensive characteristics of tree species in a lowland tropical forest. *Ecol. Monogr.* 54:209-233.

Coley, P.D., J.P. Bryant, and F.S. Chapin. (1985). Resource availability and plant antiherbivore defence. *Science* 230:895-899.

Cooke, R.C. and A.D.M. Rayner. (1984). *Ecology of Saprotrophic Fungi*. Longmans, London.

Cragg, J.B. (1981). Preface. In: *Dynamic Properties of Forest Ecosystems*, D.E. Reichle, ed. Cambridge University Press, Cambridge.

DeAngelis, D.L. (1975). Stability and connectance in food web models. *Ecology* 56:238-243.

Elton, C.S. (1966). *The Pattern of Animal Communities.* Methuen, London.

French, D.D. (1981). Multivariate comparisons of IBP Tundra Biome site characteristics. In: *Tundra Ecosystems: A Comparative Analysis,* L.C. Bliss, O.W. Heal, and J.J. Moore, eds. Cambridge University Press, Cambridge, pp. 47–75.

Gimingham, C.H. (1960). Biological flora of the British Isles. *Calluna vulgaris* L. Hull. *J. Ecol.* 48:455–483.

Goodall, D.W. and R.A. Perry. (1981). *Arid-Land Ecosystems: Their Structure, Functioning and Management.* Cambridge University Press, Cambridge.

Grace, J. and T.C. Marks. (1978). Physiological aspects of bog production at Moor House. In: *Production Ecology of British Moors and Montane Grasslands,* O.W. Heal and D.F. Perkins, eds. Springer-Verlag, London, pp. 38–51.

Grime, J.P. (1979). *Plant Strategies and Vegetation Processes.* Wiley, Chichester.

Grime, J.P. (1987). Dominant and subordinate components of plant communities: implications for succession, stability and diversity. In: *Colonization, Succession and Stability,* A. Gray, P. Edwards and M. Crawley, eds. Blackwell Scientific Publications, Oxford, pp. 413–428.

Grime, J.P. and J.G. Hodgson. (1987). Botanical contributions to contemporary ecological theory. *New Phytologist* 106 (Suppl.):283–295.

Grime, J.P., S.F. MacPherson-Stewart, and R.S. Dearman. (1968). An investigation of leaf palatability using the snail *Cepaea nemoralis* L. *J. Ecol.* 56:405–420.

Heal, O.W. and S.F. MacLean. (1975). Comparative productivity in ecosystems — secondary productivity. In: *Unifying Concepts in Ecology,* W.H. van Dobben and R.H. Lowe-McConnell, eds. Junk, The Hague, pp. 89–108.

Heal, O.W. and P. Ineson. (1984). Carbon and energy flow in terrestrial ecosystems: relevance to microflora. In: *Current Perspectives in Microbial Ecology,* M.J. Klug and C.A. Reddy, eds. American Society for Microbiology, Washington, pp. 394–404.

Leps, J., J. Osbornova-Kosinova, and K. Rejmanek. (1982). Community stability, complexity and species life-history strategies. *Vegetatio* 50:53–63.

Meentemeyer, V. and B. Berg. (1986). Regional variation in rate of mass loss of *Pinus sylvestris* needle litter in Swedish pine forests as influenced by climate and litter quality. *Scand. J. For. Res.* 1:167–180.

Odum, E.P. (1969). The strategy of ecosystem development. *Science* 164:262–270.

Odum, E.P. (1983). *Basic Ecology.* Sanders, Philadelphia.

O'Neill, R.V. and D.L. DeAngelis. (1981). Comparative productivity and biomass relations of forest ecosystems. In: *Dynamic Properties of Forest Ecosystems,* D.E. Reichle, ed. Cambridge University Press, Cambridge, pp. 411–449.

Paine, R.T. (1988). Food webs: road maps of interactions or grist for theoretical development. *Ecology* 69:1648–1654.

Pimm, S.L. (1984). The complexity and stability of ecosystems. *Nature* 307:321–326.

Pimm, S.L. and J.H. Lawton. (1977). The number of trophic levels in ecological communities. *Nature* 268:329–331.

Pomeroy, L.R., E.C. Hargrove, and J.J. Alberts. (1988). The ecosystem perspective. In: *Concepts of Ecosystem Ecology,* L.R. Pomeroy and J.J. Alberts, eds. Springer-Verlag, New York, pp. 1–17.

Profitt, G.W.H. (1985). The biology and ecology of Purple-moor Grass *Molinia caerulea* L. Moench, with special reference to the root system. Ph.D. Thesis, University of Aberystwyth, Wales.

Reichle, D.E., ed. (1981). *Dynamic Properties of Forest Ecosystems.* Cambridge University Press, Cambridge.

Reynolds, C.S. (1989). Functional morphology and the adaptive strategies of freshwater phytoplankton. In: *Growth and Reproductive Strategies of Freshwater Phytoplankton*, C.D. Sandgren, ed. Cambridge University Press, Cambridge, pp. 388–426.

Rogers, R.W. (1990). Ecological strategies of lichens. *Lichenologist* 22:149–162.

Sydes, C.L. (1984). A comparative study of leaf demography in limestone grassland. *J. Ecol.* 72:331–345.

Tansley, A.G. (1935). The use and abuse of vegetational concepts and terms. *Ecology* 16:284–307.

Van Dyne, G.M. (1972). Organization and management of an integrated ecological research program. In: *Mathematical Models in Ecology*, J.N.R. Jeffers, ed. Blackwell, Oxford, pp. 111–172.

Williamson, P. (1976). Above-ground primary production of chalk grassland allowing for leaf death. *J. Ecol.* 64:1059–1075.

3
Comparing Apples with Oranges: Methods of Interecosystem Comparison

John A. Downing

Abstract

A review of the recent literature indicates that 16%–25% of current ecological research is based on interecosystem comparisons. These comparisons are made using three general approaches: *Composite ecosystem comparisons* strive to measure aggregate similarity of ecosystems; *variable-focused comparisons* strive to find how a few important characteristics vary among ecosystems; *connection-focused comparisons* concentrate on causal relationships and how mechanical function varies among ecosystems. Although ecologists make all the usual types of statistical errors, this review concentrates on five that are particularly common in comparative ecosystem analysis: (1) Results are biased by the choice of measured variables; (2) too many variables are measured, on (3) too few ecosystems; (4) interpretations are frequently based on probable coincidences; and (5) interecosystem comparisons are often informal. The consequences of these problems and remedial measures are discussed using visual and mathematical illustrations.

Introduction

Comparative ecology has been an important tool for ecologists for more than one and one-half centuries. Comparing and contrasting the composition of ecosystems led ecologists such as Lyell, von Humboldt, Haeckel, and Darwin to great new ideas. Naturalists were present on many of the major explorations during the nineteenth century, and they drew inspiration from the similarities, differences, and contrasts observed among the ecosystems that they visited. The types and quantities of data they could collect were limited, but the sharpness of the contrasts they saw, and the elegance of their analyses, led to the formation of major concepts such as succession, natural selection, and evolution as well as the disciplines of biogeography and community ecology.

Interecosystem comparisons are still an important component of ecological analysis. More than 16% of the 229 articles published in *Ecology* in 1988, 25% of the articles published in *Ecology* in 1989, and 22% of the 294 articles published in *Oecologia* in 1988 were based on interecosystem analyses. Unlike our predecessors, twentieth-century ecologists can perform interecosystem comparisons without risking their lives, but we must instead contend with huge amounts of quantitative data. Ecologists have refined and quantified their sampling methods for living components of ecosystems. We can now use satellites, remote sensing, and aerial surveys, and can apply a battery of chemical, physical, and electronic techniques for measuring a vast number of biotic and abiotic ecosystem characteristics that would have astounded our predecessors. The ease of rapid travel makes large-scale ecosystem comparisons more feasible. The huge number of ecosystem characteristics that are measured, the complexity of current theoretical constructs, and the crushing amount of data we must process make a mastery of modern statistical and numerical methods essential.

The purpose of this review is to discuss what I believe to be the most important current problems in the quantitative comparison of ecosystems. Although nineteenth-century ecologists' analyses consisted of visual or intuitive comparisons, the contemporary scientist can employ a vast array of quantitative techniques. Because the field is so wide, I discuss here only those problems that are central to the proper application of the techniques most frequently used. To determine what those techniques are, I systematically read all the 106 articles employing interecosystem comparisons published in 1988 and the first four months of 1989 in *Ecology*, and those which appeared during 1988 in *Oecologia*. Because I thought that these methods would segregate on the basis of specific research goals or objectives, I first examined the stated research goals in each of the articles. I found that there was no consensus or clear clustering of research goals (Table 3.1). Except for a small fraction that phrased testable hypotheses,

TABLE 3.1. Verbs used to describe major research goals of interecosystem comparisons published during 1988 and 1989 in *Ecology* and during 1988 in *Oecologia*.[a]

Top 10 verbs	Frequency	Next 10 verbs	Frequency	Others
"test . . . hypothesis"	15	"identify"	4	"consider"
"investigate"	11	"assess"	3	"define"
"examine"	10	"analyze"	2	"document"
"address"	8	"characterize"	2	"elucidate"
"determine"	8	"describe"	2	"highlight"
"compare"	6	"discuss"	2	"illustrate"
"explore"	6	"measure"	2	"resolve"
"report on"	5	"show"	2	"search for"
"test for"	5	"study"	2	"see if"
"evaluate"	4	"attempt to answer"	1	etc. . . .

[a]Statements of research goals were usually found in the last paragraph of the introduction. There was no correlation between the verb used and the research technique. Frequencies indicated are number in 116 articles; some articles used more than one verb.

few introductions provided clues about the scientific approach used in the research. More disturbing on philosophical grounds was the frequent use of very vague terms to describe research goals, goals that were not clarified by further discussion (Table 3.2). Because statements of research goals are the starting point of the intellectual trajectory to be followed and form the foundation of the reader's understanding of research publications, the more frequent use of concrete terms would facilitate the dissemination of ecological research.

The vagueness of stated research goals seems more a result of writing style than fuzzy thinking, because more detailed analysis of the methods and results sections of these same articles revealed three quite well-defined types of approaches to interecosystem comparison (Table 3.3):

1. *Composite ecosystem comparisons.* These analyses make simultaneous comparisons of several (often many) characteristics of the biotic community and the physical environment. Analyses strive to measure the aggregate similarity or dissimilarity of ecosystems. The convergence of patterns of similarity among environments with patterns of similarity

TABLE 3.2. Nouns used with some of "top 10 verbs" describing research objectives in interecosystem comparisons published during 1988 and 1989 in *Ecology* and during 1988 in *Oecologia*.

Top 10 verbs	Nouns
"investigate"	the nature of
	effect of
	the distribution of
	relationships between
	links to
"examine"	patterns
	pathways
	effects
"address"	factors
	questions
	hypotheses
	the role
"determine"	contributions
	the role
"explore"	patterns
	hypotheses
"report on"	relative distributions
	measurements
	observations
"test for"	the applicability
	the effect
"evaluate"	the importance
	the relative adequacy

TABLE 3.3. Frequency of publication of types of comparisons in interecosystem comparisons published during 1988 and 1989 in *Ecology* and during 1988 in *Oecologia*.

Type of comparison	*Ecology* ($n = 44$, %)	*Oecologia* ($n = 62$, %)	Total ($n = 106$, %)
Composite	14	13	14
Variable-focused	61	82	74
Connection-focused	25	5	14

among communities is used to infer the reasons for differences and similarities among the studied systems.

2. *Variable-focused ecosystem comparisons.* These analyses concentrate on one or a few characteristics of ecosystems that are considered to be of great theoretical or practical interest. Analyses strive to find if and how these characteristics vary among ecosystems, and especially what other characteristics of these ecosystems are correlated or coincident with among-ecosystem variation in these variables of interest.

3. *Connection-focused ecosystem comparisons.* These analyses concentrate on causal relationships, and how mechanical function varies among ecosystems. Such analyses are almost exclusively based on manipulation experiments, because the complexity of ecosystems makes unconfounded natural experiments rare. Interecosystem comparisons are performed on the results of these manipulations, and subsequent analyses strive to find which of the many characteristics of ecosystems are responsible for differences or similarities in responses to manipulation.

These different approaches to interecosystem comparison use radically different methods. My systematic review of more than 800 pages of such comparisons published in recent issues of *Ecology* and *Oecologia* found only 36 different categories of statistical and numerical methods. This review is probably a good reflection of current relative use of various techniques because the cumulative number of techniques "sampled" declined rapidly after 200–300 pages had been read (Fig. 3.1). Although literally thousands of methods have been devised, many are minor modifications of more general methods, and only a few broad categories of methods are in frequent use in interecosystem research.

Composite ecosystem comparisons make simultaneous comparisons of many different biological and physical characteristics; they therefore use multidimensional techniques based on ordination and similarity measurements (Table 3.4). Principal components analysis (PCA) is often used to summarize the variability of a large number of ecosystem characteristics by using a few important composite dimensions. Correspondence analysis, also known as contingency table analysis or reciprocal averaging, is very similar to PCA, but works with categorical data like presence–absence data

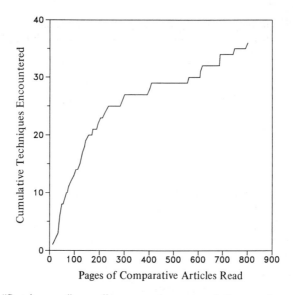

FIGURE 3.1. "Species-area" sampling curve shows cumulative number of numerical and statistical methods encountered during systematic reading of all articles about interecosystem comparisons that were published during 1988 and 1989 in *Ecology* and during 1988 in *Oecologia*. Examples of distinct methods are shown in Tables 3.4 through 4.6.

TABLE 3.4. Methods used in composite intereco-system comparisons published during 1988 and 1989 in *Ecology* and during 1988 in *Oecologia*.[a]

Method[b]	Frequency (%)
Correspondence analysis	36
Principal components analysis	29
Canonical correlation	14
Similarity indices	14

[a]Sample size is given in Table 3.3. Percentages may sum to >100% because some articles used more than one technique.
[b]Others include clustering, canonical redundancy analysis, chi-squared tests, discriminant analysis, MANOVA, dissimilarity coefficients, etc.

of various taxa. These analyses allow ecologists to find which of the ecosystems are most similar by considering all the measured ecosystem characteristics simultaneously. Canonical correlation, a generalization of multiple regression that was developed by Hotelling (1936), correlates a group of environmental characteristics with a group of measures of species abundance. Reviews of these methods are provided by Legendre and Legendre (1983) and Gower (1987).

Variable-focused ecosystem comparisons are most commonly reported in the literature by far (see Table 3.3). They differ principally from composite ecosystem comparisons in that one or a few variables are considered to be dependent variables; various methods are used to find relationships between these dependent variables (variables of interest) and measured characteristics of the ecosystems. The method most frequently used for doing this is regression analysis (see the methodological review by Draper and Smith 1981). The econometrics literature (e.g., Gujarati 1978) provides practical and readable advice about regression that is particularly useful to ecologists. One-third of all interecosystem comparisons employ some form of regression analysis (e.g., bivariate, multivariate, ANCOVA, nonlinear) (Tables 3.3 and 3.5); an understanding of these methods is therefore essential to an understanding of this burgeoning literature. Other methods frequently used include the calculation of confidence intervals and one-way to multi-way techniques for performing parametric (Steel and Torrie 1960; Sokal and Rohlf 1981) and nonparametric (Conover 1971) "analysis of variance." A technique that may show promise is identifying pairs of ecosystems that are highly similar except for some known characteristic (Crome and Richards 1988) and performing paired analyses on a series of these paired ecosystems (paired *t* test; Box et al. 1978).

Connection-focused ecosystem comparisons take their inspiration from

TABLE 3.5. Methods used in variable-focused interecosystem comparisons published during 1988 and 1989 in *Ecology* and during 1988 in *Oecologia.*[a]

Method	Frequency (%)
Regression (bivariate, multivariate, nonlinear)	41
Inspection of graphs	31
Inspection of means and confidence intervals	24
ANOVA (1- or 2-way) (parametric or nonparametric)	18
t tests	15
Correlation	13
Inspection of tables	13
Chi-squared (χ^2) analysis	13
Paired tests (parametric)	6
Nonparametric 2-random-sample tests	5
Paired tests (nonparametric)	3

[a]Sample size is given in Table 3.3. Percentages may sum to > 100% because some articles used more than one technique.

agricultural studies and therefore are often modeled on classical parametric ANOVA designs (e.g., Steel and Torrie 1960). Other techniques like
ANCOVA, MANOVA, and regression and path analysis are used, however
(Table 3.6). There appears to be little consensus about how to make comparisons *among* manipulations made in different ecosystems. Because the
roots of these studies are in agriculture, the effect of "ecosystem" is sometimes included in the design, as field effects are in agriculture. More frequently, however, differences among the ecosystems in which experiments
are performed make it necessary for manipulations to be performed slightly
differently in different ecosystems. Confusion about methods to use to
summarize differences or similarities of ecosystem responses obtained using
different experimental designs result in the frequent use of informal visual
or verbal comparisons (see Table 3.6).

Most surprising in this literature review is the frequency with which all
types of interecosystem comparisons are made by visual inspection of
graphs or tables, or by verbal discussion, without probability analysis (see
Tables 3.4–3.6). Results of ecological studies are rarely clear cut enough
that statistical analysis is not needed. In addition, interecosystem comparisons are necessarily made in many dimensions (it is the nature of the beast),
and psychologists have shown that human observers are inefficient in making multivariate judgments (Nisbett and Ross 1980). Statistical and numerical methods reduce the number of dimensions that ecologists must track
and therefore should improve the accuracy of our judgments. Finally, when
many variables are measured and compared, the probability of coincidence
is high. Statistical and numerical methods provide means of assuring ecologists that their results would not likely have occurred by chance alone.

Selected Problems Encountered in
Interecosystem Comparison

Ecologists make most of the usual types of errors in applying statistical
and numerical methods (see reviews by Green 1979; Legendre and Legendre
1983; Hurlbert 1984; Prepas 1984; Gower 1987; Legendre and Legendre
1987). These errors include improper sampling methods or designs, failure
to meet with the assumptions of analysis techniques (e.g., lack of normal-
ity, heteroscedasticity, lack of additivity, poor transformation, etc.), analysis designs inappropriate to experimental designs, pseudoreplication, etc.,
to name a few. Some are most serious in regression analysis, such as lack
of independence, autocorrelation, and collinearity (see Gujarati 1978 for
remedial measures). Most of these are not specific to interecosystem comparisons and can crop up whenever scientists let down their guard. In any
case, most of these problems are treated routinely by statistical texts, so I
do not dwell on them here. Instead, I discuss here five problems in making

TABLE 3.6. Methods used in connection-focused interecosystem comparisons published during 1988 and 1989 in *Ecology* and during 1988 in *Oecologia*.[a]

Method[b]	Frequency (%)
ANOVA (parametric)	36
Inspection of graphs	29
Inspection of tables	29
Verbal comparisons	14
Inspection of means and confidence intervals	14

[a]Sample size is given in Table 3.3. Percentages may sum to >100% because some articles used more than one technique.
[b]Others include ANCOVA, regression, MANOVA, path analysis, etc.

interecosystem comparisons that are particularly insidious, albeit elementary:

1. We bias our results by the variables we choose to measure.
2. We do not compare enough ecosystems at a time.
3. We measure too many variables in too few ecosystems.
4. We base interpretations on coincidences.
5. We often make informal interecosystem comparisons.

Problem 1. Comparing Apples with Oranges: Fruit Type, Shape, Size, or Color? A frequent criticism of students of comparative ecology is that we try to compare apples with oranges. This suggests that in comparing dissimilar ecosystems, we are trying to compare two entities that are fundamentally, or qualitatively different, like comparing two fruits of different species. I agree with the idea that ecosystems and fruits share common attributes: they are individual objects that do not coincide with each other in space, and therefore fruits are different from other fruits, and ecosystems are different from other ecosystems, by definition. Both fruits and ecosystems can be sensed and measured, that is, we can obtain data about their characteristics. It is often true that apples are more similar to other apples than they are to oranges, just as grasslands might be more similar to other grasslands than they are to forests, or lakes are to ponds, or rainforests are to deserts. This does not mean that they cannot be compared, only that they can only be compared in variables that they share. What is also true, and what is often forgotten in making interecosystem comparisons, is that the characteristics that we choose to measure on our dissimilar ecosystems will dictate the degree to which we find them different or similar. We may obtain markedly different results, depending on the characteristics we consider important.

I will illustrate this point visually, using fruits as ecosystem surrogates, to avoid getting mired in mathematics. Table 3.7 lists data on mass, size, shape, and fruit type of 26 fruits found at Montréal's famous Jean-Talon market. These fruits are far more diverse than apples and oranges, varying

TABLE 3.7. Characteristics of common fruits at Montréal's Jean-Talon market.

Fruit	Mass (g)	Length[a] (mm)	Diameter[a] (mm)	Shape	Fruit type[b]
Apple (Empire)	104.8	55.6	62.2	Comp. spheroid	Pome
Apple (Granny-S)	158.6	63.5	69.9	Comp. spheroid	Pome
Avocado	255.9	107.7	67.3	Ellipsoid	Berry
Banana	204.4	187.2	38.5	Fusiform	Berry
Cantaloupe	743.4	108.1	107.7	Spheroid	Pepo
Cucumber	347.5	205.1	51.7	Fusiform	Pepo
Eggplant	453.0	208.0	85.2	Ovoid	Berry
Grape (Italia)	7.8	26.1	22.5	Spheroid	Berry
Grape (Riber)	8.4	24.9	22.2	Spheroid	Berry
Grape (Ruby)	4.5	21.1	19.1	Ellipsoid	Berry
Grape (Thompson)	4.2	24.2	16.0	Ovoid	Berry
Grapefruit	477.0	91.5	96.5	Spheroid	Hesperidium
Green pepper	150.9	113.3	75.0	Cylindrical	Berry
Kiwi	77.0	60.3	48.6	Ellipsoid	Berry
Lemon	94.8	73.5	53.5	Ellipsoid	Hesperidium
Mango	302.7	107.3	74.8	Ellipsoid	Drupe
Melon (honey-dew)	1364.4	148.3	130.4	Spheroid	Pepo
Orange (sanguinella)	156.7	68.3	69.4	Spheroid	Hesperidium
Orange (Florida)	239.7	81.7	76.6	Ellipsoid	Hesperidium
Papaya	379.1	118.6	90.4	Pyriform	Berry
Pear (Bartlett)	151.1	79.9	63.6	Pyriform	Pome
Pear (Bosc)	130.0	93.9	56.7	Pyriform	Pome
Plum	78.5	48.7	49.6	Spheroid	Drupe
Red pepper	180.6	121.0	64.9	Cylindrical	Berry
Tangerine	123.2	53.2	64.6	Comp. spheroid	Hesperidium
Tomato	114.1	54.1	60.2	Spheroid	Berry

[a]Fruit length was measured as the greatest length parallel to the stem axis; diameter was measured as the greatest width perpendicular to this axis.
[b]Fruit types are after Hulme (1970).

in many attributes such as fruit type, shape, size, and color. Figure 3.2 (upper left) shows an ordination by fruit type, a sort of two-dimensional correspondence analysis. The apples and oranges included in this analysis find themselves rather far apart, because the apples are pomes (derived from a compound inferior ovary) while the oranges are hesperidia (modified berry with considerable development of the endocarp). The upper right of Fig. 3.2 shows the fruits ordinated by size, with one apple being between two types of oranges while a second apple is quite a bit smaller. The lower left of Fig. 3.2 shows the fruits "clustered" by similarity in shape. This time, one of the oranges falls with the spheroids, one with the ellipsoids, while both apples are compressed spheroids. Finally, in the lower right, Fig. 3.2 shows an ordination of the fruits by color and hue. The reddish apple and both the oranges group closely together, while the green apple is much more similar to a green pepper, the avocado, or the grapes than to the other apple. Both the ecosystem characteristics considered and the ecosystems chosen for study will have a great influence on the conclusions of composite ecosystem analyses employing PCA, correspondence analysis, canonical correlation, or clustering techniques. Although I believe mathematical proof to be unnecessary, such explorations have been made by numerical taxonomists studying analogous problems (Sneath and Sokal 1973; Felsenstein 1983).

The choice of variables measured is less critical to variable-focused ecosystem comparisons because the fraction of the total variance in the variable of interest that is explained by the chosen independent variables is measured directly. If variables are poorly chosen or if not enough variables are measured, this will be indicated by the significance of statistical analyses, and r^2 values of regression analyses. Figure 3.3 shows a graphical analysis of the relationship between the fresh mass of the fruits listed in Table 3.7 and their length. Because fruit length is only a rough measure of volume, and fruits vary widely in shape, the r^2 is low (.79) for the log-log relationship. Simply adding fruit diameter as a second independent variable

FIGURE 3.2. Contrasting analyses of similarities of 25 of the 26 fruits analyzed in Table 3.7. Upper left panel: "ordination" on fruit type. Clockwise from upper right: hesperidia, drupes, pomes, pepos, and berries in center. Upper right panel: "ordination" on size. Lower left panel: "ordination" on fruit shape. Top to bottom: fusiform, cylindrical, pyriform, ovoid, compressed spheroid, and spheroid. Lower right panel: similarity of color and hue. Photographs by Jean-Luc Verville of l'Université de Montréal; ordination and composition by W.L. Downing and J.A. Downing.

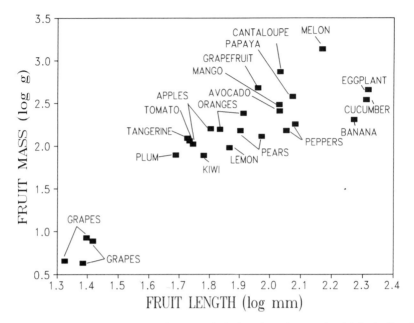

FIGURE 3.3. Log-log relationship between fruit length and mass for 26 fruits listed in Table 3.7. Regression relationship between fruit mass (M, g) and length (L, mm) is log M = −1.77 + 2.05 log L ($n = 26$; $r^2 = .79$). Multiple regression including fruit diameter (D, mm) is log M = −2.88 + 0.94 log L + 1.83 log D ($n = 26$; $R^2 = .98$).

in the regression analysis improves the R^2 (.98), and most fruits fall neatly into line. This analysis shows simultaneously how variable-focused ecosystem comparisons can be made on highly dissimilar ecosystems and the utility of residual analysis (e.g., Draper and Smith 1981) after all statistical comparisons.

Problem 2. Ecologists Do Not Compare Enough Ecosystems At a Time. The minimum number of ecosystems that must be studied in an ecosystem comparison is, of course, two. My review of the literature suggests that recent ecologists take this number of ecosystems as sufficient, rather than minimal. Almost 50% of the articles published in *Ecology* during 1988 and 1989 and in *Oecologia* during 1988 examined less than five ecosystems (median = 7; mode = 2). Fully 28% of the articles that made interecosystem comparisons studied only two ecosystems (Fig. 3.4).

Comparing only two ecosystems can tell us if the ecosystems are different, but can tell us nothing about why they are different. This is because in doing interecosystem comparisons, our replicate samples must be "ecosystems," not replicate samples taken within each ecosystem. For example, if we are interested in how variable Y varies among ecosystems, we can perform replicate measurements of Y in each of ecosystems A and B, and compare the values obtained in each, using a t test or similar nonparametric methods. This test will tell us whether or not there is a significant difference in Y between ecosystems A and B. Perhaps the reason that we chose ecosystems A and B for study was because we believed them to be similar in all attributes except for X. The t test does not show that variable Y varies with X among ecosystems, because with only two ecosystems and two characteristics measured, the probability that the high or low value of Y would be found coincident with the high or low value of X is 1 (see "interpreting coincidences"). Further, it is impossible to demonstrate that any two ecosystems differ *only* in X, because this would require an infinite number of tests with infinite replication. Incidentally, this is why manipulations are

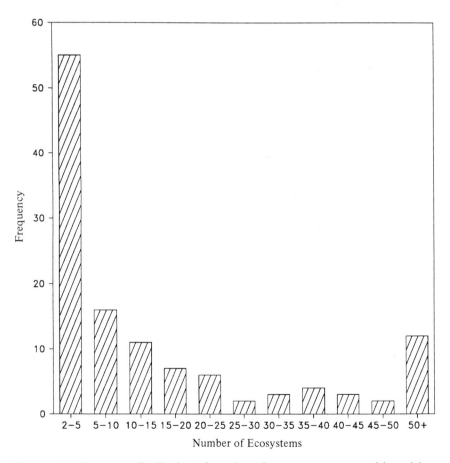

FIGURE 3.4. Frequency distribution of number of ecosystems compared in articles about interecosystem comparisons that were published during 1988 and 1989 in *Ecology* and during 1988 in *Oecologia*.

essential to connection-focused ecosystem comparisons. A real test of the hypothesis that ecosystems differing in X also differ in Y would take a minimum of three ecosystems for a correlation analysis (df = n − 2) or four ecosystems for a t test (because s^2 must be calculated within each class of X). Comparisons of two ecosystems can indeed demonstrate differences among them, but can demonstrate nothing about factors causing or correlated with these differences.

Problem 3. Too Many Variables, and Too Many Tests: Multiple Regression, Multiple Tests, and Pascal's Pyramid. The ready availability of computers has made it possible for modern ecologists to perform any of thousands of statistical and numerical analyses and to do them as many times as they like. We repeat analyses every day that would have taken years to perform (probably inaccurately!) only a few decades ago. Regression analysis is the single most frequently used method in making ecosystem comparisons (see Table 3.3). Regression analysis is employed in all its permutations: simple linear, multiple, polynomial, nonlinear, with interaction terms, with dummy variables, or in ANCOVA. Ecologists performing interecosystem comparisons measure a large number of variables on their ecosystems, frequently measuring more characteristics of ecosystems than they have ecosystems (Fig. 3.5). In addition, we are often interested in several aspects of ecosystems, so that the number of dependent variables or response variables in variable-focused analyses is sometimes superior to the number of independent variables (Fig. 3.6). If one simply performs all of the possible regression analyses on these variables and selects those with the lowest p value, one risks drawing inappropriate conclusions from multiple tests. That is, if we generate 100 sets of columns of random numbers, we will find correlations significant at $p < .05$ five times. If we perform enough tests, randomness alone will generate "highly significant" results.

This risk is particularly great where a large number of combinations of independent variables can be tried for each column of dependent variables. Let us say we are interested in predicting the variable Y in a large number of ecosystems from one variable A. Only one multiple linear regression model exists, and that is $Y = f(A)$. If we measure two independent variables, A and B, then three possible multiple linear regression models exist: $Y = f(A)$, $Y = f(B)$, and $Y = f(A,B)$. Consideration of three independent variables yields: $Y = f(A)$, $Y = f(B)$, $Y = f(C)$, $Y = f(A,B)$, $Y = f(B,C)$, $Y = f(A,C)$, and $Y = f(A,B,C)$, or seven possible models. The total number of possible combinations of independent variables for larger number of variables, as well as the relative abundance of models of different levels of complexity, can be calculated using Pascal's pyramid (Table 3.8). The total number of unique multiple regression models increases rapidly with the number of independent variables measured (Table 3.9), doubling plus 1 for every increase of 1 independent variable. At 20 independent variables, a number quite common in the ecological literature (see Fig. 3.6),

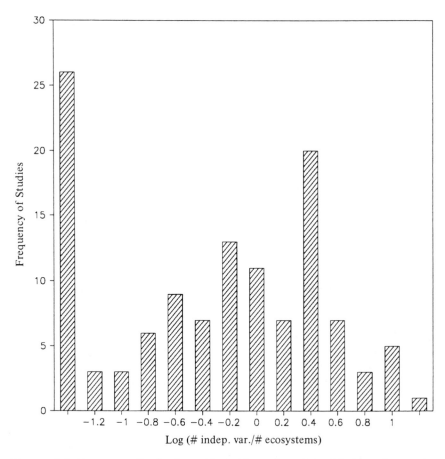

FIGURE 3.5. Frequency distribution of logarithm of number of independent variables measured divided by number of ecosystems compared in articles about interecosystem comparisons published during 1988 and 1989 in *Ecology* and during 1988 in *Oecologia*. Value of 0 indicates that as many ecosystem attributes were measured as ecosystems studied; values larger than 0 indicate that more variables were measured than ecosystems studied. An independent variable is any variable for which measurements were presented that author determined to be possibly responsible for interecosystem differences.

there are 1,048,575 unique multiple regression models possible, 52,428 of which would be statistically significant at $p = .05$ by chance alone! This would include more than 10 analyses significant at $p < .00001$. If more than 1 dependent variable is considered, then the number of possible analyses is multiplied by the number of dependent variables. If polynomial terms or interaction terms are added, the number of possible models becomes gigantic. The danger in multiple regression or in any statistical procedure in which a large number of analyses can be performed and screened is that

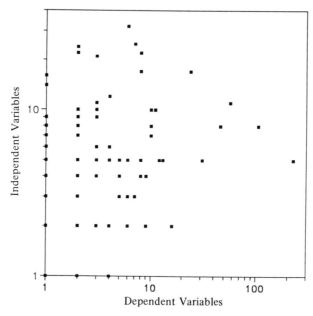

FIGURE 3.6. Relationship of number of independent variables measured to number of dependent variables measured in articles about interecosystem comparisons published during 1988 and 1989 in *Ecology* and during 1988 in *Oecologia*. Presence or absence of species were considered to be dependent variables if they were employed as such in statistical analyses. Other variables are as in Fig. 3.5.

TABLE 3.8. Number of unique multiple linear regression models that can be constructed using the number of independent variables listed in the extreme left-hand column and the number of variables in the model listed in the column heading.[a]

Number of independent variables	Number of combinations of this number of variables:									
	1	2	3	4	5	6	7	8	9	10
1	1	0	0	0	0	0	0	0	0	0
2	2	1	0	0	0	0	0	0	0	0
3	3	3	1	0	0	0	0	0	0	0
4	4	6	4	1	0	0	0	0	0	0
5	5	10	10	5	1	0	0	0	0	0
6	6	15	20	15	6	1	0	0	0	0
7	7	21	35	35	21	7	1	0	0	0
8	8	28	56	70	56	28	8	1	0	0
9	9	36	84	126	126	84	36	9	1	0
10	10	45	120	210	252	210	120	45	10	1

[a]For example, if one measures six independent variables, one can generate 20 unique multiple regression models that contain three variables. Once row one and column two are filled in, all other values can be obtained as sum of value immediately above and value immediately to its left. Total numbers of possible models (see Table 3.9) are obtained by summing model frequencies across rows.

TABLE 3.9. Total number of possible multiple regression models for different numbers of independent variables and the number of models that would be statistically significant ($p <$ .05) by chance alone.[a]

Independent variables	Models possible	Significant at .05
1	1	0.00
2	3	0.15
3	7	0.35
4	15	0.75
5	31	1.55
6	63	3.15
7	127	6.35
8	255	12.75
9	511	25.55
10	1,023	51.15
11	2,047	102.35
12	4,095	204.75
13	8,191	409.55
14	16,383	819.15
15	32,767	1,638.35
16	65,535	3,276.75
17	131,071	6,553.55
18	262,143	13,107.15
19	524,287	26,214.35
20	1,048,575	52,428.75

[a]Model frequencies are obtained by summing across rows of Table 3.8.

apparent statistical significance might cause importance to be attached to a result that was achieved by chance alone.

Remedial measures can be taken on different levels. The classical statistician's advice is to plan comparisons in advance and to make only those comparisons that are planned and for which some sort of theoretical expectation exists. This sounds narrow-minded, but limits the universe of possible comparisons, thus reducing the dilution effect of reams of ANOVAs on the few comparisons of greatest interest. The central matter is whether the significant result was "selected" or "expected" (discussed in Problem 4). Where significant results are selected, there is still some hope. Suitable techniques for correction are discussed by Cooper (1968), Miller (1977), and Kirk (1982).

Problem 4. Interpretations Are Often Based on Likely Coincidence. The aim of many interecosystem comparisons is to examine some variable of interest in more than one ecosystem, to see if it varies among these ecosystems and finally to identify characteristics of these ecosystems that may have caused differences in the variable of interest. A typical recent example of this sort studied three ecosystems (old fields) that differed in many ways such as age, history, composition, soil type, etc. The behavior of some dependent variable was examined (e.g., rates of gap formation or closure, in this case) and then reasons for interfield differences were inferred from differences in characteristics of fields. The authors found that gap creation rates seemed to be much higher in one field than the other two. They then sought ecosystem characteristics that were either greatest or lowest in the field with the higher gap creation rates and concluded that " . . . the higher disturbance rate in this field may be due to soil texture." and "A final possibility is that the differences are related to field age." A further possibility that should be considered is mere coincidence.

Falk (1981) tells the story: "When I happened to meet, while in New York, my old friend Dan from Jerusalem, on New Year's Eve and precisely at the intersection where I was staying, the amazement was overwhelming. The first question we asked each other was: 'What is the probability that this would happen?' However, we did not stop to analyze what we meant by 'this.' What precisely was the event the probability of which we wished to ascertain? I might have asked about the probability, while spending a whole year in New York, of meeting, at any time, in any part of the city, anyone from my large circle of friends and acquaintances. The probability of this event, the union of a large number of elementary events, is undoubtedly large. But instead, I tended to think of the intersection of all the components that converged at that meeting (the specific friend involved, the specified location, the precise time, etc.) and ended up with an event of minuscule probability. I would probably be just as surprised had some other combination of components from the large union taken place. The number of such combinations is immeasurably large; therefore, the probability that at least one of them will occur is close to certainty."

The question to address, therefore, in the "old fields" example given is this: "What is the probability that the lowest or highest value of at least one of the measured ecosystem characteristics will be found in the field considered to show a different value of the dependent variable?" If we had only one site descriptor (e.g., soil texture), by chance alone the probability would be $\frac{2}{3}$, because we have examined only three fields and we would notice the coincidence of either higher *or* lower values of soil texture occurring in the field with higher gap creation rates. Soil texture has a $2/n$ chance of being highest or lowest in the field with higher gap creation rates by chance alone. Considering the second variable (field age) the probability of high or low values of soil texture or field age or both occurring in the field with the higher gap creation rates is $\frac{2}{3} + \frac{2}{3} - \frac{4}{9}$ (Colquhoun 1971), or the probability of either one occurring less the probability of them both occurring: .89.

The generalized solution to this problem for larger numbers of ecosystems (n) and different numbers of independent variables (x) is straightforward. If we have n ecosystems, one of which seems different from all others in some respect, and we have estimated x independent variables, characteristics, or site descriptors on those n environments, the probability that any given one of the x independent variables will be lowest in the aberrant ecosystem is $1/n$ by chance alone. The probability that it will be lower or highest (mutually exclusive events; after Colquhoun 1971) is $2/n$. The probability of finding that at least one of the x independent variables is lowest or highest in the aberrant ecosystem is equal to the complement of the probability that such a correspondence will be found for none of the x independent variables, or:

$$1 - (1 - 2/n)^x \qquad [1]$$

Solution of this equation (Fig. 3.7) shows that in most cases there is better than a 50% chance of finding a coincident independent variable by chance alone. Search for coincident measurements only has heuristic value where many ecosystems are studied and few independent variables are observed.

Problem 5. Formalizing Interecosystem Comparisons: Secondary and Meta-Analyses in Ecological Research. There is a branch of natural science that has recently been described as being "sadly dilapidated" and "facing a crisis." It is said of this science that unlike other sciences in which "tidy,

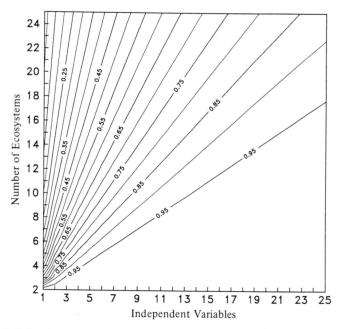

FIGURE 3.7. Probability that lowest or highest values of one or more independent variables will be found in the ecosystem judged to be different from all others in some dependent variable of particular interest. Calculated probabilities are from Equation 1.

straightforward answers to problems studied under experimental conditions are obtained in a logical, sequential fashion, building on each other," this science does not do this because the subject is "more difficult and complex to explain" and "research environments are more difficult to control." Studies in this field "not only use disparate definitions, variables, procedures, methods, samples, and so on, but their conclusions are often at odds with each other." The result is that this science yields few "acceptable results to guide public policy . . . but instead yields unending calls for further research, and the danger that funding agencies may increasingly view (this) research as muddled, unproductive, and unscientific." Comparative studies in this field "often go professionally unrewarded and are notorious for depending on the subjective judgments, preferences and biases of the reviewers; conflicting interpretations of the evidence are not uncommon, while even consistent interpretations by independent reviewers may be built on similar biases and misreadings of the literature." This science is not ecology. The preceding quotations are from Frederic Wolf's (1986) book titled *Meta-Analysis: Quantitative Methods for Research Synthesis* and concern research activity in the social and behavioral sciences.

The problem in the behavioral and social sciences, as in ecology, is that experimental analyses and observations are by necessity collected on different populations, using different designs, techniques, and methods. This has led some quantitative behavioral and social scientists to recognize three types of research. Primary analysis is the original analysis of data in a research study. Secondary analysis is the reanalysis of data for the purpose of answering the original research question with better statistical techniques or answering new questions with old data. Examples of the many recent secondary analyses of ecological data are Banse and Mosher's (1980) study of animal production, Damuth's (1981) study of the abundance of mammals, and Currie and Paquin's (1987) study of tree species richness. A further category of analysis is called meta-analysis, which is "the analysis of analyses," or the statistical analysis of a large collection of results of individual studies toward the synthesis of a general conclusion. Meta-analysis could be a rigorous alternative to the casual, narrative syntheses of the expanding research literature (Glass 1976). During the past 10 years, methods for performing meta-analysis have been developed to render research synthesis formal and objective. The use of formal meta-analysis in literature reviews can help to avoid problems such as (1) the biased selective inclusion (or exclusion) of studies; (2) differential weighting of studies and interpretations based only on subjective judgments; (3) misleading interpretations of findings; and (4) lack of consideration of characteristics of studies that may have contributed to inconsistent results (Wolf 1986).

We face many of the same problems in performing synthetic interecosystem comparisons of the results of connection-focused studies (e.g., manipulation, perturbations, etc.). Often by necessity, experimental designs and procedures differ among ecosystems. Ecologists have often been forced to compare results without rigorous test, often using informal inspection of tables, figures and narrative verbal arguments (see Tables 3.5 and 3.6). Individual ecological studies are expensive, so sample sizes (ecosystem numbers) are often low, so that the statistical power of comparisons in any one study is often weak. Although some rudimentary but exciting meta-analyses have been performed by ecologists (e.g., Connell 1983; Schoener 1983), more frequent use of the synthetic methods of meta-analysis would add rigor to our syntheses.

A wide range of synthetic methods is currently available, and the controversial nature of some of the applications of meta-analysis (e.g., Wachter 1988) is leading to rapid, critical development of better ones. The earliest methods, derived from Fisher's (1932) and Pearson's (1933) combined tests, sought to draw a quantitative consensus regarding the overall statistical significance of several independent tests. More advanced methods have been devised to examine both significance and size of effects, drawing pooled conclusions about effect size from a series of experiments. Other useful techniques test for relationships between characteristics of studies and effect size. Techniques analogous to ANOVA for testing for effects of categorical experimental characteristics, and techniques analogous to regression, for testing for continuous effects of experimental conditions, have been developed. An example use for this latter approach would be the examination of the relationship of effect size found in field fertilization experiments and soil fertility. Special attention has also been given to the combination of correlation coefficients found in disparate studies. Meta-analytical equivalents of composite ecosystem comparisons have also been developed, allowing ecologists to perform cluster analyses on sets of experiments to seek factors responsible for differences in correlations and effect sizes. Reviews of these methods are provided by Wolf (1986) and Hedges and Olkin (1985).

Conclusion

Historically, the comparative approach has been an important source of inspiration and theory to ecologists. I believe that ecosystem comparatists could have greater impact on current progress in ecology if we would phrase clearer, testable questions and pay closer attention to basic probability theory and rigorous synthetic methods. In general, ecosystem comparatists should base comparisons on a greater number of ecosystems, should make planned comparisons of them using a smaller number of variables, should be more aware that coincidences are probable when sample sizes are low, and should remember that the conclusions apply most reliably to the type of population sampled. Following these recommendations might cost more, but should offer more clear-cut solutions to ecology's important problems. The quality of the ecological theories that arise from ecosystem comparisons will be a reflection of the validity of the techniques applied: *on reconnait l'arbre à ses fruits*. Like apples and oranges, ecosystems can be compared quantitatively if ecologists are cautious about the validity of their analyses and the generality of their inferences.

Acknowledgments. I am grateful to William L. Downing for help with the ordination of fruits. Katherine A. Downing helped to collect and measure fruit samples. Simon Forget recognized Pascal's Pyramid on my full blackboard, and Alain Vaudor and Pierre Legendre suggested the use of the complement in Equation 1. Sophie Lalonde helped at several stages of analysis and preparation. Mort Strain provided useful discussions regarding coincidences and "prospecting." I am also grateful to Stephen Carpenter and an anonymous reviewer for suggestions on the manuscript.

REFERENCES

Banse, K. and S. Mosher. (1980). Adult body mass and annual production/biomass relationships of field populations. *Ecol. Monogr.* 50:355–379.

Box, G.E.P., W.G. Hunter, and J.S. Hunter. (1978). *Statistics for Experimenters*. Wiley, New York, 653 p.

Colquhoun, D. (1971). *Lectures on Biostatistics—An Introduction to Statistics with Applications in Biology and Medicine*. Clarendon Press, Oxford.

Connell, J.H. (1983). On the prevalence and relative importance of interspecific competition: evidence from field experiments. *Am. Nat.* 122:661–696.

Conover, W.J. (1971). *Practical Nonparametric Statistics*. Wiley, New York.

Cooper, D.W. (1968). The significance level in multiple tests made simultaneously. *Heredity* 23:614–617.

Crome, F.H.J. and G.C. Richards. (1988). Bats and gaps: microchiropteran community structure in a Queensland rain forest. *Ecology* 69:1960–1969.

Currie, D. and V. Paquin. (1987). Large-scale biogeographical patterns of species richness of trees. *Nature* (London) 329:326–327.

Damuth, J. (1981). Population density and body size in mammals. *Nature* (London) 290:699–700.

Draper, N. and H. Smith. (1981). *Applied Regression Analysis*, 2d Ed. Wiley, New York, 709 p.

Falk, R. (1981). On coincidences. *The Skeptical Inquirer* 6(2):18–31.

Felsenstein, J., ed. (1983). *Numerical Taxomony*. NATO Advanced Sciences Institute Series G (Ecological Sciences), No. 1. Springer-Verlag, Berlin.

Fisher, R.A. (1932). *Statistical Methods for Research Workers*, 4th Ed. Oliver and Boyd, London.

Glass, G.V. (1976). Primary, secondary, and meta-analysis of research. *Educational Researcher* 5(10):3–8.

Gower, J.C. (1987). Introduction to ordination. In: *Developments in Numerical Ecology*, P. Legendre and L. Legendre, eds. NATO Advanced Sciences Institute Series G (Ecological Sciences), No. 14. Springer-Verlag, Berlin.

Green, R.H. (1979). *Sampling Design and Statistical Methods for Environmental Biologists*. Wiley, New York.

Gujarati, D. (1978). *Basic Econometrics*. McGraw-Hill, New York.

Hedges, L.V. and I. Olkin. (1985). *Statistical Methods for Meta-analysis*. Academic Press, Orlando.

Hotelling, H. (1936). Relations between two sets of variates. *Biometrika* 28:321–377.

Hulme, A.C., ed. (1970). *The Biochemistry of Fruits and Their Products*, Vol. 1. Food Science and Technology Monographs. Academic Press, London.

Hurlbert, S.H. (1984). Pseudoreplication and the design of ecological field experiments. *Ecol. Monogr.* 54:187–211.

Kirk, R.E. (1982). *Experimental Design: Procedures for the Behavioral Sciences*, 2d Ed. Brooks/Cole, Belmont, California.

Legendre, L. and P. Legendre. (1983). *Numerical Ecology*. Developments in Environmental Modelling, 3. Elsevier, Amsterdam, 419 p.

Legendre, P. and L. Legendre, eds. (1987). *Developments in Numerical Ecology*. NATO Advanced Sciences Institute Series G (Ecological Sciences), No. 14. Springer-Verlag, Berlin.

Miller, R.G., Jr. (1977). Developments in multiple comparisons, 1966–1976. *J. Am. Stat. Assoc.* 72:779–788.

Nisbett, R. and L. Ross. (1980). *Human Inference: Strategies and Shortcomings of Social Judgment*. Century Psychology Series. Prentice-Hall, Englewoods Cliffs.

Pearson, K. (1933). On a method of determining whether a sample of size n supposed to have been drawn from a parent population having a known probability integral has probably been drawn at random. *Biometrika* 25:379–410.

Prepas, E.E. (1984). Some statistical methods for the design of experiments and analysis of samples. In: *A Manual on Methods for the Assessment of Secondary Productivity in Fresh Waters*, J.A. Downing and F.H. Rigler, eds. IBP Handbook No. 17, 2d Ed. Blackwell, Oxford, pp. 266–335.

Schoener, T.W. (1983). Field experiments on interspecific competition. *Am. Nat.* 122:240–285.

Sneath, P.H.A. and R.R. Sokal. (1973). *Numerical Taxonomy – The Principles and Practice of Numerical Classification*. W.H. Freeman, San Francisco.

Sokal, R.R. and F.J. Rohlf. (1981). *Biometry*. Freeman, San Francisco.

Steel, R.G.D. and J.H. Torrie. (1960). *Principles and Procedures of Statistics*. McGraw-Hill, New York.

Wachter, K.W. (1988). Disturbed by meta-analysis? *Science* 241:1407–1408.

Wolf, F.M. (1986). *Meta-Analysis. Quantitative Methods for Research Synthesis*. Sage University Paper series on Quantitative Applications in the Social Sciences, No. 59. Sage Publications, Beverly Hills.

4
On the Relevance of Comparative Ecology to the Larger Field of Ecology

ROBERT H. PETERS, JUAN J. ARMESTO,
BERTRAND BOEKEN, JONATHAN J. COLE,
CHARLES T. DRISCOLL, CARLOS M. DUARTE,
THOMAS M. FROST, J. PHILIP GRIME, JERZY KOLASA,
ELLIE PREPAS, AND W. GARY SPRULES

Abstract

Although all science depends on comparison, this chapter uses the term *comparative ecology* to denote a subsection of the science of ecology. In this sense, comparative ecology is distinguished quantitatively, but not qualitatively, from the rest of the science by the larger scope of its application and the greater degree of aggregation of its variables. These differences enhance problems of scale, causal attribution, imprecision, and comparability, and tend to isolate the models that emerge from comparative ecology. Nevertheless, similar problems occur throughout the science. They could be minimized if all ecological scientists would clearly identify the goals and limitations of their research. The chapter ends with a set of recommendations to help achieve this clarity of purpose and so to promote ecological science in its broadest sense.

Introduction

Comparative ecology is no more than a convenient term to represent a wide and varied collection of empirical and conceptual devices in ecology. For example, comparative ecology includes models that focus on a distinct process or attribute to describe patterns in that character among different systems (Downing, Chapter 3, this volume), such as the response of plant species to disturbance (Grime 1977; Grime et al. 1988), continental patterns in diversity (Brown 1989), the effect of depth on oceanic material fluxes (Pace et al. 1987) and of land use on phosphorus loading (Reckhow and Simpson 1980; OECD 1982), allometric rules for animal structure and function (Peters 1983; Calder 1984), the self-thinning law for plants (Yoda et al. 1963; Gorham 1979; Weller 1989), the response of different functional groups of aquatic organisms to lake trophy (Carpenter et al., Chapter 5, this volume; Peters 1986), or comparisons of different rates and biomasses of different trophic levels or functional groups (McNaughton et al., Chapter 7, this volume). Comparative ecology also includes multivariate comparisons among or between contrasted systems, such as the comparison of Mediterranean ecosystems of Mooney (1977) or the identification of chrysophyte community change across gradients in lake acidity (Smol, in press). Still other examples compare the connections and mechanisms underlying pattern, like the effect of pond permanence on the dynamics of different newt populations (Hurlbert 1970) or the component fluxes of complex food webs (Ulanowicz and Wulff, Chapter 8, this volume).

Despite obvious differences, these diverse studies share many characteristics. They tend to be synoptic analyses of many systems, although some, especially experimental studies, involve as few as two. They often involve statistical analyses of empirically determined, quantitative patterns, although the sophistication of the analyses varies and qualitative comparisons are sometimes more meaningful and appropriate (e.g., dead versus alive). Comparative analyses frequently use natural experiments to identify patterns along gradients (Vitousek and Matson, Chapter 14, this volume) or across contrasts (Smith, Chapter 13, this volume), but this does not preclude the possibility of laboratory study (Grime 1965), field experiment (Schindler 1987), or mesocosm and microcosm tests, when these are appropriate and feasible. Many comparative analyses involve simple, well-defined variables (like body mass or temperature), aggregated properties (like primary production or CO_2 flux) and relatively large systems (ecosystems, guilds, trophic levels), but exceptions to such generalizations are common. In short, neither methodology nor variable choice is sufficient to distinguish comparative ecology from the larger science of which it is part.

It is scarcely surprising that comparative ecology cannot be readily distinguished from the rest of the science. Every science obtains meaning only by comparison (Bradshaw 1987), and, in some inclusive sense, all ecology is "comparative ecology" and all science is comparative science. Although most of the authors of this chapter believe that "comparative ecology" also exists in a more restrictive sense, and that the potentialities and problems of this smaller field are not as widely recognized as they might be, we were unable to distinguish comparative ecology in a narrow sense from the larger field and eventually accepted the futility of an attempt. Nevertheless, even if comparative ecology is merely a subset of ecology, distinguished by differences in degree rather than in quality, the challenge for "comparative ecologists," intended in the restricted sense, remains to make their work relevant to the larger community. This requires clear statement of both immediate and longer term research goals, an appreciation of the overall structure of the science and the domains of comparative ecology within that framework, and a frank admission of the limits and shortcomings of ecological comparisons. Such a declaration is the purpose of this report. We begin by establishing appropriate goals for comparative ecology, then dwell on the reasons for some of the apparent idiosyncrasies of the narrower field of comparative ecology, and end with some recommendations to protect comparative ecologists from untenable excess in their positions, while rendering the goals of their research more comprehensible to their colleagues.

The report itself is a selective abstraction of the discussion group on "Legitimizing Comparative Ecology," held during the third Cary Conference (Cole, this chapter). Although this chapter cannot be described as a consensus, it seeks to reflect many of the viewpoints in that discussion and in the plenary sessions. Because the composition of the discussion groups changed during the meeting, not all participants could be included among the authors, and not all wished to be. We are nevertheless grateful to Steve Carpenter, Stuart Hurlbert, Mike Pace, Dave Tilman, Bob Ulanowicz, and, notwithstanding their anonymity, the other contributors. All were critical to the discussion that led to this chapter.

Goals for Comparison in Ecology

Every piece of science — experiment, lecture, paper, or treatise — should include an answer to the fundamental question: "What is the purpose of this work?" For the proponent, clarity of purpose directs the design of the study, the selection of appropriate methods and sites, the identification of germane observations, and the evaluation of the whole. For the granting agency, editor, and referee, clear statements of purpose are essential for critical evaluation of the intentions behind a given work and of the success of the finished piece in achieving those aims. For the potential audience, knowing the purpose of the work allows that audience to decide if its interests coincide with the material under study and therefore if the study warrants the effort required to learn and understand it. Downing (Chapter 3, this volume) has demonstrated that comparative ecosystem ecologists often identify the goal of their research only obscurely, and therefore they ought not to be surprised if the goal remains obscure to readers and critics. One result is the improper match of study and audience that surely lies behind much of the inefficiency, misunderstanding, and disappointment associated with contemporary scientific communication, including communication in comparative ecology.

The general uses to which scientific studies can be put are rather few. Most research identifies, describes, predicts, and explains patterns in nature and is usually justified as one or more of the generation of new hypotheses, the testing of existing hypotheses, or the application of these hypotheses to particular problems. This short list can serve as an aide-mémoire in the justification of research, whether or not it be intended as a contribution to comparative ecology. A writer who is considerate of the audience will usually describe the application of a research program in terms of one or more of these goals.

Much of comparative ecology is particularly relevant to the generation of hypotheses, because it often seeks to identify presumptive patterns in a range of variables for later evaluation and resolution. Among many examples from the third Cary Conference are the recycling index, a possible measure of stress sensitivity (Howarth, Chapter 9, this volume); a number of indices of total system behavior like mean food chain length, ascendancy, and throughput of energy flows (Ulanowicz and Wulff, Chapter 8, this volume); isotope ratios (Waring, Chapter 11, this volume); Redfield ratios (Smith, Chapter 13, this volume) and their counterparts on land (Vitousek and Matson, Chapter 14, this volume) and in freshwater (Caraco et al., Chapter 12, this volume). Heal and Grime (Chapter 2, this volume) pointed instead to the variables that Lindeman (1942) had suggested as valuable bases for a research program in trophic dynamics and to those that Odum (1969) identified as indicators of system maturity. Such variables are proposed as measurable system properties that are potentially meaningful because they may inform us about other properties of the system. The success of these works will therefore depend on the ability of the writers to show that information about these other properties is needed and that the proposed variables can provide this information.

Patterns that emerge from comparative ecology should also suggest new hypotheses, for the fertility of a construct is one of its chief attractions for other scientists. For example, the correlation analysis by McNaughton et al. (Chapter 7, this volume) suggests, but cannot prove, that experimental manipulation of primary production would increase secondary production by the herbivores. This hypothesis and experiment bear directly on widely discussed questions about the role of food limitation in terrestrial herbivore communities (Hairston et al. 1960; Fretwell 1987) and about the general role among ecosystems of bottom-up versus top-down control (Carpenter et al., Chapter 5, this volume; Carpenter 1988). Eventually, ecologists may wish to pursue the processes underneath these patterns with more mechanistic hypotheses, and the inspiration and development of these tests are also a legitimate goal of comparative ecology.

When comparative ecology operates on large scales and with composite variables, it can identify system properties that might never emerge from other approaches. For example, the regressions of McNaughton et al. (1989) show that the consumption of the herbivore community rises faster than primary production, so that where primary production is high, biotic interactions appear more intense. Peters (1986) reached a similar conclusion with respect to planktonic communities.

At a different level, comparative ecology can be used to test theories about regularities in the world around us and to predict where these regularities occur. For example, the seminal paper of Hairston et al. (1960; see Fretwell 1987) suggested that terrestrial herbivores are not limited by their resources because the world is green. However, the relationship between primary production and herbivore production must cast that easy generalization into doubt (McNaughton et al., Chapter 7, this volume). Such simple empirical relations can be seen as hypotheses in their own right, and because they appear to describe powerful, broadly applicable patterns in ecology, they should be rigorously tested and evaluated in their turn.

Once the patterns of comparative ecology have been identified, tested and gained some acceptance, they may serve an important role as standards of normal behavior in ecological comparisons. As standards, established patterns help us to plan our sampling campaigns by providing an informed view of what we should expect (Downing 1979), and, when the data are in hand, a knowledge of the norm is essential to any assessment. Comparative physiologists have long asked themselves the question, "compared to what?" and one answer is, "compared to average patterns generated by multispecies comparisons." Comparative ecology offers the same possibility. Strong deviations from normal patterns can be among the most informative of observations, because they lead to the consideration of other factors, both anthropogenic and natural, that influence the variable of interest; such exceptions would be invisible without the contrast provided by informed expectation.

Scientific theories, hypotheses, and models also show us which variables are most likely to describe new systems effectively, comparably, and informatively. Such information is essential to the design of effective assessment and monitoring programs, to situate and design research where logistic constraints limit realistic options to a small subset of possible measurements. In this same regard, comparative models identifying the most informative variables are instrumental in prioritizing the measurements we take in terms of the most information gained for the least expenditure. For example, Waring (Chapter 11, this volume; 1988) seeks variables that can characterize a forest in a single day's study and Fisher et al. (Chapter 10, this volume) seek variables that describe comparable patterns in streams despite the spatial and temporal diversity of running water systems.

Finally, it should be stressed that comparative ecology is not a thing unto itself, but interacts broadly with the rest of the science. Comparative analyses are a common tool in assessing ideas generated by the rest of ecology, and these other areas frequently provide the spark that fires comparative analyses. For example, Bell (1985) used a comparative analysis of the Volvoccales to test his theories regarding the evolutionary advantage of sexual reproduction. Grime (1965; Grime et al. 1988) used the comparative approach to describe the evolutionary and ecological constraints that channel biological processes along certain evolutionary paths in the hope that we may eventually create an analogue to the periodic table of the elements that describes the potentials of different plant species by their relative abilities.

Some Peculiarities of Comparative Ecology

None of the proposed goals or applications of comparative ecology should be obscure to other ecologists. There is no reason to hide these goals. However, if comparative ecologists fail to explain their goals effectively, differences in emphasis and particularities of approach can make the relationship of comparative ecology to the rest of the science difficult to discern. Figure 4.1 addresses this problem by characterizing comparative ecology with a simple scheme of three axes that order different programs of research in comparative ecology. Although this figure threatens to be a dangerous oversimplification, it can prove a useful device if not interpreted too closely.

Two of the axes in Fig. 4.1 are needed to distinguish between dependent (or response) and independent (or predictor) variables, and both represent the degree of aggregation underlying these variables. Studies relating highly specific characteristics such as individual growth to some other specific characteristic such as individual respiration rate would be placed close to the intercept. Progressively more aggregated studies might consider, for example, characteristics of populations (e.g., population density or growth rate); species (degree of r selection, average life span); guilds (rates at which leaves are shredded by stream detritivores, number of carnivorous species, proportion of coniferous trees); trophic levels (herbivore biomass, primary production); communities (biomass, diversity); or ecosystems (evapotranspiration, average phosphorus concentration of the water or soil pH). These studies would fall progressively further from the intercept and more toward the right side of Fig. 4.1. Studies comparing the aggregated properties of one axis to the more specific properties of another (e.g., an analysis of the effect of ecosystem evapotranspiration on the survivorship of an insect pest species) would lie closer to the predictor variable axis.

The third axis represents the "scope" of the study. Scope increases with the range of the phenomena under study. For example, an analysis of the relation between metal content of *Daphnia pulex* and that of its food could be addressed experimentally among test tubes, and such a study would lie close to the intercept on all three axes. However, if this relationship were considered among ponds in southern Ontario, the degree of aggregation would change little but the scope of the study would increase. Scope would increase more if both ponds and lakes were considered and still more as the geographic or taxonomic range increased. Studies of similar scope might consider more aggregative properties, for example, the metal content of all herbivores, or all animals, and these would take positions more to the right in Fig. 4.1. Such studies could be extended to higher levels of scope than the *Daphnia* work, because they could be transposed to systems where *Daphnia* cannot exist.

The examples use the hierarchy from individual to ecosystem because it

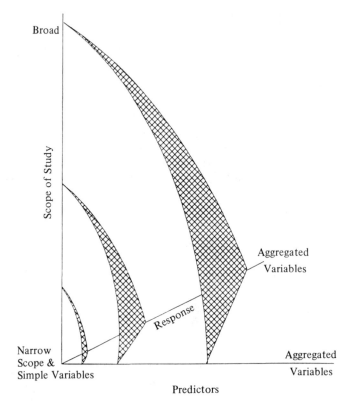

FIGURE 4.1. Schematic representation of relationships between degree of aggregation of predictor and response variables and scope of research in relationships of comparative ecology. Research that is close to the origin in that it predicts relationships among simple variables for narrowly defined phenomena is not very controversial because it involves traditional entities and topics of ecological research. However, as more phenomena are lumped together into aggregate variables describing the common property of what would usually be taken as many different entities and as wider scopes are considered, possibilities for misinterpretation, misunderstanding, and controversy increase. Planes are intended to indicate increasing risk of misunderstanding and controversy as aggregation and scope increase; they suggest that problems could arise, even if variables are simple, if the scope of the research is broad, and vice versa.

is familiar, but it is not exclusive. A variety of biological, geological, or climatic phenomena that are not readily or usefully interpretable in terms of this familiar hierarchy can still be ranged along such composite axes. For example, comparisons of different size classes, geographical extent, or temporal scales could be accommodated. This flexibility results because the axes are conceptual and not operational. As a result, any explanation of the nature of comparative ecology in these terms involves a subjective ranking of different studies in terms of scope and degree of aggregation of the variables. This subjectivity is appropriate and necessary because the misunderstandings between comparative ecologists and other practicing ecologists often reflect a failure to explain the gains and compromises required by their respective research goals at just such a personal level.

The volume contained between these three axes characterizes different studies in terms of the degree of aggregation of the dependent and independent variables and the scope of the study. In principle, a series of planes cut this volume into surfaces of increasing aggregation and scope. The difficulty our group had in distinguishing comparative ecology in a narrow sense from the larger field of ecology results from the impossibility of drawing a single plane that divided the continua in Fig. 4.1 into two disjunct groups which satisfied all participants.

In general, resistance to comparisons in ecology is less when the study considers phenomena that lie closer to the origin because these have been traditional foci of research; the more controversial studies, those that are frequently considered part of comparative ecology, lie more distally (Fig. 4.2). The planes in Fig. 4.1 are tilted to suggest isopleths of critical resistance to comparative ecology. For example, a study of wide scope involving relatively unaggregated variables may be as difficult for other ecologists to accept as one of more limited scope that involves quite aggregated characteristics.

Many of the peculiarities associated with comparative ecology derive from studies lying in the distal position of Fig. 4.1, and a major part of the challenge in explaining the relevance of one branch to another, or of legitimizing any specific study in terms of the larger science, depends on showing how studies lying at different points in this volume relate to one another. This section provides a checklist of some of these relations, interpreting them in terms of Fig. 4.1, and grouping them, largely for purposes of discussion, into four subtopics: scale, causality, imprecision, and comparability.

SCALE

Many ecological studies use small scales to establish models that can only be transposed to other problems after respecification of their parameters. Comparative analyses may be used to this end, but they may also be interpreted at other scales. Because comparative research summarizes observations from many circumstances, the summaries or synopses may describe general patterns for similar phenomena, which may extend to other levels of aggregation and scopes. Thus the positive response of limnoplankton to increased phosphorus concentrations (Pace 1986) may be taken as a small-scale model if we perceive the parameters to be so uncertain as to require redefinition for each application. Alternatively, this response may be taken as a mechanism in some larger model of lake phosphorus dynamics if we

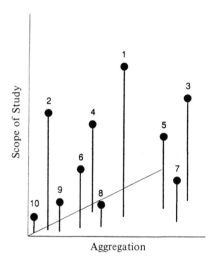

1   Species r_{max} vs Mean Size
2   Individual Metabolic Rate vs Size and Body Temp.
3   Production vs Evapotranspiration
4   Marine mortality vs Size of Life Stage
5   Lake Chlorophyll vs Phosphorus
6   Antelope Group Size vs Mean Body Size
7   Lilac Budbreak vs Climate
8   Density Dependence of Viburnum Whitefly Life Stages
9   Daphnid P Excretion vs Body Size, Temp. & Food Level
10 Descriptive Studies

FIGURE 4.2. Placement of eclectic collection of comparative studies on three-dimensional axes described in Fig. 4.1. All placements in this figure can only be suggestive because the axes are conceptual rather than operational. Identification of numbered points: 1, species r_{max} versus mean size (Blueweiss et al. 1978); 2, individual metabolic rate versus size and body temperature (Robinson et al. 1983); 3, production versus evapotranspiration (Lieth and Box 1972); 4, marine mortality versus size of life stage (McGurk 1986); 5, lake chlorophyll versus phosphorus (OECD 1982); 6, antelope group size versus mean body size (Jarman 1974); 7, lilac budbreak versus climate (Dubé and Chevrette 1978); 8, density dependence of *Viburnum* white fly life stages (Hassell et al. 1987); 9, *Daphnia P* excretion versus body size, temperature, and food level (Peters and Rigler 1973); and 10, descriptive studies.

make a causal attribution. However, as a synopsis of available information, this regression also serves as a general theory in its own right if we hypothesize that the synoptic patterns described by the mean, variance, and regression describe the likely response of other lakes. Because comparative studies at one scale need not yield patterns that apply at others, comparative ecologists should indicate their intended scale to avoid unintentional conflation of goals.

Scale effects are associated with all three axes, because scale refers to both aggregation and scope. This pervasiveness renders any treatment of "scale" necessary but easily confusing. Although many of the difficulties encountered in comparative ecology result from problems of scale, scale invokes so many problems simultaneously that the problem is better analyzed by component. Scale is therefore a generic term to refer to the distance of studies far from the origin in Fig. 4.1. It cannot be considered alone; it conditions all aspects of the problems of comparative ecology (Frost et al. 1988).

CAUSALITY

All scientists know that pattern recognition or correlation cannot show cause. This inability has many aspects. Any natural pattern is sure to correlate with more than one variable, so confounding variables make it impossible to identify any single variable as the cause of a comparative pattern. In experiments, these variables are controlled, but controls in natural experiments are almost always weak. The resultant confusion of possible causes makes both mechanistic explanation and experimental investigation of the observed patterns difficult.

Comparative ecology therefore disappoints because it is ill equipped to identify cause-and-effect relationships. This deficiency can be partly offset by large sample sizes, which establish prominent, widespread patterns and characteristics. Such extensive tests are particularly powerful in separating effects of wide significance from the many details that invariably distinguish biological entities and that confound comparisons based on limited sample sizes (Grime 1965; Clutton-Brock and Harvey 1979).

Absence of Controls. As larger and larger scales and more disparate systems are compared, the possibility of effectively isolating a single causal factor becomes increasingly remote, because the likelihood of two systems differing in only one characteristic becomes vanishingly small. Thus, large-scale comparisons are rarely, if ever, adequately controlled, and even reference systems (Likens 1985) become increasingly disparate as the scale increases.

The difficulty of finding good controls represents one of the limitations of comparative ecology and should make comparative ecologists leery of any causal attribution. This is not a fatal flaw. Despite their inadequate controls, general, empirical patterns have an advantage over closely controlled experiments. Such broad comparisons entail a fuller range of interfering factors than can ever be incorporated into simple experiments, so the imprecision in the comparisons reflects the effect of these other variables. The proportion of the total variation that is statistically explained by the comparative pattern is thus a better measure of the importance of the predictor variables than a controlled experiment can ever provide.

Human Interference. Many New World ecologists mistrust anthropogenically disturbed systems as unrepresentative of nature. The pervasiveness of contemporary man therefore renders almost all large-scale studies unnatural and unattractive to the purist. However, undisturbed systems have likely disappeared forever. Ecologists inevitably face anthropogenically perturbed systems whenever they consider questions of larger scale, especially those that represent the looming crises of global ecology. Unmeasured factors, such as human interference, can also be a virtue because their effect is likely to yield comparisons that are more robust in natural situations than comparisons obtained with strict controls. Honest assessment therefore requires comparative ecologists to acknowledge the possibility of confounding factors in their work but also to stress the advantages of applicability at scales of real interest.

Exogenous versus Endogenous Variables. The sorts of variables that are available also change with scale (Fig. 4.2). Near the origin, the predictor and response variables conform well to everyday notions of cause and effect. For example, there is a clear distinction between the organism and its environment, between predator and prey, or between parent and offspring. This ease of separation allows easy attribution of observed effects to exogenous variables, that is, to causally related variables that are clearly separable from one another. This distinction is less easy at gross scales where traditional causes and effects may be confounded within single variables. For example, the biomass of a size class may be both the cause and effect of trophic and population processes occurring within the class, and the relationship between primary production of terrestrial ecosystems and evapotranspiration (Lieth and Box 1972) is difficult to interpret causally because both variables are determined by the plants.

The difficulty of causal attribution in larger systems is related to the degree of integration of the variables. We have no difficulty in attributing the death of a prey to its predator, because the degree of integration of the prey and predator is very small. However, when integration is high, as it is in a single organism or an ecosystem, simple causal relations are very difficult to isolate, even though many characteristics covary (e.g., size and respiration rate, pulse rate and life span, evapotranspiration and productivity). The problem of causal attribution in integrated systems can therefore occur anywhere in Fig. 4.1. If the problem seems minor close to the origin, this is only because we have a better intuitive idea of appropriate entities for study close to the human scale of observation: there is little risk of confusing an organism with its environment or a predator with its prey. The situation is very different when we study larger scales, for our intuitions are less valid and our casual attributions less effective. Kolasa and Pickett (1989) have proposed use of a scale-independent theory of entities to handle this problem.

Absence of Mechanistic Explanation. Simple patterns recognized at the gross scales of the ecosystem or community can often be explained by many alternative mechanisms involving the details of the system under study. The pattern itself is often neutral in its support of these alternatives, and tests of the various mechanistic explanations would be too vast an undertaking for most laboratories. The comparative researcher is thus left to choose an inadequately supported explanation or to eschew explanation altogether. This leaves the researcher in a dilemma and, in the absence of further work in support of one mechanism or another, comparative ecologists should feel free to state their preferred explanation but also obliged to acknowledge its associated deficiencies. If mechanisms are sought, far more work is required.

Lack of Experiment. Experimental studies are so varied that the analyses of experiments in the published literature are usually inappropriate for secondary analyses (Howarth, Chapter 9, this volume). Experimental comparisons and manipulations across ecosystems are likely to prove more effective, but these are expensive and can rarely be performed extensively. As a result, the initial phases of large-scale comparative analyses are often limited to the comparison of static variables collected in punctual visits.

These practical limitations do not show that experimental comparisons are impossible at a large scale. Indeed, the most convincing evidence in comparative ecology depends on good field experiments (Smith et al. 1984; Schindler 1987) that confirmed the results of earlier and easier gradient analyses and natural experiments. The rarity of such experiments reflects only their logistic difficulty.

The high cost of experiment makes effective use of other comparative techniques essential when scale is large. Appropriate use of these alternatives in comparative analyses is one of the hallmarks of scientific wisdom, so comparative ecologists should not be timid in acknowledging necessary approximations.

IMPRECISION

Another effect of the scale dependence of sources of variation is that the residual errors around many of the patterns that emerge from large-scale analyses are relatively high. Confidence limits often cover an order of magnitude, so workers with more narrowly defined problems at lower levels of aggregation and restricted scopes are unimpressed because the unexplained error remains high relative to the total variation observed at these smaller scales.

Variance. Variables in the distal portions of Fig. 4.1 generally have larger variances, because the increase in scope and aggregation incorporates more sources of variation and so provides a fuller expression of the limits of the phenomenon. As a result, studies of wide scope or aggregation represent a wider range and larger number of factors; they identify the effects of variables that operate at larger scales but only represent less general variables as scatter or error. Because sources of variation differ across scopes and aggregations, even factors that explain a substantial proportion of the variance in large-scale investigations may fail in studies of more limited scope. Caraco et al. (Chapter 12, this volume) demonstrated this in describing the differential effects of sulfate and oxygen on the release of sedimentary phosphorus. O'Neil et al. (1986) stressed the scale dependence of correlation and causal connection as one of the distinctions of hierarchical systems.

Heterogeneity. A related problem is that those who work at lower scales may see extensive aggregation as the lumping of essentially heterogeneous phenomena. The statistical explanations offered by comparative ecologists therefore only explain variation that does not concern specialists. For example, much of the variance explained by body size in allometric regressions is effectively dismissed by specialized studies, because it is removed when the size range of the organisms under study is restricted to that of a single taxon or guild. As a result, comparative analyses often have little effect in explaining the variance that confronts colleagues working on a smaller scale.

COMPARABILITY

Any single system has an infinite number of properties and can therefore be analyzed in an infinite number of ways (Kolasa and Biesiadka 1984). As more and more systems are involved, the number of common properties necessarily declines, but the possibility for shades of meaning in the operational definition of these properties increases. The bases for comparison are therefore fewer among increasingly disparate systems, but each base itself becomes more inclusive, less rigorously defined, and less closely related to everyday experience. At larger scales, comparative ecology is restricted to a relatively small number of rather distinctive variables.

This restriction has given comparative ecology its characteristic reliance on statistical and phenomenological analyses of quantitative patterns in surrogate data culled from the literature. Because this approach may isolate comparative ecology from the rest of the science, comparative ecologists should take pains to explain that some aspects of their approach are adopted as expedients, not desiderata, and to detail restrictions inherent in this trade-off. Nevertheless, this trade-off is part of the price of generality. It is a price that many scientists willingly pay because it extends our scientific reach far beyond the particular case.

Phenomenology. Ecology has traditionally focused on interactions among individuals and populations within a single community, perhaps because this level is closest to the human scale of both the great naturalists who laid the underpinnings of the science and the everyday experience that piques our curiosity. Whatever the source of this bias, most (not all) of our theoretical tools treat phenomena near the origin in Fig. 4.1 and have less direct relevance to more distal phenomena. Deprived of these theoretical guides, comparative ecologists have come to rely more heavily on pattern recognition and simple empiricisms.

Surrogate and Simple Variables. Because only a few variables are available from the literature, because extensive analyses allow little time for intensive study within each system, and because the theoretical and conceptual underpinnings that would identify measurable variables for large-scale comparisons are poorly developed, comparative analyses rely heavily on very simple variables and better-or-worse approximations or surrogates based on whatever sparse information is available. Inevitably, these analyses are open to the criticism that the chosen variables confound different phenomena, that they increase the heterogeneity of the analyses, and that the serendipitous nature of such analyses has increased the possibility of chance correlation. As a result, any pattern identified by comparative analysis is hypothetical and should be tested thoroughly against new information. This is the nature of all scientific knowledge, but it is particularly worrisome in comparative ecology.

Secondary Analyses. The logistics and expense of large-scale comparisons have forced many comparative ecologists to rely on others' results. This permits a far more extensive database than would otherwise be the case, but restricts the scope of the study to the few variables that are routinely measured. Indeed, summaries of even the most routine measures usually require that comparative ecologists ignore substantial methodological differences.

A further disadvantage of reanalysis is that these analyses are restricted to the information that other scientists thought to be important and may incorporate their assumptions. For example, the use of species averages in allometric comparisons assumes that the species is an effective unit, but this would not be so if taxonomic splitting allowed some speciose group to bias the patterns or if the major source of variation in the pattern were more closely related to experimental technique or geographic provenance than to taxonomic position (Clutton-Brock and Harvey 1979). Grime et al. (1988) found that the predominance of morphological information biased their multivariate comparisons of British herbs and deliberately sought a more balanced set of descriptions for their analyses.

Secondary analyses offer the advantages of synoptic survey and use the storehouse of information in our libraries, but they increase error variance in the data. That uncertainty is a further price of the search for general pattern and should be frankly admitted. It should also be admitted that others, both scientists and technicians, who apply the relationship are likely to reintroduce this same uncertainty. If so, patterns emerging from secondary analysis may prove more robust when reapplied by still other workers than are relationships established in the well-defined conditions of a single expert's laboratory.

Quantification. Because the variables of interest in comparative analyses at moderate to large scales are often abstractions, and because they usually take the researcher outside the limited realms in which one can claim technical expertise, the variables are often less amenable to qualitative descriptions. For example, oligotrophy refers to low levels of productivity in aquatic systems but describes quantitatively very different phenomena to an oceanographer, alpine limnologist, or prairie limnologist. Such disparate phenomena are better compared when stripped of their verbal or conceptual context, and this is most easily achieved by quantification of the variables. Among the costs of large-scale comparisons are the loss of the information in these contexts, the inability to compare nonquantitative measures, and a sense of sterility in general patterns that necessarily lack these descriptors.

Statistical Analyses. The availability of substantial sets of numerical data leads easily to statistical comparisons. Indeed, the absence of the contextual clues and deep experience that come from defined study forces greater reliance on statistical analyses. As a result, comparative ecologists should try especially hard to reinterpret their statistics to make them transparent to other ecologists. Regrettably, there is little evidence to indicate that comparative ecologists are particularly adept at doing so.

Recommendations

The body of this chapter argues that although comparative analyses may differ from other ecological approaches, these are differences in degree not in kind. The distinction becomes more intense as scope and aggregation increase, but no sharp break separates comparative ecology from the rest of the science. Given this continuity, the cautions and recommendations that can help comparative ecologists communicate their results better to their colleagues are not bizarre or novel. They are admonitions that apply to all scientists, but which should be held in special regard by those who depart from more traditional paths of research and knowledge, because more guidance is needed in less charted territory.

A. State both short- and long-term goals of the research in concrete terms by describing what the study intends to do.
B. Relate this goal to the interests of the proposed audience and to a realistic assessment of what the research achieved.

C. Distinguish between the heuristic and predictive applications of the analytical results.

D. Be wary of causal attribution, especially if only correlational evidence is available.

E. Acknowledge the limitations associated with poor controls, unassessed factors, surrogate variables, and heterogeneous data, but also stress the compensating advantages of comparative analyses: breadth, generality, economy and timeliness.

F. Recognize that many explanations can be advanced for any set of observations and warn the reader against overenthusiasm for any particular explanation of a comparative pattern.

G. Provide experimental support for comparative patterns if possible, and if not, suggest what experiments might be done.

H. Be suspicious of extrapolation beyond the domain of the data on which empirical patterns are based, but do not despair for the potential for prediction of the synoptic description before this potential is independently assessed.

I. Admit that transfers among scales may render existing models treacherous and that such transfers must be evaluated at the scale of application.

J. Be flexible in the methods, techniques, and ideas one uses in comparison, so that whatever method is used is the one most appropriate to one's purpose.

K. Accept that comparative ecology is only a methodological variant to be used where appropriate.

In short, comparative ecologists must acknowledge that scientific knowledge is hypothetical. Comparative methods are a set of powerful techniques to describe nature, but the knowledge they provide is nonetheless flawed and fallible. Simultaneous recognition of the strengths and weaknesses of any discipline requires a blend of hubris and humility that is characteristic of all good science. In the final analysis, that must be the goal of comparative ecology.

REFERENCES

Bell, G. (1985). The origin and early evolution of germ cells as illustrated by the Volvoccales. In: *The Origin and Evolution of Sex*, H.O. Halvorson and A. Monroy, eds. A.R. Liss, New York, pp. 221–256.

Blueweiss, L., H. Fox, V. Kudzma, D. Nakashima, R. Peters, and S. Sams. (1978). Relationships between body size and some life history parameters. *Oecologia* 37: 257–272.

Bradshaw, A.D. (1987). Comparison—its scope and limits. *New Phytol.* 106 (Suppl.):3–21.

Brown, J.W. (1989). Macroecology: the division of food and space among species on continents. *Science* 243:1145–1150.

Calder, W.A. (1984). *Size, Function, and Life History.* Harvard University Press, Cambridge.

Carpenter, S.R., ed. (1988). *Complex Interactions in Lake Communities.* Springer-Verlag, New York.

Clutton-Brock, T.H. and P.H. Harvey. (1979). Comparison and adaptation. *Proc. R. Soc. Lond. B* 205:547–565.

Downing, J.A. (1979). Aggregation, transformation and the design of benthos sampling programs. *Can. J. Fish. Aquat. Sci.* 36:1454–1463.

Dubé, P.-A. and J.E. Chevrette. (1978). Phenology applied to bioclimatic zonation in Quebec. *Bull. Agric. Exp. Stn., Univ. Vermont* 684:33–42.

Fretwell, S.D. (1987). Food chain dynamics: the central theory of ecology? *Oikos* 50:291–301.

Frost, T.M., D.L. DeAngelis, S.M. Bartell, D.J. Hall, and S.H. Hurlbert. (1988). Scale in the design and interpretation of aquatic community research. In: *Complex Interaction in Lake Ecosystems*, S.R. Carpenter, ed. Springer-Verlag, New York, pp. 229–258.

Gorham, E. (1979). Shoot height, weight and standing crop in relation to density in monospecific plant stands. *Nature* (London) 279:148–150.

Grime, J.P. (1965). Comparative experiments as a key to the ecology of flowering plants. *Ecology* 46:513–515.

Grime, J.P. (1977). Evidence for the existence of three primary strategies in plants and its relevance to ecological and evolutionary theory. *Am. Nat.* 111:1169–1194.

Grime, J.P., J.G. Hodgson, and R. Hunt. (1988). *Comparative Plant Ecology.* Unwin Hyman, Boston.

Hairston, N.G., F.E. Smith, and L.B. Slobodkin. (1960). Community structure, population control, and competition. *Am. Nat.* 94:421–425.

Hassell, M.P., T.R.E. Southwood, and P.M. Reader. (1987). The dynamics of the viburnum white fly (*Aleurotrachelus jelinekii*): a case study of population regulation. *J. Anim. Ecol.* 56:283–300.

Hurlbert, S.H. (1970). The post-larval migration of the red-spotted newt, *Notophthalamus viridescens* (Rafinesque). *Copeia* 1970:515–527.

Jarman, P.J. (1974). The social organization of antelope in relation to their ecology. *Behaviour* 48:215–267.

Kolasa, J. and E. Biesiadka. (1984). Diversity concept in ecology. *Acta Biotheor.* 33:145–162.

Kolasa, J. and S.T.A. Pickett. (1989). Ecological systems and the concept of biological organization. *Proc. Natl. Acad. Sci. USA* 86:8837–8841.

Lieth, H. and E.O. Box. (1972). Evapotranspiration and primary productivity. C. W. Thornthwaite Memorial Model. *Publ. Climatol.* 25:37–46.

Likens, G.E. (1985). An experimental approach for the study of ecosystems. *J. Ecol.* 73:381–396.

Lindeman, R.L. (1942). The trophic-dynamic aspect of ecology. *Ecology* 23:399–418.

McGurk, M.D. (1986). Natural mortality of marine pelagic fish eggs and larvae: role of spatial patchiness. *Mar. Ecol. Prog. Ser.* 34:227–242.

McNaughton, S.J., M. Oesterheld, D.A. Frank, and K.J. Williams. (1989). Ecosystem-level patterns of primary productivity and herbivory in terrestrial habitats. *Nature* (London) 341:142–144.

Mooney, H.A., ed. (1977). *Convergent Evolution in Chile and California: Mediterranean Climate Ecosystems*. Dowden, Hutchinson and Ross, Stroudsburg.

Odum, E.P. (1969). The strategy of ecosystem development. *Science* 164:262–270.

OECD. (1982). Monitoring of Inland Waters (Eutrophication Control). Synthesis Report. OECD, Paris.

O'Neill, R.V., D.L. DeAngelis, J.B. Waide, and T.F.H. Allen. (1986). *A Hierarchical Concept of Ecosystems*. Princeton University Press, Princeton.

Pace, M. (1986). An empirical analysis of zooplankton community size structure across lake trophic gradients. *Limnol. Oceanogr.* 31:45–55.

Pace, M.L., G.A. Knauer, D.M. Karl, and J.H. Martin. (1987). Primary production and vertical flux in the eastern Pacific Ocean. *Nature* (London) 325:803–804.

Peters, R.H. (1983). *The Ecological Implications of Body Size*. Cambridge University Press, Cambridge.

Peters, R.H. (1986). The role of prediction in limnology. *Limnol. Oceanogr.* 31:1143–59.

Peters, R.H. and F.H. Rigler. (1973). Phosphorus excretion by *Daphnia*. *Limnol. Oceanogr.* 18:821–39.

Reckhow, K.H. and J.T. Simpson. (1980). A procedure using modeling and error analysis for the prediction of lake phosphorus concentration from land use information. *Can. J. Fish. Aquat. Sci.* 37:1439–1448.

Robinson, W.R., R.H. Peters, and J. Zimmerman. (1983). The effects of body size and temperature on metabolic rate of organisms. *Can. J. Zool.* 61:281–288.

Schindler, D.W. (1987). Detecting ecosystem responses to anthropogenic stress. *Can. J. Fish. Aquat. Sci.* 44:6–25.

Smith, V.H., F.H. Rigler, O. Choulik, M. Diamond, S. Griesbach, and D. Skraba. (1984). Effects of phosphorus fertilization on phytoplankton biomass and phosphorus retention in subarctic Québec lakes. *Verh. Int. Verh. Limnol.* 22:376–382.

Smol, J.P. (1989). Paleolimnology—Recent advances and future challenges. *Mem. Ist. Ital. Idrobiol. Dott. Marco De Marchi* 47: (in press).

Waring, R.H. (1988). Characteristics of trees predisposed to die. *BioScience* 37:569–574.

Weller, D.E. (1989). The interspecific size-density relationship among crowded plant stands and its implications for the $-3/2$ power rule of self-thinning. *Am. Nat.* 133:20–41.

Yoda, K., T. Kira, H. Ogawa, and K. Hozumi. (1963). Self-thinning in overcrowded pure stands under cultivated and natural conditions (Intraspecific competition among higher plants XI). *J. Inst. Polytech. Osaka City Univ. Ser. D. Biol.* 14:107–129.

Part II   Productivity and Trophic Structures

5
Patterns of Primary Production and Herbivory in 25 North American Lake Ecosystems

STEPHEN R. CARPENTER, THOMAS M. FROST,
JAMES F. KITCHELL, TIMOTHY K. KRATZ,
DAVID W. SCHINDLER, JOHN SHEARER,
W. GARY SPRULES, MICHAEL J. VANNI,
AND ANN P. ZIMMERMAN

Abstract

The effects of nutrients and herbivory on phytoplankton biomass and production were examined, using data from 25 lakes studied for 2 to 6 years each. Variance among lakes was substantially greater than variance among years, for all physical, chemical, phytoplankton, and zooplankton variates studied. Experimentally manipulated lakes had coefficients of variation within the range exhibited by nonmanipulated lakes. Graphical, correlative, and regression analyses illustrated the significant joint effects of both nutrients and herbivory on phytoplankton biomass and production. A Bayesian analysis of sensitivity to new information showed that the statistical models for chlorophyll are quite robust. Statistical models for primary production were deemed less conclusive, because primary production was measured in fewer lakes. We provide a list of common challenges in comparative statistical analysis of ecosystems and explain their implications for our study. The major pattern apparent in our data—that summer chlorophyll responds positively to nutrients and negatively to herbivore size—is congruent with results of whole-lake experiments in which nutrients or predators were manipulated.

Introduction

Regulation of primary production has been considered from two perspectives. The physical/chemical paradigm views primary productivity as a consequence of such factors as climate, morphometry, and nutrient supply. Organic energy flows up the food chain and ultimately determines production and biomass of upper trophic levels. This perspective was formalized in Lindeman's trophic–dynamic concept (1942), which has broadly influenced research on all ecosystem types. The International Biological Program emphasized physicochemical controls and upward transport in food chains (Brylinsky and Mann 1973; LeCren and Lowe-McConnell 1981). In lakes, the comparative studies pioneered by Vollenweider (1968) and ecosystem experiments at the Experimental Lakes Area (Schindler 1988) firmly established the major role of phosphorus in lake productivity. The physical/chemical paradigm has dominated limnology for most of its history (Persson et al. 1988).

The predation paradigm derives from community experiments on keystone predators (Paine 1966) and trophic cascades (Paine 1980). In terrestrial ecosystems, numerous examples of herbivore effects on ecosystem processes have been documented (Naiman 1988). In lakes, predator effects were described by Hrbacek et al. (1961) and formalized in the size-efficiency hypothesis (Brooks and Dodson 1965), which has prompted a large body of research on pelagic communities (Stein et al. 1988). The physical/chemical and predator literatures mainly serve ecosystem and community limnologists, respectively, and until recently have been largely independent.

Recent debates about the physical/chemical and predator perspectives in limnology have centered on control of phytoplankton biomass and productivity. Shapiro et al. (1975) argued that removal of planktivorous fishes from lakes would increase herbivory and reduce algal biomass. This conjecture stimulated considerable research (Riemann and Sondergaard 1986; Kerfoot and Sih 1987; Carpenter 1988a; Northcote 1988). Predator effects were demonstrated by whole-lake experiments in which fish community manipulations had substantial effects on phytoplankton biomass and productivity (Henrikson et al. 1980; Shapiro and Wright 1984; Carpenter et al. 1987; Carpenter and Kitchell 1988).

Experimental evidence has shifted debate from the existence of predator effects on phytoplankton to the relative magnitudes and interactions of predator and physical/chemical effects. Model analyses indicate that physical/chemical and predation effects on phytoplankton act at different time scales (Carpenter and Kitchell 1987; Carpenter 1988b). Hypotheses of physical/chemical and predator control of primary producers are therefore complementary, not contradictory (Carpenter et al. 1985; Bartell et al. 1988; Carpenter and Kitchell 1988).

The time is right for a comparative analysis of predator and physical/chemical correlates of primary production in lakes. Correlations of physical/chemical and producer variates have been analyzed exhaustively (Nicholls and Dillon 1978), but only a few comparative studies have simultaneously examined physico-chemical, primary producer, and predator variates (Mills and Schiavone 1982; Pace 1984). Direct measurements of primary production were not available to these researchers, and relatively few physical/chemical and predator variates were measured. Most comparative studies have derived data from samples over less than 1 year. It may be important to examine multiyear data to assess the range of dynamics possible in each lake, and to compare interyear and interlake variability.

For this chapter, we amassed multiyear data from 25 lakes representing a wide trophic gradient and several of the major lake districts of North America. We considered more physical/chemical and herbivore variates than any previous comparative study. The data are capable of corroborating any of several competing models to explain phytoplankton biomass and productivity: purely physical/chemical models, purely predator-driven models, or models of joint predator-resource effects such as the trophic cascade model (Carpenter and Kitchell 1988). A major value of the study is the search for pattern across a wide range of lake types. This breadth provides a check against purely local or short-term patterns that might arise in a study of more limited scope. Secondarily, the study allowed us to compare measurements of the same variates by several different research groups in different locales.

Specific objectives of this chapter are to:

1. compare the relative magnitudes of interannual and interlake variances in physical, chemical, primary producer, and herbivore variates;
2. compare interannual variances of all variates in experimentally manipulated and nonmanipulated lakes;
3. compare variability and covariability of herbivores and primary producers, while also accounting for physical/chemical factors that may affect both trophic levels;
4. compare the capacities of physical/chemical factors, herbivores, and their joint action to forecast primary production.

Lakes and Limnological Methods

Between 2 and 6 years of data were available for each of the 25 lakes studied, a total of 90 lake-years (Table 5.1). Latitudes span nearly 7°. Geochemical environments range from the base-poor Canadian Shield to calcareous southern Wisconsin. Morphometries include shallow lakes that scarcely stratify; deep dimictic lakes with long fetch; and deep, sheltered lakes that mix no more than once per annum. Certain lakes have been subjected to experimental manipulations, including eutrophication (Schindler 1974; Shearer et al. 1987; Malley et al. 1988), acidification (Watras and Frost 1989), and fish community changes (Carpenter et al. 1987; Vanni et al., unpublished).

Variances, covariances, and correlations are notoriously scale dependent (Chatfield 1980). We chose the pelagic zones of lakes as a natural spatial scale. Pragmatic constraints forced some arbitrary decisions about temporal scale. We focused on the summer stratified season, defined as 1 June through 31 August. This decision imposed a uniform temporal scale that is an important one in the annual metabolism of north temperate lake ecosystems. During the summer period, sampling frequencies ranged from weekly

TABLE 5.1. Lakes compared in this study.[a]

Lake	Project	Years	ZBAR	ZMAX	NLAT
Paul	CTI	1984–1988	5.0	15.0	46°15'
Peter	CTI	1984–1988	8.3	19.3	46°15'
Tuesday	CTI	1984–1988	10.0	18.5	46°15'
226 NE Basin	ELA	1973, 1974, 1977, 1978	5.7	14.7	49°41'
226 SW Basin	ELA	1973, 1974, 1977, 1978	6.3	11.6	49°41'
227	ELA	1973, 1974	4.4	10	49°41'
221	ELA	1985–1987	2.1	5.7	49°41'
White	LEWG	1980–1981	4.3	9.7	44°50'
Blue-Chalk	LEWG	1980–1981	9.4	21.9	45°12'
Crosson	LEWG	1980–1981	8.1	23.5	45° 5'
Mountain	LEWG	1980–1981	13.4	31.4	45°59'
Plastic	LEWG	1980–1981	8.1	16.8	45°11'
Three-Mile	LEWG	1980–1981	3.5	11.0	45°10'
King	LEWG	1980–1981	2.6	5.8	45°53'
Ruth	LEWG	1980–1981	5.7	14.6	46° 1'
Young	LEWG	1980 1981	4.3	10.4	44°43'
Big Musky	LTER	1982–1987	7.5	21.4	46°
Crystal Bog	LTER	1982–1987	1.7	2.5	46°
Crystal	LTER	1982–1987	10.4	20.4	46°
Sparkling	LTER	1982–1987	10.9	20	46°
Trout	LTER	1982–1987	14.6	35.7	46°
Little Rock (acidified basin)	LRLEAP	1984–1987	3.9	10.3	46°
Little Rock (reference basin)	LRLEAP	1984–1987	3.1	6.5	46°
Mendota	MSN	1979, 1980, 1987, 1988	12.4	25.3	43° 4'
Wingra	MSN	1972, 1973	2.4	6.1	43° 4'

[a]For each lake, the project, years analyzed in this study, maximum depth (ZMAX, m), mean depth (ZBAR, m), and north latitude (NLAT) are presented. See text for explanation of whole-ecosystem experiments in certain lakes. Project codes: CTI, Cascading Trophic Interactions; ELA, Experimental Lakes Area; LEWG, Lake Ecosystems Working Group, University of Toronto; LTER, Northern Lakes Long-Term Ecological Research; LRLEAP, Little Rock Lake Experimental Acidification Program; MSN, Madison, Wisconsin, lakes.

to monthly, so each value in the data set derived from 3 to 14 sampling events.

We made no formal attempt to cross-calibrate methods. However, protocols followed by the different laboratories were very similar, in part because of long-standing interactions among the research groups. Three of the projects (Cascading Trophic Interactions, Northern Lakes LTER, Little Rock Lake Experimental Acidification Program) have already compared their methods explicitly (Carpenter et al. 1989). Because each laboratory has already published its methods in detail, only brief summaries are presented here. We elaborate only in cases in which different research groups followed different procedures.

SAMPLING PROCEDURES

All lakes were sampled at central, deep-water stations. Lakes in the Cascading Trophic Interactions (CTI) project were sampled weekly. Lakes at the Experimental Lakes Area (ELA) were sampled every 1 or 2 weeks. All variates were available for all years from both CTI and ELA lakes. Lakes studied by the Lake Ecosystem Working Group (LEWG) were sampled every 2 to 4 weeks. All variates except primary production and cladoceran biomasses and lengths were measured in the LEWG lakes. Cladocerans were included in total crustacean biomasses and lengths from LEWG systems. Long-Term Ecological Research (LTER) program lakes were sampled biweekly. All variates were available, with these exceptions: nutrients, 1982–1985; primary production in Big Muskellunge and Crystal Bog Lakes, and in other lakes, 1982–1984; and zooplankton in 1982 and 1987. Both basins of Little Rock Lake were sampled biweekly. All variates were available from the Little Rock Lake Experimental Acidification Project (LRLEAP) except for primary production in 1984 and 1985. Lake Mendota was sampled weekly. All variates except primary production were determined. Lake Wingra was sampled weekly or biweekly. All variates except mean length for all zooplankton were available for Lake Wingra.

PHYSICAL VARIATES

Lake mean depth and its ratio to maximum depth can be correlated with primary production (Fee 1979; Carpenter 1983). Thermocline depth was included as an index of mixing and photic zone depth as an index of light penetration. Mean epilimnetic temperatures and north latitudes were also included in the data set as indicators of climate regime.

The inflection points of temperature profiles were determined by eye and assumed equal to thermocline depth. Photic zone depth was taken as the depth of 1% surface irradiance, determined using submersible quantum sensors. In Lake Wingra only, Secchi depth was converted to photic zone depths using empirical ratios calculated from unpublished data of S.R. Carpenter and extinction coefficients reported by Titus (1977). In some lakes, irradiance exceeded 1% at the maximum depth, so photic zone depth was equal to maximum depth.

CHEMICAL VARIATES

Total nitrogen and phosphorus were measured on dates as near as possible to spring overturn. Spring concentrations are close correlates of annual nutrient loading (Dillon and Rigler 1974, 1975). During summer, virtually all N and P are in algal biomass or soon will be, so total N, total P, and chlorophyll measure the same thing. This circularity makes it impossible to compare food chain and nutrient effects on algae using summer nutrient data. In any event, spring total P and annual mean total P are highly correlated ($r = .96$; Nicholls and Dillon 1978). Spring total P probably underrepresents loading to Lakes 226 NE and 227, where phosphorus was added continually through the summer (Schindler et al. 1978; Shearer et al. 1987).

Nutrient concentrations were determined on integrated water column samples (LEWG) or on samples pooled from several depths (CTI, ELA, LTER, LRLEAP, MSN). Similar methods were used by all laboratories (Prentki et al. 1977; Stainton et al. 1977; Zimmerman et al. 1983; Brock 1985; Elser et al. 1988).

PRODUCER VARIATES

Chlorophyll *a* concentration, corrected for pheopigments, was determined as a measure of algal biomass. Concentrations were determined on integrated water column samples (ELA in 1977 and later years, LEWG, LTER, LRLEAP, Lake Mendota 1987-1988) or on samples from several depths (CTI, ELA 1973-1974, Lake Wingra, Lake Mendota 1979-1980). Several extraction procedures were used, and samples were read on either spectrophotometers or calibrated fluorometers (Koonce 1972; Prentki et al. 1977; Stainton et al. 1977; Zimmerman et al. 1983; Brock 1985; Carpenter et al. 1986). Although different extraction procedures may increase error variance in our data (Stainton et al. 1977; Nicholls and Dillon 1978; Carpenter et al. 1986), the fact that all laboratories calibrated their procedures using commercial chlorophyll standards suggests that the data are comparable.

Primary production was determined using radiocarbon incubations and integrated across depths and time using numerical models that accounted for vertical inhomogeneities and fluctuations in irradiance (Fee 1973). Similar procedures were used at each site, except that incubators were used at ELA, LTER, and LRLEAP while in situ profiles were used at CTI and Lake Wingra (Koonce 1972; Prentki et al. 1977; Shearer et al. 1985; Carpenter et al. 1986). Possible errors in these measurements have been considered by Bower et al. (1987). When used with numerical models, both in situ and incubator profiles yielded production estimates close to those obtained by whole-lake radiocarbon addition (Bower et al. 1987). All production values in this chapter were derived from similar numerical models and are therefore deemed comparable.

HERBIVORE VARIATES

Biomasses and lengths of herbivores were determined in three categories: all zooplankton, all crustaceans, and all cladocerans. Length was included because of the established relationships of particle size distributions and pelagic system function (Sprules and Munawar 1986) and because grazing rates are size dependent (Peters and Downing 1984). Size-selective predation (Brooks and Dodson 1965) and size-dependent nutrient recycling are central in food chain effects on phytoplankton (Kitchell et al. 1979; Carpenter and Kitchell 1984, 1988). Crustacean herbivores, especially the large cladocerans, are pivotal in food chain responses (Carpenter et al. 1987; Carpenter and Kitchell 1988).

Zooplankton were sampled using vertical net hauls in CTI lakes, LEWG lakes, and Lake Mendota in 1979 and 1980 (Sprules et al. 1983; Brock 1985; Carpenter et al. 1987). Metered nets were used in LEWG lakes (Sprules et al. 1983). In other lakes, net efficiencies were determined by comparing vertical hauls with profiles obtained by Schindler traps or Van Dorn bottles. In Lake Mendota in 1987 and 1988, zooplankton were collected by both vertical hauls and horizontal tows of a metered Clarke–Bumpus sampler (C. Luecke and M. Vanni, unpublished data). In ELA, LTER, Little Rock, and Wingra Lakes, vertical profiles of zooplankton were collected using Schindler traps (Bartell and Kitchell 1978; Chang et al. 1980, 1984; Frost and Montz 1988; Malley et al. 1988). In Lake Wingra, the crustacean densities and lengths reported by Bartell and Kitchell (1978) were converted to biomass using the regressions of Peters and Downing (1984). In Lake Mendota, biomass were taken from Pedros-Alio (1981) for 1979–1980. For 1987–1988, biomass were calculated from lengths using the regressions of Downing and Rigler (1984) and Lynch et al. (1986). Biomass calculation procedures for CTI, ELA, and LEWG lakes are given by Carpenter et al. (1987), Malley et al. (1988), and Sprules et al. (1983). For LTER and LRLEAP lakes, length–weight regressions were measured directly (T. Frost, unpublished data) or taken from Downing and Rigler (1984).

Statistical Approach and Methods

COMPARISONS OF VARIABILITY

One-way ANOVA (analysis of variance) was used to compare interlake variance with that among years within lakes: lakes were used as categories, years as "replicates." We do not imply that years are the proper source of error variance for interlake comparison and attempt no significance tests; ANOVA simply provides a convenient framework for calculating and comparing the variances. These variances are distinctly different from those used by Kratz et al. (1987), who obtained estimates of interlake and interyear variance from two-way ANOVA of data completely cross-classified by lake and year. In our data, cross-classification does not obtain, and an independent estimator of the interannual variance is not possible.

Coefficients of variation (CV, standard deviation/mean) were used to compare variability among lakes and among variates. Because the CV is dimensionless, comparisons among variates do not depend on the units. For comparisons of manipulated and nonmanipulated lakes, we regarded Peter, Tuesday, 226 NE basin, and Little Rock acidified basin as manipulated systems. For the purposes of this analysis, ELA lakes 227 and 221 were not regarded as manipulated lakes because the years we analyzed do not include transient dynamics induced by perturbation.

COVARIATION

We examined covariation graphically and employed significance tests selectively. On the basis of analysis of residuals from ANOVA, common log transformations were applied to total N and P, producer variates, and zooplankton biomasses. Any attempt to dissect this data set by exhaustive significance tests on correlation coefficients is futile. Among the 18 basic variates, there are 153 pairwise correlations. Using the Bonferroni criterion for multiple tests, we would be forced to use a nominal significance level of .0003 to maintain an experiment-specific error rate of .05. We avoided this restrictive criterion by minimizing significance tests.

Because years may be autocorrelated within lakes, correlations based on lake-years may have inflated degrees of freedom. Therefore, we used averages of all years for each lake in correlations tested for significance so that each lake contributed only one observation to the analysis.

MODEL COMPARISONS

We used multiple regression to compare three models for each of four dependent variates (epilimnetic chlorophyll and primary production; photic zone chlorophyll and primary production). The pure physical/chemical model used mean depth, thermocline depth, temperature, total N, and total P as predictors; the pure herbivory model used lengths and biomasses of total zooplankton, crustaceans, and cladocerans as predictors; and the mixed model used both sets of predictors. Stepwise multiple regression (entry and removal probabilities = .15) was used to remove redundant predictors and yield final models containing similar numbers of predictor variates.

Because not all variates were measured in every lake every year, certain lake-years can be used for some models but not all. When all possible lake-years are analyzed, the number of observations differs among models and could affect the interpretation of results. To check this possibility, we analyzed two versions of the data set. In the first version (Maximum N), all possible lake-years were used to fit each model. In the second version (Equal N), lake-years with missing data were eliminated, so that N was identical for each model.

There are important restrictions on the interpretation of stepwise multiple regressions using nonexperimental, colinear data (Box 1966; Box et al. 1978; Draper and Smith 1981; Hocking 1983). First, the P values employed in stepwise regression cannot be interpreted as probabilities of type 1 error (Draper and Smith 1981; Hocking 1983). Second, omission of a predictor from a stepwise regression equation does not imply that the predictor is unrelated to the dependent variate (Box et al. 1978). More often than not, predictors omitted from stepwise regression equations simply have high correlations with other predictors that happened to be included in the equations.

Because of these restrictions, we avoided significance tests and attached no importance to which predictors ended up in the models. Our analysis rests on comparison of models that contain different subsets of predictors prior to regression. The best model will have the largest R^2 and the smallest error mean square. The error standard deviation S (square root of the error mean square) is the average deviation of the data from the fitted model. Because our dependent variates were log transformed, error factors of 10^S equal the average ratios of observations and predictions.

Error variances and R^2 values compare models quantitatively but do not tell us if the differences among models are so great that we should prefer one model and exclude the others. To address this question, we considered the sensitivity of the models to new data. Suppose that the ranking of models does not change when we add a new data point that is inconsistent with the best model but consistent with the other models. Then we have strong evidence in favor of the best model, and many observations inconsistent with it will be needed to change our minds. Conversely, suppose that the ranking of models changes when we add a new data point that is inconsistent with the best model but consistent with the other models. Then we have little confidence in the ranking of models, because a modest amount of new data could change the ranking.

The sensitivity of the model comparison to new data can be determined using Bayesian statistics (Walters 1986). The probability that model M_i is correct, given our data Y_1 and a hypothetical new data point Y_2 is

$$P(M_i \mid Y_1, Y_2) = L(M_i \mid Y_1) * L(M_i \mid Y_2) / T \qquad [1]$$

where L is the likelihood function and T is simply the sum of the $L(M_i \mid Y_1) * L(M_i \mid Y_2)$ over all models (Box and Tiao 1973). Equation 1 assumes that all models had equal probabilities of being correct before data set Y_1 was analyzed, and that the standardized likelihood measures the probability that model M_i is correct after analyzing Y_1 and before analyzing Y_2 (Box and Tiao 1973). The likelihood function is calculated from the residuals of the regressions using the normal function (Box and Tiao 1973). For residuals centered around zero, the likelihood is

$$L = exp(-N/2) / [(2 \pi S^2)^{0.5}]^N \qquad [2]$$

where N is the number of observations and S is the standard deviation of the residuals. The likelihood decreases steeply and nonlinearly as the standard deviation of the residuals increases (Fig. 5.1). L also depends on N. In the range of our data (S between 0.18 and 0.42; N between 35 and 75), L is much more sensitive to variation in S than it is to variation in N.

To determine the sensitivity of our analysis to new information, we calculated probabilities of each model (Equation 1), given a hypothetical new

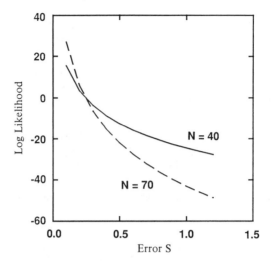

FIGURE 5.1. Natural log of likelihood function versus error standard deviation for models fit to 40 (solid line) or 70 (dashed line) observations.

observation that deviated from the best model but fit both of the other models. The deviation from the best model ranged from 0 to 3 SD. This procedure is a severe challenge to the best-fitting model, because an actual new observation would probably deviate from all models to some degree.

Results

INTERLAKE AND INTERYEAR VARIATION

For all variates measured, interlake variance was substantially greater than variance among years (Table 5.2). Many of the distributions were skewed, with the mean falling nearer the minimum value than the maximum value. In the case of TN, TP, algal variates, and zooplankton biomasses, common log transformation normalized distributions in subsequent analyses.

Ranges of variates in our data set were similar to those reported in other comparative studies (Table 5.3). For lognormally distributed variates, the ratio of the maximum to the minimum may be more appropriate than the range as a basis for comparison. The range of total P in our data exceeds that in the 12 phosphorus-chlorophyll studies reviewed by Nichols and Dillon (1978) and that in the studies by Mills and Schiavone (1982) and Pace (1984). The studies by Schindler (1978) and Canfield and Bachmann (1981) included extremely eutrophic systems with total P and chlorophyll values that exceeded the range of our data. The eutrophic Balkan fishponds where predator control of phytoplankton was first described also have extremely high total P and chlorophyll concentrations. Schindler (1978) found 150-

TABLE 5.2. Variables studied with acronyms, ranges, means, and ratios of interlake to interyear variance.[a]

Variable	Acronym	Minimum	Maximum	Mean	Variance ratio
Thermocline depth	THERM	1.6	11.5	5.6	64.3
Photic zone depth	PHOT	1.4	26.8	9.0	90.2
Epilimnion temperature	TEMP	16.5	26.0	21.0	3.85
Spring total nitrogen	TN	50	1,440	400	9.84
Spring total phosphorus	TP	1.0	173	19.4	9.45
Epilimnion chlorophyll	EPICHL	0.9	84.1	8.36	24.5
Photic zone chlorophyll	PHOCHL	0.8	87.4	9.54	18.9
Epilimnion primary production	EPIPPR	25.1	986	182	11.7
Photic zone primary production	PHOPPR	20.3	1,055	133	23.9
Zooplankton biomass	TZBIO	61	11,190	2,002	9.17
Crustacean biomass	CRUSBIO	4.9	10,721	1,829	10.2
Cladoceran biomass	CLADBIO	0.68	5,430	833	10.9
Zooplankton length	TZLEN	0.100	0.635	0.244	10.1
Crustacean length	CRUSLEN	0.197	1.38	0.48	6.86
Cladoceran length	CLADLEN	0.27	1.43	0.75	6.62

[a]Variance ratios were calculated from one-way ANOVAs that yielded mean squares among lakes and among years within lakes. Units: depths, m; temperature, °C; nutrients and chlorophyll, $\mu g/l$; primary production, mg C m^{-3} d^{-1}; herbivore biomasses as dry weights, mg m^{-2}; herbivore lengths, mm.

TABLE 5.3. Ranges of total phosphorus concentration ($\mu g/l$), chlorophyll concentration ($\mu g/l$), zooplankton biomass (dry mass, $\mu g/l$), and zooplankton length (mm) from selected multilake comparison studies.

Total P	Chlorophyll	Zooplankton Biomass	Zooplankton Length	Reference
5–85	1–85	—	—	Nichols and Dillon 1978[a]
3–316	0.2–200			Schindler 1978[b]
5–1,200	0.8–500	—	—	Canfield and Bachman 1981
—	0.6–158	9–3,800	—	McCauley and Kalff 1981
8–105	2–110	46–790	0.5–1.5	Mills and Schiavone 1982
3–56	1–29	66–373	0.37–0.86	Pace 1984[c]

[a]Based on 12 separate multilake comparisons involving 7 to 143 lakes each.
[b]Range of annual primary production: 3.2–500 g C m^{-2} yr^{-1}.
[c]Zooplankton length based on cladocerans only.

fold variation in annual primary production for 60 lakes. For summer only, our data exhibit 500-fold variation. The 180-fold range of zooplankton biomass in our data is less than the 420-fold range reported by McCauley and Kalff (1981) but greater than the ranges reported by Mills and Schiavone (1982) and Pace (1984).

Physical variates and zooplankton lengths had relatively low coefficients of variation, while algal variates and zooplankton biomass had relatively high CV (Fig. 5.2). Nutrients were intermediate in variability. In most of the lakes, TN and TP behaved as physical variates, with CV < 0.3. In a few lakes, however, TN and TP were as variable as biotic variates.

Manipulated lakes were not distinctly more variable from year to year than nonmanipulated lakes (Fig. 5.2). Because of transient dynamics induced by manipulation, the manipulated lakes had somewhat larger than average CV, but the values are within the range exhibited by nonmanipulated lakes.

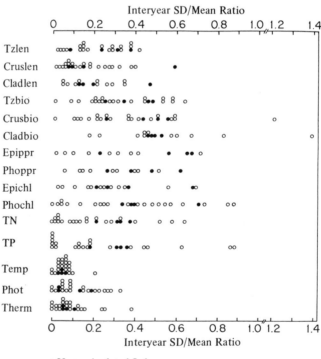

○ Unmanipulated Lake
● Manipulated Lake

FIGURE 5.2. Interannual coefficient of variation (standard deviation to mean ratio) for each variate (see Table 5.2 for acronyms). Each dot represents a lake; solid dots are manipulated lakes.

The largest manipulation effects were seen in Lake 226, eutrophied by successive additions of P, N, and inorganic C (Schindler 1974), and in Tuesday Lake, where planktivorous fishes were replaced by piscivorous ones (Carpenter and Kitchell 1988). Algal responses in Lake 226 and Tuesday Lake were about the same (Figure 5.3). In Lake 226, nutrient buildup from successive enrichments had substantial effects on chlorophyll and primary production, independent of changes in herbivore biomass. In Tuesday Lake, increased herbivore biomass in the second year (after planktivore removal) caused the major reduction in primary production, and the system then appeared to stabilize. In Tuesday Lake, spring total P actually increased during the study, from about 5 μg/l premanipulation to 14–18 μg/l post manipulation.

PATTERNS OF COVARIATION

Strong positive relationships were evident between nutrients and algal variates, consistent with many comparative studies (see Table 5.3) and whole-lake experiments (Schindler 1988). Spring total P had the strongest relationships and was the only physical/chemical variate with obvious relationships to all algal variates. For example, photic zone chlorophyll and total P are highly correlated (Fig. 5.4; $r = .636$, $p < .001$). The extreme lakes in Fig. 5.4 are not experimental ones. The outlier on the far left of Fig. 5.4 is Lake 221, which has very high chlorophyll concentrations for its phosphorus level. Planktivorous fishes are abundant in Lake 221 and may explain its deviation from the scatterplot of other lakes.

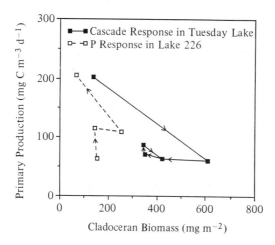

FIGURE 5.3. Trajectories of response to manipulation in Tuesday Lake (fish manipulation caused increased herbivory and decreased primary production; solid symbols and lines) and Lake 226 (nutrient addition caused increased primary production; open symbols and lines).

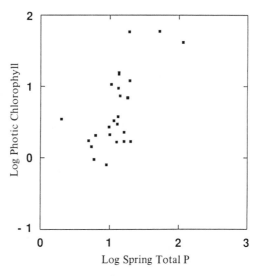

FIGURE 5.4. Log photic zone chlorophyll concentration during summer (μg/l) versus log total P concentration (μg/l) near spring overturn.

Negative relationships occurred between herbivore and algal variates, consistent with whole-lake experiments (Carpenter and Kitchell 1988) and some comparative studies (Mills and Schiavone 1982; Pace 1984). In contrast to Pace (1984), we found negative correlations between zooplankton biomass and algal variates. Because the sign and magnitude of correlations between herbivore and algal biomass are functions of the sampling interval (Carpenter and Kitchell 1987, 1988), we attach no biological significance to this difference. It is more productive to search for correlations of algal variates with herbivore size or partial correlations between algal and herbivore variates after nutrient effects have been removed (Pace 1984; Carpenter and Kitchell 1988).

Relationships between algal variates and herbivore size were not affected by the units of the algal variates (Fig. 5.5). For example, one might expect biomass or production per unit area to be less sensitive to herbivory than biomass or production per unit volume, because photic zone thickness is partially controlled by algal biomass. However, areal photic zone chlorophyll is also negatively related to zooplankton length (Fig. 5.5B; $r = -.635, p < .001$). Relativizing algal variates to total P also had little effect on covariation (Fig. 5.5C; $r = -.708, p < .001$). The extreme points in Fig. 5.5 are not manipulated lakes.

Partial correlations of herbivore and algal variates (with ZBAR, THERM, TEMP, TN, and TP effects removed) were mostly large and negative (Table 5.4). Despite the very restrictive Bonferroni criteria, seven correlations are significant, all of them negative. Photic zone chlorophyll and primary production have about the same correlations with herbivore variates, whether they are expressed on volumetric or areal bases.

The duality of control implied by these data is best visualized in three dimensions (Fig. 5.6). High chlorophyll is associated with high total P and small zooplankton. Low chlorophyll occurs with low total P and small zooplankton, or with high total P and large zooplankton. While total P and zooplankton length are not strictly orthogonal, they have no significant relationship ($r = -.135, p = .550$).

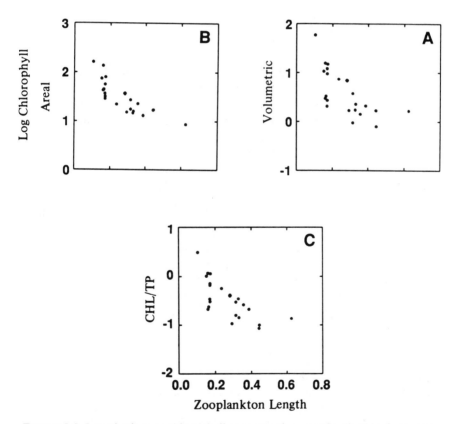

FIGURE 5.5. Log photic zone chlorophyll concentration, standardized in three ways, versus mean zooplankton length: (A) volumetric concentration (μg/l); (B) areal density (mg m^{-2}); (C) ratio of volumetric chlorophyll concentration to volumetric total P concentration.

TABLE 5.4. Partial correlations of herbivore and primary producer variates after effects of physical/chemical variates are removed.[a]

ALGAL VAR	TZBIO	CRUSBIO	CLADBIO	TZLEN	CRUSLEN	CLADLEN
EPICHL	−.631	−.784[b]	−.876	−.656	.088	−.301
PHOCHL	−.652[b]	−.654[b]	−.548	−.867[c]	−.707	−.477
PHOCHL*	−.647	−.626	−.613	−.802	−.700	−.343
EPIPPR	−.637	−.730	−.764[d]	−.732	−.674	.394
PHOPPR	−.627	−.685	−.713	−.852[b]	−.791[d]	.027
PHOPPR*	−.794	−.858	−.905	−.633	−.602	.117

[a]All algal variates and zooplankton biomasses were log transformed before analysis. Variable names are given in Table 5.2. PHOCHL* and PHOPPR* are photic zone chlorophyll and primary production, respectively, converted to areal units (mg m^{-2} and mg C m^{-2} d^{-1}, respectively). Significance tests were performed for volumetric algal variates, and p values were corrected by the Bonferroni criterion for multiple tests.
[b]$p < 0.01$.
[c]$p < 0.001$.
[d]$p < 0.05$.

COMPARISON OF REGRESSION MODELS

Satisfactory multiple regressions were obtained for all models and all dependent variates. Residual plots and tests for overly influential observations (Draper and Smith 1981) revealed no obvious problems with the fits.

The mixed model (both physicochemical and herbivore variates) fit the data best for all dependent variates, using both Maximum N and Equal N data sets (Tables 5.5, 5.6). R^2 was consistently highest, and the error SD consistently lowest, for the mixed model. Variance explained by regression was highest for epilimnetic and photic zone chlorophyll, and photic zone primary production was highest using the Maximum N data set. Less variance was explained by regression for epilimnetic primary production and photic zone primary production using the Equal N data set.

Regression equations are best compared on the basis of how large an error one can expect from them (Draper and Smith 1981). When the dependent variate is log transformed, the error factor 10^S (where S is the error SD) relates the regression errors to the original units. For example, an error factor of 2 means that we expect average errors of twice or half of the predictions.

Error factors for our regressions range from about 1.2 to 2.6 (Fig. 5.7). The error factor for the mixed model is distinctly lower than those of the other models for chlorophyll in both data sets. The model error factors are more similar for the primary production regressions.

Analysis for sensitivity to new information followed two distinctly different patterns. We present one example of each pattern (Fig. 5.8). Analyses that would be unaffected even by highly aberrant new data are exemplified by photic zone chlorophyll from the Maximum N data set (Fig. 5.8A). Over the full range of deviations studied, the mixed model has the highest probability of being correct by a wide margin. All the chlorophyll analyses were virtually identical: the mixed model was superior by a wide margin, for all deviations considered. Analyses that were sensitive to new data are exemplified by epilimnetic production from the Equal N data set (Fig. 5.8B). The model probabilities of being correct change substantially in the presence of new data that deviate from the best model. All the productivity analyses showed similar sensitivity to new information.

Discussion

We have demonstrated the value of comparative analyses at two distinct levels. On one level, cross-system comparisons revealed new patterns in the relationships among nutrients, algae, and herbivores that have significant

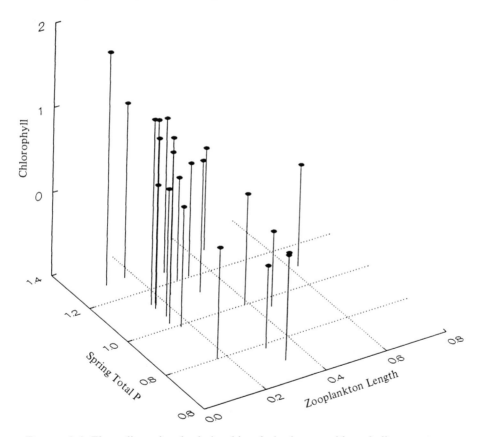

FIGURE 5.6. Three-dimensional relationship of photic zone chlorophyll concentration during summer (μg/l), total phosphorus concentration near spring overturn (μg/l), and zooplankton length during summer (mm). Note common log scales for chlorophyll and phosphorus.

TABLE 5.5. Comparison of physical/chemical, herbivore, and mixed (i.e., physical/chemical plus herbivore) models for algal variates based on all possible lake-years.[a]

Dependent variable	Model	Independent variables	N	R^2	Error SD
EPICHL	PHYS/CHEM	ZBAR, THERM, TN, TP	52	0.732	0.259
	HERBIVORE	CRUSBIO, CLADBIO, CLADLEN	57	0.687	0.222
	MIXED	THERM, TEMP, TN, TZBIO, CRUSBIO, CLADLEN	41	0.842	0.179
PHOCHL	PHYS/CHEM	THERM, TN, TP	68	0.396	0.406
	HERBIVORE	TZBIO, TZLEN	75	0.482	0.326
	MIXED	ZBAR, THERM, TZLEN	75	0.678	0.259
EPIPPR	PHYS/CHEM	THERM, TN, TP	40	0.434	0.288
	HERBIVORE	CRUSBIO, TZLEN, CLADLEN	37	0.406	0.276
	MIXED	THERM, TZBIO, CRUSBIO, CLADLEN	39	0.523	0.271
PHOPPR	PHYS/CHEM	THERM, TEMP, TN, TP	40	0.638	0.232
	HERBIVORE	TZBIO, CRUSBIO	39	0.582	0.249
	MIXED	TEMP, TP, CLADBIO, CLADLEN	37	0.721	0.211

[a]Algal and zooplankton biomass variates were log transformed. Candidate predictors were selected to maximize N, in view of missing values. Data were then subjected to stepwise multiple regression. Independent variates in the final model are presented: N (number of lake years), R^2 (coefficient of determination), and error standard deviation (square root of the error mean square). Variable names are explained in Table 5.2.

TABLE 5.6. Comparison of physical/chemical, herbivore, and mixed (i.e., physical/chemical plus herbivore) models for algal variates.[a]

Dependent variable	Model	Independent variables	N	R^2	Error SD
EPICHL	PHYS/CHEM	THERM, TEMP, TN	41	0.729	0.224
	HERBIVORE	CRUSBIO, CLADBIO, CLADLEN	41	0.715	0.230
	MIXED	THERM, TEMP, TN, TZBIO, CRUSBIO, CLADLEN	41	0.842	0.179
PHOCHL	PHYS/CHEM	THERM, ZBAR	59	0.243	0.421
	HERBIVORE	CRUSBIO, TZLEN	59	0.525	0.333
	MIXED	ZBAR, THERM, TZLEN	59	0.706	0.264
EPIPPR	PHYS/CHEM	THERM, TN	35	0.309	0.298
	HERBIVORE	CRUSBIO	35	0.348	0.284
	MIXED	THERM, TZBIO, CRUSBIO, CLADLEN	35	0.431	0.279
PHOPPR	PHYS/CHEM	THERM, TN	35	0.360	0.246
	HERBIVORE	CRUSBIO, CLADLEN	35	0.462	0.226
	MIXED	TEMP, TP, CRUSBIO, CLADLEN	35	0.578	0.204

[a]Lake-years for analysis were limited to those for which all predictors had been measured, so that N was the same for all models for each dependent variate. Otherwise, analysis was same as that presented in Table 5.4.

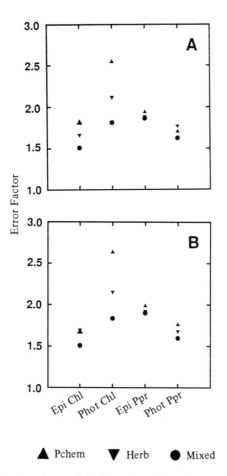

FIGURE 5.7. Error factors for models fit to four dependent variates for (A) maximum N and (B) equal N data sets. Dependent variates are eplimnetic chlorophyll (EPI CHL), photic zone chlorophyll (PHOT CHL), epilimnetic primary production (EPI PPR), and photic zone primary production (PHOT PPR). Symbols: upward-pointing triangle, physical/chemical model; downward-pointing triangle, herbivore model; circle, mixed model. See Tables 5.5 and 5.6 for further information about models.

implications for functioning of lake ecosystems. On another level, our analysis synthesized and reconciled whole-lake experiments with comparative limnology. Responses of lakes to nutrient and food chain perturbations are consistent when viewed as trajectories of similar systems in different dimensions. The comparative approach places results of whole-system experiments in the more general context of natural systems. The broad scope of natural interlake variability exceeds both interyear variability and variability induced by ecosystem experimentation. Ecosystem experiments are

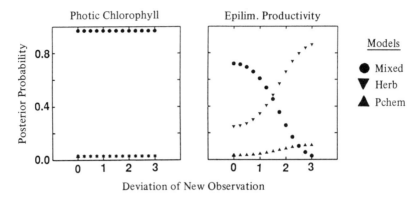

FIGURE 5.8. Sensitivity of model comparisons for photic zone chlorophyll and epilimnetic productivity to new information. Posterior probability of each model is plotted versus deviation of hypothetical new observation from mixed model. Deviation is expressed as multiples of error SD. Symbols as in Figure 7.7. Stars denote points where symbols for physical/chemical and herbivore models overlap.

effective for studying responses to perturbation and establishing dynamic mechanisms (Likens 1985; Walters 1986; Carpenter 1988a; Schindler 1988). They avoid the statistical difficulties of comparative studies (see following), while posing their own unique statistical challenges (Walters 1986; Carpenter 1988a; Carpenter et al. 1989). We have found ecosystem experiments and comparisons to be complementary, mutually reinforcing approaches.

Our remaining points are in two categories. First, we discuss statistical issues with general implications for comparative ecosystem studies. Second, we discuss ecological conclusions germane to the functioning of lake ecosystems.

STATISTICAL EVALUATION

Some statisticians' feelings about multiple regression are similar to those of some physicists about the atomic bomb: now that they see how people are using it, they wish they had never thought of it. Wilkinson (1988, p. 481) commented that "stepwise regression is probably the most abused computerized statistical technique ever devised." Box (1966; Box et al. 1978) presented powerful arguments against regression analysis of nonexperimental data and concluded that "to find out what happens to a system when you interfere with it, you have to interfere with it (not just passively observe it)." Hocking (1983) cautioned that "fitting of equations to observational data (as opposed to data from carefully designed experiments) is, at best, a risky business." Hocking (1983) and Box (1980) noted that Bayesian approaches may be more satisfactory.

In view of these warnings, it is wise to assess our study in light of a list of common problems with regression studies of nonexperimental data (Table 5.7). We believe that our methods are sufficiently consistent for a valid analysis. The range of predictors is similar to the range known from temperate lakes. Semiconfounding (or multicolinearity) is present, but we avoided its effects by (a) avoiding comparative significance tests of predictors and (b) ignoring which predictors entered the multiple regressions. Nonsense correlation is not an issue, because our variates were selected with specific questions in mind, and other studies have linked them mechanistically to algal biomass and productivity in lakes. Serial correlation among years probably occurs within lakes, but we avoided its effects on significance tests by averaging each lake's data over all years.

Items 6 and 7 (Table 5.7) cause scale effects in variances, covariances, correlations, and regression coefficients: their magnitudes or signs change as a function of the time span over which data are averaged (Chatfield 1980; Carpenter and Kitchell 1987, 1988; Carpenter 1988b). Our results are specific to the scale chosen: the summer stratified season. This choice is easily justified. It is an important time period for ecosystem metabolism and for human usage and management problems. Physical/chemical factors clearly have strong effects at that scale. Theory and experiments also suggest substantial food web effects on algae during summer stratification (Carpenter and Kitchell 1987, 1988). In eutrophic lakes, herbivore influence may be greater during the spring clear-water phase than in summer (Sommer et al. 1986). Thus, our analyses of summer data may underestimate the magnitude of herbivore effects during the ice-free season, or on an annual basis.

The practical consequence of these caveats (Table 5.7) is that most ecological regression studies are asymmetric in the following sense. We assume that items 1, 2, 4, and 5 can be eliminated by judicious data selections.

TABLE 5.7. Difficulties in regression analysis of nonexperimental data.[a]

1. Inconsistent data: the data are influenced by systematic differences in methods used at different locales or times.
2. Limited range of predictor variables: the range of one or more predictors in the data set is much less than the natural or typical range; the effects of these predictors are therefore underestimated.
3. Semiconfounding of predictor variables: two or more predictors are colinear; their independent effects cannot be assessed.
4. Nonsense correlation: two variates that are not causally related have a high correlation because a third, unmeasured variate affects them both.
5. Serially correlated errors: observations from nearby locales or times are not independent; the degrees of freedom are fewer than they appear to be.
6. Dynamic relationships: recent but unknown historical events affect the apparent relationships among variates.
7. Feedbacks: correlations among variates are dictated by compensatory interactions and reveal little or nothing about the mechanistic relationships.

[a]Source: Box et al. 1978.

Then positive results (significant correlations or convincing scatterplots) imply some underlying relationship that may have predictive value. In contrast, negative results (nonsignificant correlations, variates omitted from multiple regressions, shotgun scatterplots) reveal nothing about potential relationships among the variates. If orthogonal data were collected (item 3) or the samples were scaled appropriately (items 6 and 7), significant correlations might be obtained for the same variates. Because negative results are uninformative, multiple regression alone cannot be used to compare competing hypotheses or models.

Our approach does compare competing hypotheses by carefully considering variates and specifying models before analysis, as advocated by Box (1980), Hocking (1983), Walters (1986), and Wilkinson (1988). Multiple alternative models may be a fruitful and powerful approach for future comparative studies. In many cases, such models can be compared adequately using R^2 or error mean squares. The sequential Bayesian approaches advocated by Box and Tiao (1973), Box (1980), and Walters (1986) provide substantially enhanced model comparisons. Bayesian methods can also determine what experiments can discriminate rival models with the fewest measurements (Walters 1986). Another strength of the Bayesian framework is its capacity to incorporate new information in sequential research programs (Box and Tiao 1973; Walters 1986). A disadvantage of Bayesian analysis is the requirement that each model be assigned a probability of being correct before analysis. However, in most applications the prior probability has negligible effects on the outcome once a modest amount of data is collected (Box and Tiao 1973; Walters 1986).

The main disadvantage of our approach is that the models are hierarchic: the mixed model is fitted with the combined predictors of the herbivore and physical/chemical models. Therefore, the fit of the mixed model can be no worse than the better of the other two models. One must then ask if the mixed model represents an improvement over the alternatives. Our analysis of the potential effect of new information addresses that question. Rival models with overlapping sets of predictors are probably inevitable in multifactor analyses of complex, interacting systems. We have illustrated one possible approach to comparing them. However, experiments designed to discriminate alternative models may be the most satisfactory approach (Box 1966; Walters 1986).

ECOLOGICAL IMPLICATIONS

These data corroborate the important effects of physical/chemical factors and herbivory on phytoplankton biomass and productivity in lakes. The nutrient effects in our data echo a large literature that includes both comparative and experimental studies. There is no question that nutrients, especially phosphorus, have powerful effects on algal biomass and productivity (Schindler 1977, 1988). However, our results are not consistent with the view that physical/chemical factors alone control lake productivity (Harris 1986).

Graphical analyses and correlations suggest that herbivore and physical/chemical effects are about the same magnitude in this data set. However, a rigorous comparison of variances attributable to physical/chemical and herbivore effects requires data in which herbivore and physical/chemical variates are truly independent. Large-scale experiments may be the best way of obtaining such orthogonal data.

In the whole-lake experiments we analyzed, magnitudes of algal response to phosphorus additions and food web manipulations are about equal. Relative to the range of interannual variability evident in nonmanipulated lakes, the manipulations employed in these experiments are not extreme or unusual. This fact refutes the criticism that whole-lake experiments have employed unrealistically large perturbations (Crowder et al. 1988). It does not contradict the conclusion that substantial, sustained manipulations are needed for successful whole-lake experiments (Carpenter 1989), but does show that successful manipulations are within the range of natural variation.

The mixed models that predict chlorophyll from both physical/chemical and herbivore variates are clearly superior to the alternatives. The mixed models for primary production are also superior to purely physical/chemical or purely herbivore models, but the differences among models are less clear cut. Given data from more lakes, we could compare models for primary production more definitively. The most informative data points for discriminating the models will come from lakes that represent extreme conditions: fishless versus planktivore-rich systems spanning a wide range of nutrient levels.

This analysis supports the concept that phytoplankton biomass and production in lakes are jointly controlled by nutrient and food chain processes (Carpenter et al. 1985; McQueen et al. 1986; Carpenter and Kitchell 1988). However, the data contain no evidence that the magnitude of herbivore effects changes with nutrient level (McQueen et al. 1986). Rather, herbivore effects appear to explain residuals from regressions of phytoplankton variates on physical/chemical variates. The converse is also true: physical/chemical variation can explain residuals from regressions of phytoplankton variates on herbivore variates. These patterns are consistent with whole-lake experiments (Henrikson et al. 1980; Shapiro and Wright 1984; Carpenter and Kitchell 1988) and paleolimnological studies (Kitchell and Carpenter 1987; Leavitt et al. 1989) that document joint control of phytoplankton by herbivory and abiotic factors.

Production rates and chlorophyll concentrations respond conservatively to perturbations of lake ecosystems other than nutrient loading (Schindler 1988). Food web regulation must involve strong interactions (Paine 1980) to manifest sustained changes in ecosystem processes (Carpenter and Kitchell 1988). These compensatory response capacities of communities and ecosystems are similarly challenged by chemical and food chain perturbations. Delimiting the mechanistic and stochastic components of system response capacity is both an opportunity and a necessity if ecologists are to develop practical and predictive tools for resource management. At the least, a better understanding of abiotic and food chain processes will help identify indicators and causative components of community and ecosystem dynamics. These become the basis for predictions and tests in systems likely to be affected by a changing global climate.

Acknowledgments. We thank J.J. Magnuson and J. Paloheimo for helpful advice on this study. P. Chang provided data on zooplankton in Lakes 226 and 227. J. Temte and Y. Rentmeester helped compile Lake Mendota data. K. McTigue prepared the figures. The Cascading Trophic Interactions project is supported by the U.S. National Science Foundation (grants BSR 83-08918, 86-04996, and 86-06271). Research at the Experimental Lakes Area is funded through the Canadian Department of Fisheries and Oceans. The Northern Lakes Long-Term Ecological Research Site is supported by U.S.N.S.F. grants DEB-80-12313 and BSR 85-14330. Cooperative agreement #CR-812216-01-0 with the U.S. Environmental Protection Agency funds the Little Rock Lake Experimental Acidification Project. The Lake Mendota Program is supported by the Wisconsin DNR (NRG-90785).

REFERENCES

Bartell, S.M. and J.F. Kitchell. (1978). Seasonal impact of planktivory on phosphorus release by Lake Wingra zooplankton. *Verh. Int. Verein. Limnol.* 20:466–474.
Bartell, S.M., A.L. Brenkert, R.V. O'Neill, and R.H. Gardner. (1988). Temporal variation in regulation of production in a pelagic food web model. In: *Complex Interactions in Lake Communities*, S.R. Carpenter, ed. Springer-Verlag, New York, pp. 101–118.
Box, G.E.P. (1966). Use and abuse of regression. *Technometrics* 8:625–629.
Box, G.E.P. (1980). Sampling and Bayes' inference in scientific modeling and robustness. *J. R. Stat. Soc. Ser. A* 143:383–430.
Box, G.E.P. and G.C. Tiao. (1973). *Bayesian Inference in Statistical Analysis.* Addison-Wesley, Reading, Massachusetts.
Box, G.E.P., W.G. Hunter, and J.S. Hunter. (1978). *Statistics for Experimenters.* Wiley, New York.
Bower, P.M., C.A. Kelly, E.J. Fee, J.A. Shearer, D.R. DeClerq, and D.W. Schindler. (1987). Simultaneous measurement of primary production by whole-lake and bottle radiocarbon additions. *Limnol. Oceanogr.* 32:299–312.
Brock, T.D. (1985). *A Eutrophic Lake: Lake Mendota, Wisconsin.* New York: Springer-Verlag.
Brooks, J.L. and S.I. Dodson. (1965). Predation, body size, and composition of plankton. *Science* 150:28–35.
Brylinsky, M. and K.H. Mann. (1973). An analysis of factors governing productivity in lakes and reservoirs. *Limnol. Oceanogr.* 18:1–14.

Canfield, D.E. and R.W. Bachman. (1981). Prediction of total phosphorus concentrations, chlorophyll *a*, and Secchi depths in natural and artificial lakes. *Can. J. Fish. Aquat. Sci.* 38:414–423.

Carpenter, S.R. (1983). Lake geometry: implications for production and sediment accretion rates. *J. Theor. Biol.* 105:273–286.

Carpenter, S.R., ed. (1988a). *Complex Interactions in Lake Communities*. New York: Springer-Verlag.

Carpenter, S.R. (1988b). Transmission of variance through lake food webs. In: *Complex Interactions in Lake Communities* (S.R. Carpenter, ed.). Springer-Verlag, New York, pp. 119–138.

Carpenter, S.R. (1989). Replication and treatment strength in whole-lake experiments. *Ecology* 70:453–463.

Carpenter, S.R. and J.F. Kitchell. (1984). Plankton community structure and limnetic primary production. *Am. Nat.* 124:159–172.

Carpenter, S.R. and J.F. Kitchell. (1987). The temporal scale of variance in limnetic primary production. *Am. Nat.* 129:417–433.

Carpenter, S.R. and J.F. Kitchell. (1988). Consumer control of lake productivity. *BioScience* 38:764–769.

Carpenter, S.R., J.J. Elser, and M.M. Elser. (1986). Chlorophyll production, degradation, and sedimentation: implications for paleolimnology. *Limnol. Oceanogr.* 31:112–124.

Carpenter, S.R., J.F. Kitchell, and J.R. Hodgson. (1985). Cascading trophic interactions and lake productivity. *BioScience* 35:634–639.

Carpenter, S.R., T.M. Frost, D. Heisey, and T.K. Kratz. Randomized intervention analysis and the interpretation of whole ecosystem experiments. *Ecology* 70 (in press).

Carpenter, S.R., J.F. Kitchell, J.R. Hodgson, P.A. Cochran, J.J. Elser, M.M. Elser, D.M. Lodge, D. Kretchmer, X. He, and C.N. von Ende. (1987). Regulation of lake primary productivity by food web structure. *Ecology* 68:1863–1876.

Chang, P.S.S., D.F. Malley, W.J. Findlay, W. Mueller, and R.T. Barnes. (1980). Species composition and seasonal abundance of zooplankton in Lake 227, Experimental Lakes Area, northwestern Ontario, 1969–1978. *Can. Data Rep. Fish. Aquat. Sci.* 182:1–101.

Chang, P.S.S., D.F. Malley, W.J. Findlay, R.T. Barnes. (1984). Zooplankton in Lake 226, Experimental Lakes Area, northwestern Ontario, 1971–1978 data. *Can. Data Rep. Fish. Aquat. Sci.* No 484. 208 pp.

Chatfield, C. (1980). *The Analysis of Time Series*. New York: Halsted Press.

Crowder, L.B., R.W. Drenner, W.C. Kerfoot, D.J. McQueen, E.L. Mills, U. Sommer, C.N. Spencer, and M.J. Vanni. (1988). Food web interactions in lakes. In: *Complex Interactions in Lake Communities*, S.R. Carpenter, ed. Springer-Verlag, New York, pp. 141–160.

Dillon, P.J. and F.H. Rigler. (1974). The phosphorus-chlorophyll relationship in lakes. *Limnol. Oceanogr.* 19:767–773.

Dillon, P.J. and F.H. Rigler. (1975). A simple method for predicting the capacity of a lake for development based on lake trophic status. *J. Fish. Res. Bd. Can.* 32:1519–1531.

Downing, J.A. and F.H. Rigler. (1984). *A Manual on Methods for the Assessment of Secondary Productivity in Fresh Waters*. Blackwell, Oxford.

Draper, N. and H. Smith. (1981). *Applied Regression Analysis*, 2nd Ed. Wiley, New York.

Elser, J.J., M.M. Elser, N.A. MacKay, and S.R. Carpenter. (1988). Zooplankton-mediated transitions between N and P limited algal growth. *Limnol. Oceanogr.* 33:1–14.

Fee, E.J. (1973). A numerical model for determining integral primary production and its application to Lake Michigan. *J. Fish. Res. Bd. Can.* 30:1447–1468.

Fee, E.J. (1979). A relation between lake morphometry and primary productivity and its use in interpreting whole-lake eutrophication experiments. *Limnol. Oceanogr.* 24:401–416.

Frost, T.M. and P.K. Montz. (1988). Early zooplankton response to experimental acidification in Little Rock Lake, Wisconsin, USA. *Verh. Int. Verein. Limnol.* 23:2279–2285.

Harris, G.P. (1986). *Phytoplankton Ecology*. Chapman and Hall, London.

Henrikson, L., Nyman, H.G., Oscarson, H.G., and J.A.E. Stenson. (1980). Trophic changes without changes in external nutrient loading. *Hydrobiologia* 68: 257–263.

Hocking, R.R. (1983). Developments in linear regression methodology: 1959–1982. *Technometrics* 25:219–230.

Hrbacek, J.M., Dvorakova, M., Korinek, V., and L. Prochazkova. (1961). Demonstration of the effect of the fish stock on the species composition of zooplankton and the intensity of metabolism of the whole plankton assemblage. *Verh. Int. Verein. Limnol.* 14:192–195.

Kerfoot, W.C. and A. Sih, eds. (1987). *Predation: Direct and Indirect Impacts on Aquatic Communities*. University Press of New England, Hanover, New Hampshire.

Kitchell, J.F. and S.R. Carpenter. (1987). Piscivores, planktivores, fossils and phorbins. In: *Predation: Direct and Indirect Impacts on Aquatic Communities*, W.C. Kerfoot and A. Sih, eds. University Press of New England, Hanover, New Hampshire, pp. 132–146.

Kitchell, J.F., R.V. O'Neill, D. Webb, G.W. Gallepp, S.M. Bartell, J.F. Koonce, and B.S. Ausmus. (1979). Consumer regulation of nutrient cycling. *BioScience* 29:28–34.

Koonce, J.F. (1972). Phytoplankton Succession and a Dynamic Model of Algal Growth and Nutrient Uptake. Ph.D. Dissertation, University Wisconsin-Madison.

Kratz, T.K., T.M. Frost, and J.J. Magnuson. (1987). Inferences from spatial and temporal variability in ecosystems: analyses of long-term zooplankton data from a set of lakes. *Am. Nat.* 129:830–846.

Leavitt, P.R., S.R. Carpenter, and J.F. Kitchell. Whole lake experiments: the annual record of fossil pigments and zooplankton. *Limnol. Oceanogr.* 34 (in press).

LeCren, E.D. and R. Lowe-McConnell, eds. (1981). *Functioning of Freshwater Ecosystems*. Cambridge University Press, London.

Likens, G.E. (1985). An experimental approach for the study of ecosystems. *J. Ecol.* 73:381–396.

Lindeman, R.L. (1942). The trophic-dynamic aspect of ecology. *Ecology* 23:399–418.

Lynch, M., L.J. Weider, and W. Lampert. (1986). Measurement of the carbon balance in *Daphnia. Limnol. Oceanogr.* 31:17–33.

Malley D.F., P.S.S. Chang, and D.W. Schindler. (1988). Decline of zooplankton populations following eutrophication of Lake 227, Experimental Lakes Area, Ontario: 1969–1974. *Can. Tech. Rep. Fish. Aquat. Sci.* 1619:1–25.

McCauley, E. and J. Kalff. (1981). Empirical relationships between phytoplankton and zooplankton biomass in lakes. *Can. J. Fish. Aquat. Sci.* 38:458–463.

McQueen, D.J., J.R. Post, and E.L. Mills. (1986). Trophic relationships in freshwater pelagic ecosystems. *Can. J. Fish. Aquat. Sci.* 43:1571–1581.

Mills, E.L. and A. Schiavone. (1982). Evaluation of fish communities through assessment of zooplankton populations and measures of lake productivity. *North Am. J. Fish. Manage.* 2:14–27.

Naiman, R.J. (1988). Animal influences on ecosystem dynamics. *BioScience* 38: 750–752.

Nicholls, K.H. and P.J. Dillon. (1978). An evaluation of phosphorus-chlorophyll-phytoplankton relationships for lakes. *Int. Rev. Gesamten Hydrobiol.* 63:141–154.

Northcote, T.G. (1988). Fish in the structure and function of freshwater ecosystems: a "top-down" view. *Can. J. Fish. Aquat. Sci.* 45:361–379.

Pace, M.L. (1984). Zooplankton community structure, but not biomass, influences the phosphorus-chlorophyll *a* relationship. *Can. J. Fish. Aquat. Sci.* 41:1089–1096.

Paine, R.T. (1966). Food web complexity and species diversity. *Am. Nat.* 100:65–75.

Paine, R.T. (1980). Food webs, linkage interaction strength, and community infrastructure. *J. Anim. Ecol.* 49:667–685.

Pedros-Alio, C. (1981). Ecology of Heterotrophic Bacteria in the Epilimnion of Eutrophic Lake Mendota, Wisconsin. Ph.D. Dissertation, University of Wisconsin-Madison.

Persson, L., G. Andersson, S.F. Hamrin, and L. Johansson. (1988). Predator regulation and primary production along the productivity gradient of temperate lake ecosystems. In: *Complex Interactions in Lake Communities*, S.R. Carpenter, ed. Springer-Verlag, New York, pp. 45–68.

Peters, R.H. and J.A. Downing. (1984). Empirical analysis of zooplankton filtering and feeding rates. *Limnol. Oceanogr.* 29:763–784.

Prentki, R.T., D.S. Rogers, V.J. Watson, P.R. Weiler, and O.L. Loucks. (1977). Summary tables of Lake Wingra basin data. *Univ. Wis. Inst. Environ. Stud. Rep.* 85:1–89.

Riemann, B. and M. Sondergaard. (1986). *Carbon Dynamics of Eutrophic, Temperate Lakes*. Elsevier, Amsterdam.

Schindler, D.W. (1974). Eutrophication and recovery in experimental lakes: implications for lake management. *Science* 184:897–899.

Schindler, D.W. (1977). The evolution of phosphorus limitation in lakes. *Science* 195:260–262.

Schindler, D.W. (1978). Factors regulating phytoplankton production and standing crop in the world's lakes. *Limnol. Oceanogr.* 23:478–486.

Schindler, D.W. (1988). Experimental studies of chemical stressors on whole-lake ecosystems. *Verh. Int. Verein. Limnol.* 23:11–41.

Schindler, D.W., E.J. Fee, and T. Ruszczynski. (1978). Phosphorus input and its consequences for phytoplankton standing crop and production in the Experimental Lakes Area and similar lakes. *J. Fish. Res. Bd. Can.* 35:190–196.

Shapiro, J. and D.I. Wright. (1984). Lake restoration by biomanipulation: Round Lake, Minnesota: the first two years. *Freshwater Biol.* 14:371–383.

Shapiro, J., V. Lamarra, and M. Lynch. (1975). Biomanipulation: an ecosystem approach to lake restoration. In: *Proceedings of a Symposium on Water Quality Management Through Biological Control*, P.L. Brezonik and J.L. Fox, eds. University of Florida, Gainesville, pp. 85–96.

Shearer, J.A., E.J. Fee, E.R. DeBruyn, and D.R. DeClerq. (1987). Phytoplankton productivity changes in a small double-basin lake in response to termination of experimental fertilization. *Can. J. Fish. Aquat. Sci.* 44:47–54.

Shearer, J.A., E.R. DeBruyn, D.R. DeClerq, D.W. Schindler, and E.J. Fee. (1985). Manual of phytoplankton primary production methodology. *Can. Tech. Rep. Fish. Aquat. Sci.* 1341:1–58.

Sommer, U., Z.M. Gliwicz, W. Lampert, and A. Duncan. (1986). The PEG model of seasonal succession of planktonic events in freshwaters. *Arch. Hydrobiol.* 106: 433–471.

Sprules, W.G. and M. Munawar. (1986). Plankton size spectra in relation to ecosystem productivity, size, and perturbation. *Can. J. Fish. Aquat. Sci.* 43:1789–1794.

Sprules, W.G., J.M. Casselman, and B.J. Shuter. (1983). Size distribution of pelagic particles in lakes. *Can. J. Fish. Aquat. Sci.* 40:1761–1769.

Stainton, M.P., M.J. Capel, and F.A.J. Armstrong. (1977). The Chemical Analysis of Fresh Water, 2nd Ed. *Can. Fish. Mar. Serv. Misc. Spec. Pub.* 25.

Stein, R.A., S.T. Threlkeld, C.D. Sandgren, W.G. Sprules, L. Persson, E.E. Werner, W.E. Neill, and S.I. Dodson. (1988). Size-structured interactions in lake communities. In: *Complex Interactions in Lake Communities*, S.R. Carpenter, ed. Springer-Verlag, New York, pp. 161–180.

Titus, J.E. 1977. The Comparative Physiological Ecology of Three Submersed Macrophytes. Ph.D. Dissertation, University of Wisconsin-Madison.

Vollenweider, R.A. (1968). Scientific Fundamentals of the Eutrophication of Lakes and Flowing Waters, with Particular Reference to Nitrogen and Phosphorus as Factors in Eutrophication. Organization for Economic Cooperation and Development, Paris.

Walters, C. (1986). *Adaptive Management of Renewable Resources*. MacMillan, New York.

Watras, C. and T.M. Frost. (1989). Little Rock Lake: perspectives on an experimental ecosystem approach to seepage lake acidification. *Arch. Environ. Contam. Toxicol.* 18:157–165.

Wilkinson, L. (1988). *SYSTAT: The System for Statistics*. Systat Inc., Evanston, Illinois.

Zimmerman, A.P., K.M. Noble, M.A. Gates, and J.E. Paloheimo. (1983). Physico-chemical typologies of south-central Ontario lakes. *Can. J. Fish. Aquat. Sci.* 40:1788–1803.

APPENDIX. Means of annual values for variates analyzed, by lake. See text for acronyms, units, and methods of measurement. Missing values are denoted by dots.

Lake	NLAT	ZBAR	ZMAX	TN	TP
Big Muskellunge	46.000	7.500	21.400	420.000	12.800
Blue Chalk	45.200	9.000	22.000	348.000	9.000
Crosson	45.080	8.000	23.000	405.000	10.000
Crystal	46.000	10.400	20.400	135.250	6.300

Crystal Bog	46.000	1.700	2.500	833.000	19.400
King	45.880	3.000	6.000	395.000	12.500
Little Rock (Ref)	46.000	3.100	6.500	343.000	13.250
Little Rock (Trt)	46.000	3.900	10.300	300.000	13.500
Lake 221	49.700	2.100	5.700	699.333	2.000
Lake 226 NE	49.680	5.700	14.700	451.000	13.500
Lake 226 SW	49.680	6.300	11.600	426.750	10.500
Lake 227	49.680	4.400	10.000	762.000	19.000
Mendota	43.070	12.400	25.300	1144.750	115.000
Mountain	45.980	13.000	31.000	422.500	6.000
Paul	46.250	5.000	15.000	69.420	18.040
Peter	46.250	8.300	19.300	66.160	13.100
Plastic	45.180	8.000	16.000	285.000	5.500
Ruth	46.020	6.000	15.000	340.000	5.000
Sparkling	46.000	10.900	20.000	213.050	11.400
Three Mile	45.170	3.000	11.000	542.500	16.000
Trout	46.000	14.600	35.700	183.050	9.750
Tuesday	46.250	10.000	18.500	108.300	14.040
White	44.830	4.000	10.000	357.500	16.000
Wingra	43.070	2.400	6.100	891.500	52.250
Young	44.720	4.000	10.000	477.500	20.000

Lake	THERM	PHOT	TEMP	EPICHL	PHOCHL
Big Muskellunge	9.367	14.500	20.517	2.333	3.333
Blue Chalk	7.750	23.800	20.750	.	0.800
Crosson	5.100	6.200	21.850	.	2.350
Crystal	8.383	17.800	20.500	1.350	2.117
Crystal Bog	1.833	2.417	21.967	9.717	12.267
King	2.750	5.250	20.650	.	1.750
Little Rock (Ref)	5.850	6.000	21.800	3.625	11.800
Little Rock (Trt)	6.650	8.725	21.575	2.375	16.300
Lake 221	3.300	3.367	19.967	3.200	3.467
Lake 226 NE	4.200	5.450	20.025	14.150	15.650
Lake 226 SW	4.400	7.000	19.900	6.725	11.400
Lake 227	2.700	2.800	20.100	60.300	63.150
Mendota	10.025	4.895	22.600	22.150	41.350
Mountain	8.850	18.550	20.550	.	0.950
Paul	3.540	5.300	21.500	3.158	8.162
Peter	4.168	7.298	21.720	2.048	3.936
Plastic	6.600	22.100	20.500	.	1.450
Ruth	6.500	8.450	20.350	.	1.750
Sparkling	7.667	13.650	20.650	1.483	3.567
Three Mile	5.550	7.200	21.100	.	2.900
Trout	10.000	11.883	19.467	2.500	2.717
Tuesday	2.524	3.028	21.160	5.464	9.070
White	5.000	9.650	19.300	.	1.850
Wingra	2.500	1.400	25.950	59.050	59.050
Young	6.100	12.600	21.650	.	1.750

Lake	EPIPPR	PHOPPR	TZBIO	CRUSBIO	CLADBIO
Big Muskellunge	.	.	1674.000	1523.000	823.800
Blue Chalk	.	.	4061.500	3678.500	.
Crosson	.	.	2112.500	1920.500	.
Crystal	46.650	39.850	1945.960	1865.500	159.720

(continued)

Lake	EPIPPR	PHOPPR	TZBIO	CRUSBIO	CLADBIO
Crystal Bog	.	.	717.000	707.750	430.250
King	.	.	1209.500	1127.000	.
Little Rock (Ref)	67.000	107.800	914.000	875.000	284.250
Little Rock (Trt)	115.600	100.500	2283.000	2233.000	1031.750
Lake 221	63.333	55.333	.	.	.
Lake 226 NE	181.500	123.500	470.075	416.050	155.575
Lake 226 SW	80.250	62.000	382.225	305.525	83.725
Lake 227	764.000	503.500	70.750	38.550	22.240
Mendota	.	.	5132.000	4349.250	2322.500
Mountain	.	.	1291.000	983.000	.
Paul	142.620	86.300	1510.000	1411.600	885.600
Peter	95.360	55.600	1167.200	1111.800	614.000
Plastic	.	.	1924.000	1707.000	.
Ruth	.	.	1400.500	988.000	.
Sparkling	50.000	37.833	3115.680	2941.960	890.860
Three Mile	.	.	1564.500	1199.500	.
Trout	62.200	48.800	6720.320	6458.360	3173.800
Tuesday	120.620	96.720	1051.200	945.800	373.600
White	.	.	1008.500	784.500	.
Wingra	506.452	866.129	3435.000	463.000	289.000
Young	.	.	3384.500	2749.000	.

Lake	TZLEN	CLADLEN	CRUSLEN
Big Muskellunge	0.159	0.294	0.662
Blue Chalk	0.445	0.616	.
Crosson	0.387	0.553	.
Crystal	0.169	0.389	0.774
Crystal Bog	0.170	0.311	0.671
King	0.625	1.122	.
Little Rock (Ref)	0.171	0.368	0.550
Little Rock (Trt)	0.160	0.395	0.648
Lake 221	.	.	.
Lake 226 NE	0.173	0.468	0.656
Lake 226 SW	0.152	0.381	0.533
Lake 227	0.105	0.270	0.526
Mendota	.	0.759	1.095
Mountain	0.314	0.516	.
Paul	0.281	0.466	1.132
Peter	0.314	0.543	1.180
Plastic	0.357	0.515	.
Ruth	0.328	0.587	.
Sparkling	0.163	0.319	0.562
Three Mile	0.331	0.682	.
Trout	0.170	0.353	0.616
Tuesday	0.234	0.498	0.773
White	0.292	0.472	.
Wingra	.	0.495	0.490
Young	0.443	0.906	.

6
Relationships of Primary and Secondary Production in Lakes and Marine Ecosystems

LAWRENCE R. POMEROY

Abstract

Interactions of primary and secondary production involve processes that occur in food webs. Controls may be top down (predator control) or bottom up (nutrient limitation), and separating them may be difficult. Predator-prey interactions have been examined for both macroorganisms and microorganisms, but total food webs that include both have only rarely been described, and the descriptions are never really complete. It is apparent that both kinds of controls have effects that may be direct, indirect, or interactive. The incompleteness of many earlier studies, particularly with regard to microbial processes, limits our ability to interpret and model ecosystem function.

Nutrient limitation affects plants primarily and the whole food web secondarily. We recognize characteristic communities that have strong external nutrient supplies and ones in which recycling is the predominant source. Nutrient limitation takes several forms. Nutrient supply may limit the currently dominant population; it may limit the potential net primary production, or it may limit gross primary production in excess of respiration. While nutrient supply may limit total ecosystem production, it also influences community structure in ways that can have far-reaching consequences. Just as standing stocks of biomass are not a strong predictor of production, standing stocks of nutrients are not a strong predictor of nutrient limitation.

Even in communities lacking significant inputs of nutrients, some organisms may be well supplied. Microzones in the plankton probably expose phytoplankton to nutrient concentrations that vary by orders of magnitude. Experimental manipulation of lakes has demonstrated both top-down and bottom-up controls on community structure. All populations, from bacteria to fishes, are affected. The ocean contains a number of mesoscale and regional inputs of nutrients that produce localized, transitory blooms of phytoplankton in an otherwise oligotrophic sea. Because of their short-term nature, blooms tend to be out of synchrony with zooplankton and are limited by depletion of the nutrient supply rather than by grazing. The ocean is thus bottom-up controlled in localized blooms and may be top-down controlled elsewhere. As in most terrestrial systems, only a small fraction of the production of seagrass beds and salt marshes is utilized directly by grazers. However, the major producers on coral reefs, macroalgae growing between the corals, are heavily grazed. This affects community structure but not community productivity.

Using the existing empirical database and modeling approaches, we can make some preliminary generalizations about the interactions of primary and secondary production in aquatic and marine ecosystems. Interactions between bottom-up and top-down controls occur. To improve our understanding of this important aspect to ecosystem function, we need research that integrates ecosystem and population approaches.

Introduction

The interactions of primary and secondary producers are a central consideration for both population and ecosystem ecologists, but no unified body of theory and conclusions has emerged that cuts across the hierarchical levels from species populations to ecosystems. Indeed, modelers at the population and ecosystem levels often have conflicting views of food chain relationships. This is in part a result of the bias of the former toward top-down, predator-mediated interpretations and of the latter toward bottom-up, nutrient-limited interpretations of system function. A real synthesis is difficult, simply because the problem is one that bridges a hierarchical no-man's land across which both methods and goals change radically. Population ecologists normally deal with relatively simple, definable systems and make models of high predictive value, while ecosystem ecologists deal with systems of high complexity and make models that are descriptive but highly condensed and heuristic rather than predictive. O'Neill et al. (1986) treated this problem in detail and provided examples.

The interactions of primary and secondary production are the processes that occur in food webs. It is probably fair to say that no complex, natural food web has been fully described. In every case there is inadequate descriptive knowledge of ecosystem processes, not only quantification but simple description of what species populations are present and what they eat. The difficulty of empirical analysis makes food webs a fertile field for modelers. Because food web models describe complex systems, they suffer from the condensation problems that traditionally plague ecosystem modelers. Recent empirical research tells us that a quantitatively large part of ecosystem energetics is nearly always microbial in both terrestrial and aquatic systems, and the so-called decomposer box in food web models must be unpacked to understand producer-consumer interactions. The unpacking is now well underway in most pursuits of ecosystem energetics, but the realization of its consequences does not seem to have impressed population ecologists and food web modelers. For example, Paine (1988) takes food chain theorists to task for modeling "incomplete" food webs taken from the literature, but he never mentions microorganisms as being among the underrepresented components.

Population interactions, although important, do not always control the interactions of primary and secondary producers. I examine the evidence for various controls on food chain structure and function: predator-prey interactions, nutrient limitations, and physical factors. We hope eventually to find general principles concerning population regulation that will apply in all communities and ecosystems—terrestrial, freshwater, and marine. Several investigators have proposed hypotheses of food web dynamics. I compare what we know about the range of aquatic systems, attempting to test some of those hypotheses with available data. Ideally, one should set out to do this by collecting comparable data in identical ways across the full range, from lakes to oceans. In practice, very different approaches are often used, and this is sometimes dictated by the logistical problems of sampling. Lakes, except for the largest, are more readily accessible than oceans. It is necessary to recognize the limitations on hypothesis testing that results from the analysis of diverse and often fragmentary information.

Predator–Prey Interactions

In a seminal paper, Hairston et al. (1960) pointed out some interactions between ecosystem energetics and population regulation. Their suggestion that decomposers as a group must be food limited was ahead of its time. The implications of that were rediscovered independently by marine micro-biologists later on and are discussed further on. But the speculation of Hairston et al. that grazers normally are limited by predation, not re-sources, set off a wave of activity that still rolls onward, attempting to make generalizations about interactions of productivity and food chain structure. Although they do not cite Hairston et al., one of the early, significant validations of the ideas of Hairston et al. was the observation of Brooks and Dodson (1965) that lakes containing populations of the alewife, a planktivorous fish, had distinctive assemblages of grazing zooplankton. In those lakes with alewife populations the dominant zooplankton were small species with a modal length of 0.3 mm, while in adjacent lakes with-out alewife populations, the dominant zooplankton were large species with a modal length of 0.8 mm.

In a reexamination of the concepts of Hairston et al. (1960), Fretwell (1977) extracted four potentially testable hypotheses.

1. *Populations are limited by either food or predation.* Apparently the assumption is that environmental stresses are always secondary or affect all players equally. Abiotic factors may limit populations.

2. *A population limited by predation does not limit the population on which it feeds.*
3. *A population limited by food limits the population on which it feeds.*
4. *Trophic levels are populations.* In fact, Hairston et al. stated that most trophic levels contain more than one population. Other interpretations of trophic levels abound (e.g., Wiegert and Owen 1971; Pomeroy and Wiebe 1988; Burns 1989; Moloney and Field 1991), but I will not discuss this question further.

Fretwell adds two further assumptions of his own that were not contained in the original concepts of Hairston et al.:

5. *Food chains increase in length as primary production increases.*
6. *Food chains get longer as ecological efficiency increases.*

Based on these six assumptions, Fretwell postulated that the original Hairston et al. hypothesis that plants are resource limited applies to food chains with an odd number of links—1, 3, 5. In food chains with an even number of links—2, 4, 6—the plants are grazer limited. There is no immediate test for this hypothesis. Oksanen et al. (1981) further speculated that the pressure of herbivory on primary producers should be most severe in unproductive environments. With increasing primary productivity, they said that the role of predation on herbivore regulation should increase and the impact of herbivory on plant populations should decrease. However, they added, "In very productive systems increase in herbivory pressure is again probable, at least in aquatic systems." These speculations of Fretwell and of Oksanen et al. will be discussed later in this chapter as they apply to lakes and marine systems.

GRAZERS AND PREDATORS IN MICROBIAL FOOD CHAINS

From Elton onward, studies of food chains focused on eukaryotic and usually macroscopic organisms that ecologists could see and touch. In fact, more than half of marine primary production in lakes and oceans is by organisms in the size range 0.5-5 μm (Platt et al. 1983; Probyn 1985; Stockner and Porter 1988). Such small organisms are utilized primarily by small grazers, usually protozoans (Sheldon et al. 1977), and so the ocean contains a complex microbial food web, based in part on 0.2- to 2.0-μm picoautotrophs (Sieburth et al. 1977a) and in part on heterotrophic bacteria in the lower part of that size range. Heterotrophic bacteria, probably the most numerous organisms on the planet (median frequency of 10^5 ml^{-1} of seawater $\approx 10^{29}$ cells in the ocean), use resources that are partly autotrophic and partly heterotrophic in origin (Eppley et al. 1981), partly dissolved and partly particulate (Cho and Azam 1988). Bacterial production is correlated with primary production and averages 30% of primary production (Cole et al. 1988). Because they multiply rapidly, heterotrophic bacteria compete effectively with zooplankton in consuming organic matter made available by crashing phytoplankton blooms. Painting (1989) demonstrated in the South African upwellings that because the life history of the dominant grazing zooplankton was longer than the phytoplankton bloom cycle in an upwelling, bacteria are major consumers of primary production. The bacteria are preyed on primarily by flagellates, but also by ciliates and pelagic tunicates. Recycling of nutrients by the protozoans extends phytoplankton production, while the production of protozoans provides a supplementary food source for crustacean zooplankton after the peak of phytoplankton production has passed.

Surveys in the ocean at all depths and latitudes find bacterial numbers that vary by just three orders of magnitude, 1×10^4 to 1×10^7 ml^{-1}. While it has been suggested by Fenchel (1982) that the upper limit is controlled by grazing flagellates, the work of Proctor and Fuhrman (1990) raises the possibility that marine heterotrophic bacteria are controlled at high population density by virus infections. According to Hairston et al. (1960) the heterotrophic marine bacteria should be resource limited, but there appears to be an interaction between resource supply and predation or parasitism similar to that modeled by O'Brien (1974) for phytoplankton. Most of the dissolved organic matter in the sea is refractory, and bacterial utilization of the more refractory fraction is relatively slow and inefficient (Pett 1989). The more labile dissolved materials are usually very dilute (nanomolar to picomolar), which puts a premium on small size and maximum surface-to-volume ratio. When phytoplankton blooms produce high concentrations of labile dissolved organic matter, large bacteria (> 1 μm) with higher half-saturation constants, k_s, overgrow the small bacteria (0.1–0.3 μm), and numbers may exceed 10^6 ml^{-1} before flagellate grazing catches up. At 10^4 ml^{-1} they have a rarity refuge from flagellate predators and viruses, so the flagellates crash when bacteria have been reduced to that level of abundance (Fenchel 1982).

Communities shift up and down metabolically between different quasi-stable states as the primary production changes. Coexisting, competing food chains co-occur on the basis of a changing mixture of primary and secondary dissolved and particulate organic resources. Population cycles occur in a few days in moving, mixing water. Usually it is all but impossible to follow the fate of such a community in the sea, because the community is constantly being changed by the regime of eddy diffusion. Mere description of the processes continues to be a daunting and costly task, and quantification is still problematic. In such a system, modeling energy flow and population interactions becomes an important recourse, and now probably there is enough basic description of the system to permit meaningful modeling insights.

Nutrient Limitation

The supply of essential elements, especially N and P, is often viewed as a widespread and significant limitation to populations and communities of plants, applying Liebig's elemental analysis of plants and soils to ecosystem processes. However, communities do not respond to changes in nutrient supply in the same way as do single species populations. The response of a population to nutrient concentration can be described by the Monod application of Michaelis–Menten enzyme kinetics to whole organisms (Dugdale 1967; O'Brien 1974; Tilman 1982). Growth rate of the population is proportional to the concentration of the limiting nutrient up to a saturating concentration. Above that, increased nutrient causes no further increase in population growth rate. Dugdale (1967) further assumed, apparently correctly, that all species present at one time in a phytoplankton community would have nearly the same half-saturation constant, k_s. He further predicted correctly that species adapted to low-nutrient waters would have a low k_s, and that implies a low saturating uptake velocity, V_{max}. The effect of a change in a limiting nutrient will depend on the magnitude of the change relative to the k_s of the dominant species populations. When the change does not go far below k_s or exceed saturation, changes in the growth rate of the dominant populations occur. When the change does go far below k_s or above saturation, succession of the dominant species to a group with appropriately lower or higher nutrient saturation will occur. Using a dynamic simulation model of population growth under Michaelis–Menten nutrient limitation, O'Brien (1974) showed that species with a k_s much greater than the concentration of the limiting nutrient will grow more slowly than populations having a lower k_s, and at equal grazing pressure the species with a lower k_s will be dominant. The model also predicts that these species will be smaller in size, as has been found empirically by Eppley et al. (1969).

Interpretations of phytoplankton population dynamics may lead to confusion about when nutrient limitation has occurred. Howarth (1988) pointed out that there are at last three different definitions of nutrient limitation in general use:

1. *The limitation of growth of currently dominant populations.* Changes in populations will be of the Michaelis–Menten type that occur below the saturation limits of extant dominant populations.
2. *Limitation of the potential rate of net primary production.* Changes in populations will go beyond the limits of the extant dominant populations, leading to changes in species dominance.
3. *Limitation of net ecosystem production, that is, gross primary production in excess of total ecosystem respiration.* These population changes are expressed as an effect on net increase in organic content or output of the ecosystem.

Modeling by both O'Brien (1974) and Tilman (1982) makes no distinction between types 1 and 2 but suggests a continuum of species of differing k_s. Probably that is an accurate representation of nature, but groups of species within such a continuum can be seen to dominate communities at given concentrations of nutrients. Indeed, there may not be a clear separation between type 1 and type 2 nutrient limitation, but it is important to recognize the different processes that are involved. Subtle changes in nutrients may lead to subtle or even not-so-subtle changes in species dominance. Major changes in the nutrient regimen, such as those seen in spring blooms or mesoscale features of the ocean, cause type 2 changes.

Situations exist in both lakes and oceans in which the current population is growing near its optimal rate in the presence of concentrations of PO_4, and NH_4 so dilute that they challenge the skill of the analyst to detect them (Goldman et al. 1979). In these conditions, the smallest picoplankton with low k_s are dominant. An increase in nutrient concentration will permit larger organisms with higher k_s to grow as well. According to Moloney's (1988) model, however, even under these conditions the picoplankton are fastest growing, and are limited not by nutrients but by predators. When this is carried to the extreme and the density of phytoplankton is such that some other factor, such as light or carbon, becomes limiting, we have eutrophication, which may be natural (Netarts Bay, Oregon, or Walvis Bay, Namibia) or anthropogenic (certain lakes, rivers, and estuaries). By the type 2 definition, most communities of phytoplankton are nutrient limited. Many lakes provide good examples, because they are frequently limited in the type 2 sense by available phosphorus. However, Brylinsky (1980) showed that the largest partial correlation with lake primary production was with latitude, that is, solar insolation, and all partial correlations accounted for 50% of total variation. His finding suggested that in many cases neither nutrients nor grazers, but rather a physical factor, controls primary production. Whether such a phytoplankton community is type 1 limited is a more difficult question. In their quasi-steady-state, low-nutrient condition, the populations of dominant phytoplankton are relatively stable and do not crash. The O'Brien model would interpret this as a balance between growth rate and grazing rate: both top-down and bottom-up controls are effective. When nutrients are increased, populations with a higher k_s become dominant, a type 2 nutrient effect. Typically, such a change happens so quickly that grazers cannot use all or even most of the growth. As modeled by O'Brien, this leads to a depletion of the limiting nutrient and a population crash (type 1 nutrient effect). For example, in the spring bloom in Conception Bay, Newfoundland, Douglas et al. (1986) estimated that <5% of the phytoplankton production was eaten by zooplankton during the bloom.

Most modeling is done on the assumption that the limiting nutrient is well mixed in the water, and the mean concentration — the concentration found in water analyses — represents the real limit to which individual phytoplankters are subjected. Several lines of evidence now suggest that there are microzones in most natural waters that expose individual phytoplankters to nutrient concentrations that differ from the mean concentration by orders of magnitude. Many species are adapted to take advantage of temporally transitory nutrient pulses (e.g., Caperon and Meyer 1972). Lehman and Scavia (1982a,b) demonstrated in laboratory experiments that zooplankton excretion produced micropatches of phosphate that are taken up selectively by individual phytoplankters which happened to be in the line of fire.

That experimental observation has been challenged on several grounds. First, the laboratory experiment did not duplicate commonly occurring levels of eddy turbulence that would tend to mix away local concentrations of nutrients in a matter of seconds. This criticism is based on mean intensities of eddy diffusion, not what is happening on a scale of micrometers, and no satisfactory empirical test has yet been performed (Lehman 1984). Second, it has been argued that organisms near enough to a zooplankter to benefit from excretory largess will be eaten. Feeding studies have shown crustacean zooplankton to be highly selective feeders, but also sloppy and careless (Koehl 1984; Strickler 1984). Not all suitable organisms are captured successfully. Many organisms are not selected because of size or other attributes which the zooplankter does not find to be acceptable. Many experimental studies of phytoplankton, going back to Caperon and Meyer (1972), show that phytoplankton adapted to oligotrophic conditions often have k_s well above expected ambient nutrient concentrations and are able to take up a cell quota of nutrients in as little as a few seconds. No one seems to have modeled the growth of a phytoplankton population with this kind of differential, cell-by-cell nutrient source.

Another kind of microzone is the organic aggregate, a conglomerate of dead and living material of diameters < 10 μm to > 10 cm, accumulated through the action of turbulent shear on sticky bacterial extracellular exopolymers (McCave 1984; Biddanda 1988). Alldredge (1979) showed that many aggregates contain higher concentrations of PO_4 and NH_4 than bulk water and also contain healthy phytoplankton. A third kind of microzone is the interior of a zooplankter's gut. Nonselective feeders frequently ingest phytoplankton they cannot digest. Porter (1976) has shown that many of them survive the trip down a zooplankter's alimentary canal and may indeed profit from the experience.

There are three-dimensional mosaics of nutrient supply not only on the space scale of millimeters and time scale of seconds but also on larger scales. In any water body there are cellular circulation patterns that concentrate many buoyant or motile phytoplankton (Ryther 1955; Margalef 1956; Pomeroy et al. 1956). We may not see higher nutrient concentrations in such small-scale convergences, but they are regions of improved diffusive exchange, because all the water in the mixed layer passes through them in a relatively short time. During calm conditions, local nutrient concentrations can build up in zooplankton patches and dense schools of fishes. The thermocline in many lakes and in many parts of the ocean occurs near the light compensation depth, and pulses of nutrient-rich water are advected upward through the thermocline by billow turbulence (Woods and Wiley 1972), internal waves (Armi and Farmer 1985), and solitons (McGillivary 1988).

These conditions provide pulses of nutrients to vertically and horizontally limited patches of phytoplankton, often of species adapted to the low-light regime of the lower mixed layer and thermocline. Mesoscale (10–100 km) nutrient patchiness is well known, involving jet-stream intrusions (Atkinson et al. 1985), river plumes (Edmund et al. 1981), warm-core rings (McCarthy and Nevins 1986), and other oceanic fronts. This level of patchiness involves type 2 nutrient limitation.

MINERALIZATION BY MICROBIAL FOOD CHAINS

Until recently, microorganisms were relegated to a catchall role of "decomposers," yet even ecosystem-level studies of energy flux usually ignored microbial metabolism. Evidence is accumulating that much respiratory activity in the sea is microbial (Pomeroy and Johannes 1968; Sieburth 1977; Azam et al. 1983), and that microbial respiration often equals or exceeds primary production (Pomeroy and Johannes 1968; Sieburth et al. 1977b; Gieskes et al. 1979; Smith and Mackenzie 1987; Griffith 1989). To the extent that heterotrophic microorganisms consume the primary production, they are cycling nutrients rapidly and will tend to sustain primary production. The finding of Zahary (1988) that respiration equivalent to 50% of primary production occurs in just the upper 10 m of the Hartbeesport Reservoir (South Africa) suggests that high respiratory rates may occur in lakes as well as in the sea.

INTERACTIONS OF GRAZING AND NUTRIENT LIMITATION

O'Brien's (1974) model of phytoplankton population regulation showed interaction between grazing rate and nutrient limitation in determining population growth rate and the size of a steady-state population. Type 1 nutrient limitations appear to be modulated by the effect of grazers; type 2 nutrient limitations sometimes interact with grazing pressure and sometimes not. Phytoplankton blooms that result from an infusion of nutrients often result in changes in biomass (measured as chlorophyll) of two orders of magnitude. Both in O'Brien's models and in nature, this usually terminates with nutrient depletion and a population crash. Bacteria, at least in the sea, appear to be limited either by grazing pressure or viruses above 10^7 ml^{-1}, and their blooms are terminated by depletion of excess labile substrate without any surplus bacterial biomass remaining. However, careful confirmation in the sea and comparative studies in lakes and estuaries, where bacterial numbers may rise above 10^7 ml^{-1}, are needed.

The O'Brien model suggested that multiple limiting nutrients did little to change the outcome, because one nutrient would always determine population response. However, Tilman (1982) pointed out that because of different requirements for various essential chemical elements by phytoplankton species, sequential limitation by two elements would produce different sequences of species succession, depending on which nutrients were limiting.

Both top-down and bottom-up effects have been demonstrated in a variety of natural and experimental systems. One way to evaluate the various scenarios is to compare existing descriptive data with predictions generated by various models of the interactions of primary and secondary producers. Such modeling is a popular indoor sport for ecologists and one with severe limitations. A major limitation is the extent, quality, and completeness of the information available, as previously noted. Another is the validity of post hoc interpretations of any given set of observations. I review some examples with the warning that the approach only serves to suggest where further effort toward validation might be well placed. Both bottom-up and top-down controls and potential interactions will be considered.

A Comparison of Aquatic Systems

Having reviewed what we know and what is speculated about the interactions between primary and secondary production in general terms, I now attempt to examine how these processes may vary across the range of aquatic systems from ponds to oceans. As I have pointed out, such comparisons are hampered by the lack of uniformity in data gathering, dictated at least in part by the physical size of the system.

LAKES

Hairston et al. (1960) stated that their conclusions applied primarily to terrestrial communities, perhaps because of then-existing misconceptions about aquatic community and ecosystem structure and processes. Lakes are instructive systems in that they are aquatic islands in a terrestrial sea, harboring a limited array of species populations that arrange themselves in a small number of configurations. Thus, natural experiments occur, and intentional experiments are more tractable and repeatable than they would be in most marine systems. Variations in abundance of carnivores in lakes can have major, readily demonstrable effects on the standing stocks and productivity of phytoplankton (Brooks and Dodson 1965; Carpenter et al. 1985; Carpenter, Chapter 5, this volume). Bottom-up effects also can be demonstrated in lakes. Nutrient supply to the phytoplankton can induce population changes that affect the species composition and abundance of the carnivores (Stockner and Porter 1988). The relationship between size of predator and prey becomes a key factor in aquatic systems. In British Columbia lakes, Stockner and Shortreed (1985) found that the presence of micro- and nanoplankton, with *Daphnia* as the dominant grazer, led to efficient transfer of energy to sockeye salmon smolts. Extremely oligotrophic coastal lakes lacked *Daphnia*, had longer and more varied food chains based on picoplankton, and produced fewer, smaller salmon. Carefully calculated fertilization changed the food web and improved salmon production. Recognition of a microbial food web and its effect as an energy sink led to a manipulated change in population structure with benefit to a fishery (Stockner and Porter 1988). While this suggests bottom-up effects translated from producer to predator, Kerfoot and DeMott (1984), using replicated enclosures in Lake Mitchell, Vermont, found a different set of population responses to fertilization. Addition of fertilizer to increase available nitrogen and phosphorus by an order of magnitude caused an increase in *Daphnia* fecundity, but only a twofold increase in chlorophyll. *Daphnia* controlled the phytoplankton population.

If trout, a keystone predator, were also added, however, *Daphnia* numbers decreased and alternate food chains of small grazers (rotifers, ciliates) appeared. These were eaten by a predatory zooplankter, *Mesocyclops*, which was in turn eaten by the trout. If this food web was fertilized heavily, an algal bloom occurred, because some algal species were too large to be consumed by the smaller grazers, and top-down control of producers was lost. In this simple system, some of Fretwell's (1977) predictions of top-down control were validated. However, comparing the Kerfoot and Demott experimental results with the Stockner and Shortreed empirical ones, the loss of grazer control of the primary producers in lake enclosures may have been an artifact of the simple structure of the enclosures, which offered no refuges and had no extensive feeding guilds of zooplankton to utilize changing resources. So even lakes, with few species of grazers and predators, deviate from model systems because of matches and mismatches between predator and prey size and fecundity. Microbial components of the food web are integral to this process and cannot be ignored (Porter et al. 1988). Top-down effects in lakes can influence populations of zooplankton (Brooks and Dodson 1965) or bacteria (Riemann and Sondergaard 1986), and bacteria can compete very successfully with eukaryotes for resources that potentially either can use.

PELAGIC MARINE ECOSYSTEMS

Energy flux through pelagic marine food webs is a continuing subject of controversy. Those who measure heterotrophic processes find respiratory rates as high as or higher than reported rates of photosynthesis. Rates of photosynthesis estimated by bottle methods are lower than estimates based on oxygen accumulation below the seasonal thermocline. Satellite imagery presents a picture of mesoscale, transitory high-production events in eddies and fronts in a low-production ocean. Although PO_4, NO_3, and NH_4 are present in ocean surface water in very small amounts, the normal, midocean phytoplankton community has a low k_t, and there is evidence that this community is not nutrient limited (Goldman et al. 1979). As we have seen, there appears to be an interaction between nutrient uptake, growth, and grazing like that in O'Brien's (1974) simulation. When inputs of nutrients occur, they stimulate growth of larger phytoplankton species with higher

k_s, which temporarily overgrow the normal picoplankton flora and then crash when nutrients become depleted. Phytoplankton blooms of that sort occur in all parts of the ocean, for various reasons. In temperate and polar seas and lakes, blooms are strictly seasonal. Although the causal mechanisms of spring phytoplankton blooms of temperate lakes and oceans are not understood fully, they are basically responses to increased light in the presence of excess nutrients and increasing stability of the euphotic zone (Gran and Braarud 1935; Sverdrup 1953).

In the subtropics and tropics, blooms are sporadic, forced by physical events that inject a pulse of nutrients into the euphotic zone. Some tropical lakes have a seasonal change in stability in response to changes in the wind regime (Talling 1966). Phytoplankton growth declines sharply when the nutrients become limiting. At this point, only a very small fraction of the bloom has been consumed by grazers, and much of it is sedimented out of the euphotic zone and lost to the benthos or to microbial consumers along the way (Smetacek et al. 1978; Douglas et al. 1986; Wassman 1987; Grebmeier et al. 1988). So the spring bloom, while it lasts, behaves like the terrestrial systems of Hairston et al. (1960) and is strictly resource limited. Grazer populations have been depleted during the winter—not only by predators, but by other winter stresses as well. When the spring bloom arrives, the grazing zooplankton go into a reproductive mode, but they do not complete a generation in time to become resource limited during the bloom. Only in the post-bloom conditions of nutrient-limited low-primary production do they become resource limited. The same sequence of events occurs in low latitudes in a bolus of nutrient-rich water brought into the euphotic zone in a coastal upwelling or a jet-stream intrusion, but in those cases an order of magnitude more of the primary production appears to be used in the short term by the pelagic consumers (Pomeroy et al. 1987). This disparity needs further scrutiny and confirmation.

A likely example of a more stable resource limitation of grazers in the marine plankton is found in the Gulf of Alaska in the subarctic eastern Pacific Ocean. No spring bloom of phytoplankton occurs there. It had been postulated that there is sufficient photosynthesis through the winter so that large, mesozooplankton species are able to survive winter in sufficient numbers to prevent the initiation of a spring bloom. A recent modeling analysis suggests that mesoplankton cannot exert the necessary control but that microplankton may do so (Frost 1987). As is so often the case, validation requires better empirical information, in this instance on winter rates of photosynthesis and abundance of the smaller microplankton. Meanwhile, Evans and Parslow (1985) have suggested a purely physical limit, the presence throughout the year of a shallow mixed layer.

In the tropics, blooms occur under several sets of circumstances, sometimes but not always involving a temporary population imbalance. In their initial bloom phase, the coastal upwellings off Peru and South Africa are systems depauperate in grazers. The water has recently upwelled, and grazers have been seeded in recently and have not had an opportunity to reproduce (Peterson et al. 1988). Around the edges, in the water that has been at the surface for weeks and where grazers have been able to reproduce, their populations can be high. Painting (1989) found that zooplankton grazers were resource limited in the final stages of succession in upwelled water. The dinoflagellate blooms of more stable tropical or temperate waters probably owe their existence at least in part to their motility as well as reproduction. Physical circulation has concentrated motile dinoflagellates out of a larger body of water over a short period of time, and the growth of grazers has not caught up with them (Ryther 1955; Margalef 1956; Pomeroy et al. 1956).

The most extensive blooms in tropical oceans are of *Oscillatoria* (*Trichodesmium*), which occur over thousands of square kilometers, from the Coral Sea to the Sargasso Sea (Carpenter 1983). The bundles of trichomes are large enough to be seen individually with the unaided eye, so the individual plant unit that is presented to potential grazers is larger by orders of magnitude than any other common pelagic plant except *Sargassum*. Most zooplankton see *Oscillatoria*, if they see it at all, as a part of the landscape rather than as something to eat. Also, being blue-green bacteria, *Oscillatoria* are likely to be well equipped with allelopathic defenses. The fate of these blooms, so far as we know, is to deplete nutrient resources, die, and go into the microbial food web. Because *Oscillatoria* is a sporadic bloomer, no specialist grazer can develop by using it as an exclusive resource, and no other resource in that size category occurs in that ecosystem.

Because of the difficulty of following events in pelagic communities in the sea, modeling is an important adjunct to empirical observations. A recent marine model by Moloney (1988) brings together aspects of nutrient limitation and food chain dynamics and reaches a number of provocative conclusions. Discarding the trophic-level concept as unworkable for the marine plankton, Moloney developed a generic, size-based model of nitrogen and carbon of planktonic communities, using the trophic continuum concept of Cousins (1985). She varied nutrients to simulate various specific planktonic community types, and a stable, low-nutrient community is compared with a coastal upwelling community. Three size categories of phytoplankton are distinguished as the base of five food chains, ranging from a two-link chain from microphytoplankton to fishes to a seven-link chain based on primary dissolved organic carbon (released from phytoplankton of all sizes) and consisting of heterotrophic bacteria, flagellates, ciliates, zooplankton, and fishes. Because these are size based, each food chain may be composed of any of a number of species. Moreover, a number of other food chains will be present: detritus-based ones, nonselective particle feeding by net-casters, consumption of gametes and zooplankton by protozoans, and parasitism. Nevertheless, Moloney's model is the most comprehensive to date, and it probably gives us a realistic insight of how such systems work.

Several important conclusions emerged from Moloney's (1988) model. In both stable, oligotrophic and unstable, eutrophic planktonic communities, most energy is dissipated through short food chains. Long food chains are more predominant, however, in the oligotrophic community, contrary to Fretwell's (1977) and Pimm's (1982) speculations, because most of the primary production is by picoplankton that can be consumed only by other microorganisms. Thus, most zooplankton and all nekton are dependent on secondary food sources. In productive upwellings, food chains are mostly very short, with fishes and zooplankton grazing directly on net phytoplankton. Efficiency is further improved, because search time for food is reduced in the upwellings. Respiratory losses are high in all planktonic communities, however, because in the turbulent, changeable communities there are many mismatches between production and consumption resulting from different generation times. Resources, even when present, are inefficiently utilized, pointing again to the dominance of microbial processes that are ready to assimilate any organic matter, dissolved or particulate. Many concurrent, parallel food chains occur, with much prey switching and changing predator dominance.

Microbial dominance of the ocean and high microbial respiratory rates mean that plant nutrients are being recycled rapidly. An important result of rapid microbial utilization of primary production, and very little storage of it, is the support of continued phytoplankton growth by regenerated PO_4 and NH_4. It is estimated that more than 90% of primary production in central ocean gyres is supported by recycled nutrients, and even in coastal upwellings 50%–60% of production is supported by recycled nutrients (Chavez and Barber 1987). Some nitrogen and phosphorus are lost with sinking fecal pellets or dead plankton, but relatively little leaves the upper mixed layer (cf. Lande and Wood 1987). From an ecosystem perspective, this is an important attribute of the ocean system, one that shapes planktonic species and communities. Evolution toward smallness permits the phytoplankton and bacteria to remain suspended in the water at very little energetic cost and to grow essentially without resource limitation, even though their storage reserves are probably good for about 1 day.

When vertical advection brings a pulse of nutrients into the euphotic zone of the ocean, the first response is by the picoplankton, who are soon overgrown by larger, faster-growing species (Moloney 1988). This transitory pulse of picoplankton found by Moloney in her model has been seen in the Peru upwelling, but observations like those of Semina (1972) have usually been ignored and assumed to be a mistake. Long sets of sequential observations at sea are so difficult and costly that they are attempted rarely, and what we have to work with is a series of snapshots of the system.

SHALLOW BENTHIC MARINE ECOSYSTEMS

Coral reefs and seagrass beds provide a contrast in food chain control, although both are very productive. Several species of sea grasses, from subpolar waters to the tropics, appear to share the characteristic of underutilization by grazers, although major grazers, such as geese and turtles, exist. The limiting resource appears to be primarily space or light. The grass beds are limited to shallow, soft-bottom environments protected from extremes of waves, tides, and ice. Sea grasses are important structural components of their communities, like typical terrestrial plant communities.

In contrast, the structure of coral reefs is mainly corals which, although they contain symbiotic dinoflagellates and are net autotrophs, are not the principal autotrophic populations. The quantitatively dominant autotrophs are red and green algae and blue-green bacteria that form a thin lawn over the rock between the coral heads (Johannes et al. 1972; Smith and Marsh 1973), sometimes supplemented by the calcareous red algae *Porolithon* and *Lithothamnion* (Marsh 1970). Tropical reefs are perhaps the most productive community in the world, and although they occur in nutrient-depleted ocean water, they are not resource limited, because large volumes of water flow across the reefs and the flux of nutrients therefore is not limiting (Pomeroy et al. 1974). Schools of grazing fishes keep the algal lawn cropped so closely that it escaped the detection of ecologists until upstream-downstream measurements showed that active primary producers must be present on what had been viewed as bare reef rock (Johannes et al. 1972). Exclosure experiments verified that luxuriant turfs of algae grow when grazing pressure is removed. Does this mean, following Fretwell (1977), that the seagrass food chain has an odd number of links and the coral reef food chain an even number of links? In a system with a food web composed of thousands of food chains of varying length, it is not likely, nor does it seem to be an interesting question.

Superficially, salt marshes are intertidal grasslands. The grasses or sedges that dominate these systems are tough and unpalatable. The usual grazers are present in about the usual abundance: chewing and sucking insects and mammalian grazers such as deer, where they are not excluded by human activity around the periphery. The difference between salt marshes and terrestrial grasslands is that marine processes infiltrate the marshes. Seawater brings abundant estuarine phytoplankton that are consumed by filter-feeding invertebrates. Seawater also brings nutrients to support growth of benthic microflora, mainly diatoms, in the surface layer of intertidal sediments. Several lines of evidence suggest that these algal producers support a considerable part of the truly marine food chains in salt marshes. Invertebrates among the marsh grass, such as the mussel *Geukensia*, have stable isotope signatures similar to the grass, while invertebrates in or near the tidal creeks have isotope signatures more like phyoplankton (Peterson et al. 1985). The grass also has an important structural function, for without the grass, the intertidal zone would be an unstable mud bank.

Discussion

The most significant bottom-up controls for ocean communities or ecosystems are mesoscale and regional events, with a periodicity of days to a year, that contribute limiting nutrients to the euphotic zone and set the level of primary production. Lakes are systems with seasonal or stochastic mixing events, depending on morphology and latitude. Very large lakes have some of the mesoscale features of oceans, but even the largest are only comparable physically to epicontinental seas, like the Tongue of the Ocean in the Bahamas. The ocean is composed of a set of large, oligotrophic basins with productive mesoscale and regional features embedded in them. The oceans have been misinterpreted, because until very recently it was difficult to find and sample the significant mesoscale features. Partly as a result of this sampling bias, primary and secondary productivity may have been underestimated. By changing the level of primary production, infusions of limiting nutrients in lakes or in local features of oceans influence the entire food web.

The dominance of extremely small primary producers has a strong effect on the size of the prey populations and in turn on the success of grazers and on the efficiency of transfer of energy to terminal consumers. Long microbial food chains characteristic of oligotrophic conditions are inherently inefficient energy transfer systems. However, eutrophic, mesoscale or regional features, such as eddies and upwellings, have short, efficient food chains, but do not remain stable long enough to support even the larger grazing zooplankton (Moloney 1988). Stable systems, such as that of the subarctic eastern Pacific Ocean, are exceptional (Evans and Parslow 1985).

Historically, the central gyres of the ocean and the summer epilimnia of lakes have been viewed as environments inhabited by nutrient-stressed phytoplankton. Now it appears likely that phytoplankton selected for low k_s grow at rates within their linear range of response to a limiting nutrient. They are also being grazed at some finite rate. Their growth occurs within limits set by the interaction of both grazing and nutrient supply. Changes in either will at first change the production of currently dominant species and over time may lead to dominance by other phytoplankton species. Such changes occur as seasonal or sporadic whole-lake events and as mesoscale or regional events in the ocean. In the usual course of events, they are initiated by infusion of limiting nutrient, not by changes in keystone carnivores, but subsequent grazer and predator effects can modify the structure of the food web, the energy flux, and the rate of a nutrient cycling.

Ecosystem ecologists and population ecologists look at the interaction between primary and secondary production differently and ask different questions about it. Although ecosystem processes are literally the sum of population processes, it soon becomes clear that questions about overall system behavior are not best approached by summing population processes. Yet, qualitative changes in community structure have major effects on the net flux of energy and materials, and only through description of these changes will we understand what happened. Populations of macroorganisms can be identified, separated, dissected. Some of these population processes "show through" at the ecosystem level [e.g., O'Neill et al. (1986)], while others do not. Thus, the ecosystem ecologists should be more conversant than they usually are with theory, principles, and data concerning populations in an ecosystem under study. Acquisition of such knowledge is possible in a lake, but very troublesome in a coral reef or the central ocean gyres. For both, much of the population data we should like to have are lacking, and, as a result, comparisons across ecosystems are limited. To make matters more difficult, the ocean is not an easily accessible system for population studies, except for the intertidal margins. Changes in microbial communities are very rapid and probably patchy. The species of heterotrophic bacteria and some of the small flagellates usually cannot be identified by rapid morphological examination, so the kinds of population analyses done routinely with macroorganisms are still uncommon for bacteria.

Shallow benthic marine ecosystems differ in significant ways from the pelagic and from one another. They are in between the pelagic and the terrestrial, in that they have a stable substratum exposed to sunlight but are subaquatic and even have significant seston components. Most shallow, benthic systems do not experience the short-term cycles of nutrient abundance and depletion of the deep-water pelagic systems, because they have larger, more stable reservoirs in sediments or biomass. Departures from quasi-steady state tend to be in the form of small-scale mosaics resulting from physical disturbance. Such events are comparable to events in a forest but occur on a much shorter time scale. Most are open systems with very short recycling times (Howarth, Chapter 9, this volume). The microbial food web within benthic systems is concentrated in the sediments, where sediments are present. Where sediments are largely absent, as in many coral reefs, microbial processes may have relatively little impact on system processes, except as they contribute to downstream pelagic systems.

As population ecologists have been drawn toward ecosystem-level questions involving energetics and food chains, they have come up against problems familiar to ecosystem ecologists: system complexity, fragmentary data sets, and multiple or changing limits. Extant theories of food chain control of communities are indeed helpful in understanding events seen in nature, but by themselves they do not provide complete predictive ability. The current theories of food chain length are difficult to test with incomplete data from natural communities. They cannot be tested by uncritical recourse to the literature. All published descriptions of food chains, even those of very simple systems such as tree holes and phytotelmata, are incomplete. Various short segments can be identified, but the microbial components are still poorly described and are largely absent from most literature more than a few years old. Microbial food chains cannot be dismissed as a completely separate decomposer pathway. Bacteria, and in terrestrial systems both bacteria and fungi, compete with metazoans for many resources, frequently with great success. They are in turn consumed in some metazoan food chains. We have seen that there are problems with most current theories of the control of food chain length by either grazers or primary productivity in aquatic and marine ecosystems. Even in lakes, with relatively low species diversity, the size continuum influences the structure of the productivity continuum, and it is influenced by limiting nutrient supply.

Ecology is a young science that deals with complex systems on a wide spectrum of scales of time and space. We still receive frequent, major surprises as description of ecosystems improves. It is important to admit, indeed to emphasize that description—dare we say, natural history?—is still an important component of ecosystem research. In older fields such as physics, chemistry, or biology, scientists are working with a well-established body of data. Areas of the unknown always remain around the perimeter, and there may still be surprises, but major ones are going to be rare especially in physics and chemistry. Creating theory in ecosystem ecology, however, is a bit like creating theory in chemistry without a periodic table. We have not yet described the system fully and, given its complexity, we have to understand that at some level we may have great difficulty describing it. Obviously, one of the great challenges to theoreticians in any field is generalization through effective simplification or condensation. It remains a great challenge in ecosystem analysis, both for the theoretician and the empiricist.

Acknowledgments. This chapter was revised substantially on the basis of comments during the Cary Conference. David Tilman and Michael Pace were especially helpful. Research cited from my laboratory was supported by the U.S. National Science Foundation and the U.S. Department of Energy.

REFERENCES

Alldredge A.L. (1979). The chemical composition of macroscopic aggregates in two neritic seas. *Limnol. Oceanogr.* 24:855–866.

Armi L. and D. Farmer. (1985). The internal hydraulics of the Strait of Gibraltar and associated sills and narrows. *Oceanol. Acta* 8:37–46.

Atkinson L.P., D.W. Menzel, and K.A. Bush. (1985). *Oceanography of the Southeastern Continental Shelf.* American Geophysical Union, Washington, D.C.

Azam F., T. Fenchel, J.G. Field, J.S. Gray, L.A. Meyer-Reil, and F. Thingstad. (1983). The ecological role of water-column microbes in the sea. *Mar. Ecol. Prog. Ser.* 10:257–263.

Biddanda, B.A. (1988). Microbial aggregaton and degradation of phytoplankton-derived detritus in seawater. II. Microbial metabolism. *Mar. Ecol. Prog. Ser.* 42: 89–95.

Brooks, J.L. and S.I. Dodson. (1965). Predation, body size, and composition of plankton. *Science* 150:28–35.

Brylinsky, M. (1980). Estimating the productivity of lakes and reservoirs. In: *The Functioning of Freshwater Ecosystems*, E.D. Le Cren and R.H. Lowe, eds. Cambridge University Press, Cambridge, pp. 411–447.

Burns, T.P. (1989). Lindeman's contradiction and the trophic structure of ecosystems. *Ecology* 70:1355–1362.

Caperon, J. and J. Meyer. (1972). Nitrogen-limited growth of marine phytoplankton. II. Uptake kinetics and their role in nutrient limited growth of phytoplankton. *Deep-Sea Res.* 19:619–632.

Carpenter, E.J. (1983). Nitrogen fixation by marine Oscillatoria (*Trichodesmium*) in the world's oceans. In: *Nitrogen in the Marine Environment*, Academic Press, London, pp. 65–103.

Carpenter, S.R., J.F. Kitchell, J.R. Hodgson. (1985). Cascading trophic interactions and lake productivity. *BioScience* 35:634–639.

Chavez, F.P. and R.T. Barber. (1987). An estimate of new production in the equatorial Pacific. *Deep-Sea Res.* 34A:1229–1243.

Cho, B.C. and F. Azam. (1988). Major role of bacteria in biogeochemical fluxes in the ocean's interior. *Nature* (London) 332:441–443.

Cole, J.J., S. Findlay, and M.L. Pace. (1988). Bacterial production in fresh and saltwater ecosystems: a cross-system overview. *Mar. Ecol. Prog. Ser.* 43:1–10.

Cousins, S.H. (1985). The trophic continuum in marine ecosystems: structure and equations for a predictive model. *Ecosystem Theory for Biological Oceanography. Can. Bull. Fish. Aquat. Sci.* 76–93.

Douglas, D.J., D. Deibel, R.J. Thompson, P.C. Griffith, and L.R. Pomeroy. (1986). COPE 86: Short-term temperature response of heterotrophic and photosynthetic cold ocean microplankton. *EOS* 67:977.

Dugdale, R.C. (1967). Nutrient limitation in the sea: dynamics, identification, and significance. *Limnol. Oceanogr.* 12:685–695.

Edmund, J.M., E.A. Boyle, B. Grant, and R.F. Stallard. (1981). The chemical mass balance in the Amazon plume. I. The nutrients. *Deep-Sea Res.* 28A:1339–1371.

Eppley, R.W., J.N. Rogers, and J.J. McCarthy. (1969). Half-saturation constants for uptake of nitrate and ammonia by marine phytoplankton. *Limnol. Oceanogr.* 14:912–920.

Eppley, R.W., S.G. Horrigan, J.A. Fuhrman, E.R. Brooks, C.C. Price, and K. Sellner. (1981). Origins of dissolved organic matter in southern California coastal waters: experiments on the role of zooplankton. *Mar. Ecol. Prog. Ser.* 6:149–159.

Evans, G.T. and Parslow, J.S. (1985). A model of annual plankton cycles. *Biol. Oceanogr.* 3:327–347.

Fenchel, T. (1982). Ecology of heterotrophic microflagellates. IV. Quantitative occurrence and importance as bacterial consumers. *Mar. Ecol. Prog. Ser.* 9:35–42.

Fretwell, S.D. (1977). The regulation of plant communities by food chains exploiting them. *Perspect. Biol. Med.* 20:169–185.

Frost, B.W. (1987). Grazing control of phytoplankton stock in the open subarctic Pacific Ocean: a model assessing the role of mesoplankton, particularly the large Calanoid copepods *Neocalanus* spp. *Mar. Ecol. Prog. Ser.* 39:49–68.

Gieskes, W.W., G.W. Kraay, and M.S. Baars. (1979). Current ^{14}C methods for measuring primary production—gross underestimates in oceanic waters. *Nature* (London) 13:58–78.

Goldman, J.C., J.J. McCarthy, and D.G. Peavey. (1979). Growth rate influence on the chemical composition of phytoplankton in oceanic waters. *Nature* (London) 279:210–215.

Gran, H.H. and T. Braarud. (1935). A quantitative study of the phytoplankton in the Bay of Fundy and the Gulf of Maine including observations on hydrography, chemistry, and turbidity. *J. Biol. Bd. Can.* 1:219–467.

Grebmeier, J.M., C.P. McRoy, and H.M. Feder. (1988). Pelagic-benthic coupling on the shelf of the northern Bering Sea and Chuckchi Sea. I. Food supply source and benthic biomass. *Mar. Ecol. Prog. Ser.* 48:57–67.

Griffith, P.C. (1989). Pelagic Community Respiration in the Continental Shelf Waters of the South Atlantic Bight. Ph.D. Dissertation, University of Georgia, Athens, Georgia.

Hairston, N.G., F.E. Smith, and L.B. Slobodkin. (1960). Community structure, population control and competition. *Am. Nat.* 94:421–425.

Howarth, R.W. (1988). Nutrient limitation of net primary production in marine ecosystems. *Annu. Rev. Ecol. Syst.* 19:89–110.

Johannes, R.E., J. Alberts, C. D'Elia, R.A. Kinzie, L.R. Pomeroy, W. Sottile, W. Wiebe, J.A. Marsh, Jr., P. Helfrich, J. Maragos, J. Meyer, S. Smith, D. Crabtree, A. Roth, L.R. McCloskey, S. Betzer, N. Marshall, M.E.Q. Pilson, G. Telek, R.I. Clutter, W.D. DuPaul, K.L. Webb, and J.M. Wells, Jr. (1972). The metabolism of some coral reef communities: a team study of nutrient and energy flux at Eniwetok. *BioScience* 22:541–543.

Kerfoot, W.C. and W.R. DeMott. (1984). Food web dynamics: Dependent chains and vaulting. In: *Trophic Interactions within Aquatic Ecosystems*, American Association for the Advancement of Science, Washington, D.C., pp. 347–382.

Koehl, M.A.R. (1984). Mechanisms of particle capture by copepods at low Reynolds numbers: possible modes of selective feeding. In: *Trophic Interactions within Aquatic Ecosystems*. American Association for the Advancement of Science, Washington, D.C., pp. 135–166.

Lande, R. and A.M. Wood. (1987). Suspension times of particles in the upper ocean. *Deep-Sea Res.* 34:61–72.

Lehman, J.T. 1984. Grazing, nutrient release, and their impacts on the structure of phytoplankton communities. In: *Trophic Interactions within Aquatic Ecosystems*, American Association for the Advancement of Science, Washington, D.C., pp. 49–72.

Lehman, J.T. and D. Scavia. (1982a). Microscale nutrient patches produced by zooplankton. *Proc. Natl. Acad. Sci. USA* 79:5001–5005.

Lehman, J.T. and D. Scavia. (1982b). Microscale patchiness of nutrients in plankton communities. *Science* 216:729–730.

Margalef, R. (1956). Estructura y dinámica de la ⟨purga de mar⟩ en la Ría de Vigo. *Invest. Pesq.* 5:113–134.

Marsh, J.A. (1970). Primary productivity of reef-building calcareous red algae. *Ecology* 51:255–263.

McCarthy, J.J. and J.L. Nevins. (1986). Utilization of nitrogen and phosphorus by primary production in warm-core ring 82-B following deep convective mixing. *Deep-Sea Res.* 33A:1773–1788.

McCave, I.N. (1984). Size spectra and aggregation of suspended particles in the deep ocean. *Deep-Sea Res.* 31A:329–352.

McGillivary, P.A. (1988). Oceanic and atmospheric forcing: implications for particle entrainment in the Santa Barbara Channel. *EOS* 69:1093.

Moloney, C.L. (1988). A Size-Based Model of Carbon and Nitrogen Flows in Plankton Communities. Ph.D. Dissertation, University of Cape Town, Cape Town, South Africa.

Moloney, C.L. and J.G. Field. (1991). Modelling carbon and nitrogen flows in a microbial plankton community. In: *Protozoa and Their Role in Marine Processes*. Springer-Verlag, Heidelberg, pp. 443–474.

O'Brien, W.J. (1974). The dynamics of nutrient limitation of phytoplankton algae: a model reconsidered. *Ecology* 55:135–141.

Oksanen, L., S.D. Fretwell, J. Arruda, and P. Niemela. (1981). Exploitation ecosystems in gradients of primary productivity. *Am. Nat.* 118:240–261.

O'Neill, R.V., D.L. DeAngelis, and T.F.H. Allen. (1986). *A Hierarchical Concept of Ecosystems*. Princeton University Press, Princeton, New Jersey.

Paine, R.T. (1988). Food webs: road maps of interaction or grist for theoretical development? *Ecology* 69:1648–1654.

Painting, S.J. (1989). Bacterioplankton Dynamics in the Southern Benguela Upwelling Region. Ph.D. Dissertation, University of Cape Town, Cape Town, South Africa.

Peterson, B.J., R.W. Howarth, and R.H. Garritt. (1985). Multiple stable isotopes used to trace the flow of organic matter in estuarine food webs. *Science* 227:1351–1363.

Peterson, W.T., D.F. Arcos, G.B. McManus, H. Dan, D. Bellatoni, T. Johnson, and P. Tiselius. (1988). The nearshore zone during coastal upwelling: daily variability and coupling between primary and secondary production off central Chile. *Prog. Oceanogr.* 20:1–40.

Pett, R.J. (1989). Kinetics of microbial mineralization of organic carbon from detrital *Skeletonema costatum* cells. *Mar. Ecol. Prog. Ser.* 52:123–128.

Pimm, S.L. (1982). *Food Webs*. Chapman and Hall, London.

Platt, T., D.V. Subba Rao, and B. Irwin. (1983). Photosynthesis of picoplankton in the oligotrophic ocean. *Nature* (London) 301:702–704.

Pomeroy, L.R. and R.E. Johannes. (1968). Respiration of ultraplankton in the upper 500 meters of the ocean. *Deep-Sea Res.* 15:381–391.

Pomeroy, L.R. and W.J. Wiebe. (1988). Energetics of microbial food webs. *Hydrobiologia* 159:7–18.

Pomeroy, L.R., H.H. Haskin, and R.A. Ragotzkie. (1956). Observations on dinoflagellate blooms. *Limnol. Oceanogr.* 1:54–60.

Pomeroy, L.R., G. Paffenhöfer, and J.A. Yoder. 1987. Summer upwelling on the southeastern Continental Shelf of the U.S.A. during 1981. Interactions of phytoplankton, zooplankton and microorganisms. *Prog. Oceanogr.* 19:353–372.

Pomeroy, L.R., M.E.Q. Pilson, and W.J. Wiebe. (1974). Tracer studies of the exchange of phosphorus between reef water and organisms on the windward reef of Eniwetok Atoll. In: *Proceedings of the 2nd International Symposium on Coral Reefs*. Great Barrier Reef Committee, Brisbane, pp. 87–96.

Porter, K.G. (1976). Enhancement of algal growth and productivity by grazing zooplankton. *Science* 192:1332–1334.

Porter, K.G., H. Paerl, R. Hodson, M. Pace, J. Priscu, B. Riemann, D. Scavia, and J. Stockner. (1988). Microbial interactions in lake food webs. In: *Complex Interactions in Lake Communities*. Springer-Verlag, New York, pp. 209–227.

Probyn, T.A. (1985). Nitrogen uptake by size-fractionated phytoplankton populations in the southern Benguela upwelling system. *Mar. Ecol. Prog. Ser.* 22:249–258.

Proctor, L.M. and J.A. Fuhrman. (1990). Viral mortality of marine bacteria and cyanobacteria. *Nature* (London) 343:60–62.

Richards, F.A., ed. (1981). *Coastal Upwelling*. American Geophysical Union, Washington, D.C.

Riemann, B. and M. Sondergaard. (1986). Regulation of bacterial secondary production in two eutrophic lakes and in experimental enclosures. *J. Plankton Res.* 8:519–536.

Ryther, J.H. (1955). Ecology of marine autotrophic dinoflagellates with reference to red water conditions. In: *The Luminescence of Biological Systems*. American Association for the Advancement of Science, Washington, D.C., pp. 387–414.

Semina, H. J. (1972). The size of phytoplankton cells in the Pacific Ocean. *Int. Rev. Gesamten Hydrobiol.* 57:177–205.

Sheldon, R.W., W.H. Sutcliffe, and M. A. Paranjape. (1977). Structure of pelagic food chain and relationship between plankton and fish production. *J. Fish. Res. Bd. Can.* 34:2344–2353.

Sieburth, J.M. (1977). International Helgoland Symposium: convener's report on the informal sessions on biomass and productivity of microorganisms in planktonic ecosystems. *Helgol. Wiss. Meeresunters.* 30:697–704.

Sieburth, J.M., K.M. Johnson, C.M. Burney, and D.M. Lavoie. (1977a). Estimation of in situ rates of heterotrophy using diurnal changes in organic matter and growth rates of picoplankton in diffusion culture. *Helgol. Wiss. Meeresunters.* 30:565–574.

Sieburth, J.M., V.B. Smetacek, and J. Lenz. (1977b). Pelagic ecosystem structure: heterotrophic compartments of the picoplankton and their relationship to plankton size fractions. *Limnol. Oceanogr.* 23:1256–1263.

Smetacek, V.B., K. von Brockel, B. Zeitschel, and W. Zenk. (1978). Sedimentation of particulate matter during a phytoplankton spring bloom in relation to the hydrographic regime. *Mar. Biol.* 47:211–226.

Smith, S.V. and F.T. Mackenzie. (1987). The ocean as a net heterotrophic system: implications from the carbon biogeochemical cycle. *Global Biogeochem. Cycles* 1:187–198.

Smith, S.V. and J.A. Marsh. (1973). Organic carbon production and consumption on the windward reef flat of Eniwetok Atoll. *Limnol. Oceanogr.* 18:953–961.

Stockner, J.G. and K.G. Porter. (1988). Microbial food webs in freshwater planktonic ecosystems. In: *Complex Interactons in Lake Communities*. Springer-Verlag, New York, pp. 69–83.

Stockner, J.G. and K.S. Shortreed. (1985). Whole-lake fertilization experiments in coastal British Columbia lakes: empirical relationships between nutrient inputs and phytoplankton biomass and production. *Can. J. Fish. Aquat. Sci.* 42:649–658.

Strickler, J.R. (1984). Sticky water: A selective force in copepod evolution. In: *Trophic Interactions within Aquatic Ecosystems*, American Association for the Advancement of Science, Washington, D.C., pp. 287–295.

Sverdrup, H.U. (1953). On conditions for the vernal blooming of phytoplankton. *J. Cons. Int. Explor. Mer.* 18:287–295.

Talling, J.F. (1966). The annual cycle of stratification and phytoplankton growth in Lake Victoria (East Africa). *Int. Rev. Gesamten Hydrobiol.* 51:545–621.

Tilman, D. (1982). *Resource Competition and Community Structure*. Princeton University Press, Princeton, New Jersey.

Wassman, P. (1987). Sedimentation of organic matter and silicate out of the euphotic zone of the Barents Sea. *EOS* 68:1728.

Wiegert, R.G. and D.F. Owen. (1971). Trophic structure, available resources, and population density in terrestrial vs. aquatic ecosystems. *J. Theor. Biol.* 30:69–81.

Woods, J.D. and R.L. Wiley. (1972). Billow turbulence and ocean microstructure. *Deep-Sea Res.* 19:87–121.

Zahary, T. (1988). Plankton respiration in a hypertrophic lake. *Arch. Hydrobiol. Beih. Ergebn. Limnol.* 31:151.

7
Primary and Secondary Production in Terrestrial Ecosystems

SAMUEL J. MCNAUGHTON, MARTIN OESTERHELD,
DOUGLAS A. FRANK, AND KEVIN J. WILLIAMS

Abstract

Ecosystem-level values of net primary productivity and herbivore biomass, consumption, and secondary productivity in terrestrial ecosystems were assembled from the literature. Data on belowground processes and trophic levels higher than herbivores were too rare in the literature to warrant a comparative analysis. All herbivore trophic-level properties were positively correlated with net primary productivity. Different ecosystem types were located at different positions on a common line. However, rather than regarding herbivore-level properties as simple consequences of producer properties, we believe it is more appropriate to regard primary production as an integrative variable indicative of processes throughout the ecosystem. Although the data were limited for ecosystems where there are periodic herbivore "outbreaks," the data available suggest that these pulsed ecosystems may be fundamentally different from ecosystems suffering chronic, consistent levels of herbivory.

Introduction

The roots of comparative ecosystem analysis and studies of trophic organization sink deeply into the origins of ecology. Edward Forbes (1844) described the distinct ecosystems occurring at different depths in the Aegean Sea and noted the regulating influences of tides and currents, substrate, and freshwater influxes upon those ecosystems. Stephen Forbes (1887) described the trophic organization of lakes, recognizing the importance of both predation and competition as regulating processes, and characterizing such cryptic but potentially important interactions as those between a facultatively predaceous plant and predatory fish. Thus, comparative analyses and characterization of trophic relationships are among the original endeavors of ecology.

A principal objective in preparing this chapter was to assemble data that would allow us to characterize the relationships between primary and secondary productivity across a diverse array of terrestrial ecosystem types. That task proved much more difficult than we had anticipated, given the antiquity of ecology's treatment of those topics. Computer literature searches, combining the key phrases of "primary production" and "secondary production," and variants, produced little. Recent articles treating both subjects did not lead, as had been expected, to a body of papers also dealing with both. Rather, each paper existed essentially independently, with little reference to other papers combining treatments of primary and secondary productivity. Most papers treating both topics tended to emphasize the characterization of processes in a particular type of ecosystem rather than a general consideration of productivity patterns. Therefore, our search finally became a journal-by-journal and book-by-book inspection to find data sources.

This chapter is organized as a consideration of general relationships between primary and secondary productivities, emphasizing factors influencing the linkage between those flows in ecosystems. We emphasize the first two trophic levels, primary producers and herbivores, because data on higher, predator trophic levels proved too rare, and too idiosyncratically determined, to provide the basis for a comparative analysis. Similarly, we confine our consideration to aboveground data, since belowground data were too rare in the literature to provide a comparative analysis. An abbreviated treatment of the topic appeared in McNaughton et al. (1989).

Strength of Trophic Coupling: Limits on Transfer

Three major, consolidated factors that can limit secondary productivity are the amount, availability, and suitability of food. Thus, understanding differences in secondary productivity among ecosystems depends on understanding (a) primary productivity and (b) trophic transfers. Limitations of primary production have been studied at levels from the individual plant to the globe (Lieth 1975a). Variables known to be correlated with primary productivity include precipitation, photosynthetically active radiation, evapotranspiration, temperature, and length of growing season (Lieth and Box 1972; Lieth 1975a). Some of the factors associated with primary production will have little effect on consumers other than through their limitation of primary production. For example, soil moisture may have major effects on primary production but have little effect upon secondary production except through food amount.

Primary production is correlated with precipitation in temperate grasslands (Van Dyne et al. 1976), temperate forests (Sharp et al. 1975), and tropical grasslands (Murphy 1975). Evapotranspiration has been correlated with primary production on a global basis (Lieth 1975a). Therefore, secondary productivity, to the extent that it is food quantity limited, can be expected to be associated with water balance.

In addition, water availability in terrestrial ecosystems can act directly to influence secondary production through drinking water supply (McNaughton and Georgiadis 1986). If drinking water limits secondary productivity, it would not be uniformly manifested in all groups of herbivores or in all ecosystems. Many cow-sized ungulates require 20–30 l of water a day, and water availability can constrain both foraging range and food processing (Robbins 1983). Other herbivores can, or must, fulfill their water requirements solely by extracting moisture from food. These organisms may be influenced by food succulence and may have physiological specializations that increase water extraction efficiency and reduce rates of water loss.

Food availability has both spatial and temporal components. Secondary productivity may commonly be limited by food availability in the leanest rather than the best season of the year. Length of growing season is positively correlated with net primary productivity on a global basis (Lieth 1975a), and the shortage of food in unproductive periods may be a principal limit on secondary productivity. Animals have evolved an array of mechanisms to cope with short growing seasons, including hibernation and migration. Such mechanisms will reduce secondary productivity by creating periods during which maintenance costs exceed food consumption.

Food quality also can be an important limit on secondary productivity. Quality of food for consumers is, of course, a vastly complex topic that ranges in scope from caloric density to nutrient balances to the quantities and properties of secondary plant compounds that can influence herbivores and detritivores (Coley et al. 1985). A detailed explication of these limits on secondary productivity is far beyond the scope of this chapter, but we do consider the role of secondary chemicals at the ecosystem level in the concluding discussion.

In addition to factors operating to limit secondary productivity through food supply, predation and disease can also regulate food web fluxes. Field studies provide clear evidence that predation can limit herbivore populations, maintaining them below food-limited levels (Caughley et al. 1980; Erlinge 1987). Kangaroo and emu densities in Australia are substantially higher in areas protected from dingo predation by fencing (Caughley et al. 1980). This provides evidence that predation can limit biomasses of principal prey, but the role of such limitation in system-level flows is unknown. Because trophic dynamics were not measured in concert with these studies, it is possible that other herbivores may be enhanced by the absence of kangaroos.

Similarly, little is known about the role of disease as a regulator of energy and chemical fluxes through the trophic web, although circumstantial evidence suggests that it could be extremely important. Pathogens could affect the rate and pathways of energy flow through food webs, although flow through the pathogens themselves may be minor. Certainly, evidence from domesticated animals indicates that disease can play a major regulatory role (Byerly 1977; Burnridge 1982). Little is known about the effects of disease on productivity of wild animals, but the potential for large effects is suggested by epidemics. When myxomatosis decimated rabbit populations in Great Britain, shifts in vegetation dynamics of plant communities were pronounced (Thomas 1963; Ross 1982). Rinderpest, or cattle plague, had a major effect in African ecosystems throughout the twentieth century (Plowright 1982), reducing susceptible ungulate populations to fractions of their density in the absence of the disease. Thus, a single pathogen, representing a negligible amount of energy flow, disrupted entire ecosystems for an extended period.

Although diseases and predators can certainly curb prey or host populations, this may not be translated into effects on secondary production. Rather, they may shift the balance among alternative pathways in the food web; from grazing food webs to detritus food webs, or from vertebrate- to invertebrate-based grazing webs. Maintenance of similar levels of secondary production in the presence of varying levels of predation or disease may depend on the variety of alternate pathways for energy and materials flow in the ecosystem.

Data on Primary and Secondary Production

To provide a comparative analysis of the relationships between primary and secondary production in terrestrial ecosystems, a data set consisting of measures of net aboveground primary production (NAP) was compiled from the literature; it encompasses 104 sites from nine terrestrial ecosystem types (Table 7.1). We believe this is a unique compilation, because for each value of NAP there was a corresponding value for either herbivore biomass (B), consumption (C), or secondary productivity (SP). If original authors did not give energy equivalents of biomass-based measurements, we converted the data to energy, using standard conversion factors (Golley 1968). It must be emphasized that the data compiled were collected with a variety of methods and goals. Many literature values were not included because herbivore-level properties were estimated from producer data. Other studies were eliminated because they included only a portion of the herbivore trophic level or were otherwise judged to be incomplete or inappropriate for inclusion. Therefore, rather than an exhaustive compilation of all literature on primary and secondary productivity, we attempted to assemble a discriminating data set in which primary and secondary trophic-level data were high-quality measurements taken in a coordinated fashion on the same sites. The data are restricted to plants and herbivores above ground because there was insufficient data to draw conclusions about higher trophic levels or belowground components.

TABLE 7.1. Data source references.

Reference	B[a]	C[a]	SP[a]
Batzli et al. 1980	*[b]	*	*
Bray 1964		*	
Bunderson 1986	*		
Cargill and Jefferies 1984		*	
Chew and Chew 1970	*	*	*
Coe et al. 1976	*		*
Coughenour et al. 1985	*	*	*
Coupland and Van Dyne 1979		*	*
DeAngelis et al. 1981		*	
East 1984	*		
French et al. 1979	*		
Golley et al. 1962	*		*
Gosz et al. 1978	*	*	*
Grier and Logan 1977		*	
Harris et al. 1975	*	*	
Haukioja and Koponen 1975	*	*	
Leigh and Windsor 1982		*	
McNaughton 1985		*	
Menhinick 1967	*	*	*
Nielson 1978		*	
Odum 1970	*		*
Ohmart et al. 1983		*	
Pfeiffer and Wiegert 1981		*	*
Schowalter et al. 1981	*	*	
Scott et al. 1979	*	*	*
Sinclair 1975	*	*	*
Smalley 1960		*	*
Van Hook 1971	*	*	*
Wiegert and Evans 1967	*		*
Wielgolaski 1975	*	*	

[a]Abbreviations: B, biomass; C, consumption; SP, secondary productivity.
[b]Asterisks indicate herbivore trophic-level properties obtained from each reference.

NET ABOVEGROUND PRODUCTIVITY

The range of NAP was from 126 kJ m^{-2} yr^{-1} for a desert to 29,490 kJ m^{-2} yr^{-1} for a tropical forest. Mean values of NAP for different types of ecosystems (Table 7.2) fell within the ranges for similar ecosystems estimated by Lieth (1975b). Typical trends in NAP were evident in the data, with desert and tundra having low values. Ecosystems with intermediate precipitation levels (grasslands) and growing seasons (temperate forests) had intermediate values of NAP. Highest NAP values were from tropical forests and salt marshes. Particularly because of an overrepresentation of grasslands compared to forests, the NAP data used here were skewed right, so they were log transformed for all statistical tests. Analysis of variance

TABLE 7.2. Sample sizes and means for ecosystems in data sets.

Ecosystem	NAP[a]		B[a]		C[c]		SP[a]	
	Sample size	Average NAP	Sample size	Average B	Sample size	Average C	Sample size	Average SP
Combined	104[b]	9,885	51	36	69	3,604	36	33.6
Desert	6	1,686	5	4	2	134	3	0.4
Tundra	6	1,075	6	2	6	67	4	0.8
Temperate grassland	20	4,941	7	4	14	674	7	60.9
Temperate old field	6	5,657	3	9	4	397	4	43.1
Tropical grassland	50	11,272	22	34	31	7,217	14	6.3
Temperate forest	*6	21,055	*2	5	*6	267	*1	3.3
Tropical forest	4	22,589	3	70	2	2,473	0	–
Salt marsh	3	18,109	0	–	3	1,442	2	192.0
Agriculture tropical grassland	2	12,778	2	50	0	–	0	–

[a]Abbreviations: NAP, net aboveground primary production; B, biomass; C, consumption; SP, secondary production.
[b]Units are kJ m^{-2} yr^{-1} for NAP, C, and SP, and kJ m^{-2} for B. Asterisks, Means for nonoutbreak years; data set includes an additional temperate forest point measured during an outbreak of insect herbivores.

(ANOVA) revealed significant differences ($p < .00001$) in NAP among types of ecosystems using either all 104 localities or any of the consumer-based subsets.

Trophic-Level Relationships

Secondary production of the herbivore trophic level and net aboveground primary production were positively related ($r^2 = .364$, df $= 34$, $p < .0001$) by

$$\log SP = 1.10 (\log NAP) - 3.27$$

where SP and NAP are in units of kJ m^{-2} yr^{-1} and "log" indicates common logarithms (Fig. 7.1). The relationship could not be discriminated statistically from a linear pattern with the conversion factor between primary and secondary productivity ranging between 0.1% and 0.2%. However, a considerable degree of dispersion around the line was associated

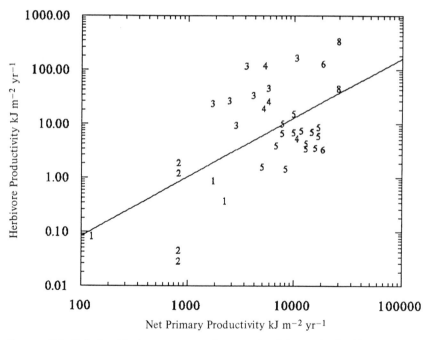

FIGURE 7.1. Relationship between secondary productivity at the herbivore trophic level and net primary productivity. Key to ecosystem types: 1, desert; 2, tundra; 3, temperate grassland; 4, temperate successional old field; 5, unmanaged tropical grassland (game parks); 6, temperate forest; 7, tropical forest; 8, salt marsh; 9, managed tropical grassland (agroecosystem).

with particular ecosystem types. Residuals for tropical grasslands were significantly lower than zero, indicating that these sites have a low SP/NAP ratio. In contrast, temperate grasslands and salt marshes fell well above the line, suggesting a high efficiency of conversion of NAP into SP. These patterns are related to the major herbivore classes in the different types of ecosystems. Ecosystems with invertebrates, principally insects, as the major herbivore class, invariably fell above the general line, while those with principally vertebrate herbivores, largely ungulates, consistently fell below the line (Fig. 7.2). Two conditions of a single temperate deciduous forest were included (Fig. 7.1): a point with low SP value corresponds to a "normal" level of herbivore activity, while the high SP point corresponds to a period of insect "outbreak." Similarly, the four tundra points represent two sites, each measured during both low and high years of rodent populations. Although NAP varied little, SP was much higher during the outbreak years. This suggests that some ecosystems may have a potential for extraordinary variation in SP, both spatially and temporally. These isolated cases, combined with widespread natural history observations of substantial herbivore fluctuations in many different types of ecosystems, suggest that fluxes into herbivore-based and detritus-based trophic webs may alternate over varying time periods in many ecosystems.

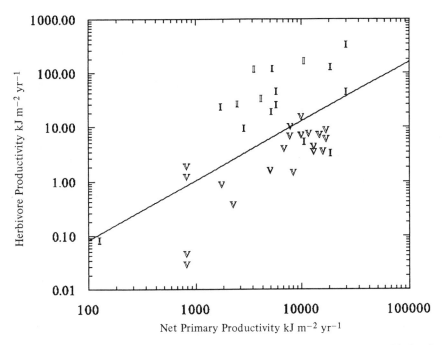

FIGURE 7.2. Relationship between secondary productivity at herbivore trophic level and net primary productivity with life form of dominant herbivores indicated as I (invertebrates) and V (vertebrates).

Herbivore biomass also was positively related ($r^2 = .579$, df $= 49$, $p < .00001$) to NAP by

$$\log B = 1.52\,(\log NAP) - 4.79$$

with units in kJ m^{-2} and kJ m^{-2} yr^{-1}, respectively (Fig. 7.3). Because the slope of this line was significantly larger than one, B has an exponential relationship to NAP. The ratio of herbivore biomass to NAP increased from around 0.0005 yr^{-1} at low NAP to 0.0025 yr^{-1} at highest NAP. In contrast to the relationship between SP and NAP, there was no evidence that any ecosystem type departed significantly from the common line (residual test, $p = .8$). Tropical grasslands and tropical forests had the highest values for herbivore biomass.

The increase in herbivore biomass with NAP (Fig. 7.3) was not translated into a proportional increase in secondary production (see Fig. 7.1). A fivefold increase in the B : NAP ratio was accompanied by just a twofold increase in the SP : NAP ratio. Therefore, as NAP and B increase, the turnover of herbivore biomass decreases. This pattern across systems was

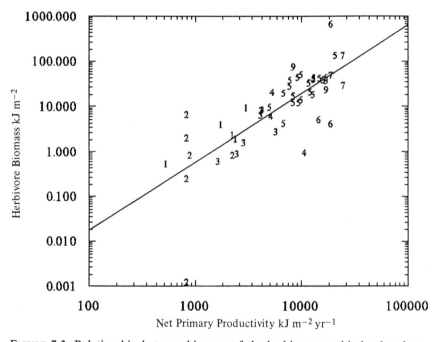

FIGURE 7.3. Relationship between biomass of the herbivore trophic level and net primary productivity. Ecosystem symbols: 1, desert; 2, tundra; 3, temperate grassland; 4, temperate successional old field; 5, unmanaged tropical grassland (game parks); 6, temperate forest; 7, tropical forest; 8, salt marsh; 9, managed tropical grassland (agroecosystem).

Reprinted by permission from Nature vol. 341 p. 143
Copyright © 1989 Macmillan Magazines Ltd.

also evident within groups of similar systems. For tropical and temperate grasslands, considered as separate groups, B increased significantly with NAP ($r^2 = .49, p < .0003$, tropics; $r^2 = .65, p < .03$, temperate), but SP did not ($r^2 = .12$ and $.48$, respectively, neither significant at $p = .05$).

Consumption by herbivores was positively related ($r^2 = .383$, df $= 67$, $p < .00001$) to NAP by

$$\log C = 1.38 (\log NAP) - 2.32$$

with units as before (Fig. 7.4A). There were some notable departures from the common line, however. Tropical grasslands were consistently above and temperate and tropical forests consistently below the line. First inspection of these departures suggested that secondary chemicals, which are more common in tree than grass foliage (Coley et al. 1985), could act as a significant regulator of herbivory. However, much of the production in forests is wood, a generally invulnerable food class for herbivores. We therefore investigated whether low consumption rates in forests simply reflected a low proportion of leaf production or were a consequence of low levels of herbivory on the foliage itself. When only foliage production (NFP) was considered for forest ecosystems, the relationship between C and NFP was highly correlated ($r^2 = .594$, df $= 73, p < .00001$) by

$$\log C = 2.04 (\log NFP) - 4.80$$

with units as before (Fig. 7.4B). Forest points appear incorporated into the same cloud of points as the herbaceous ecosystems. This indicates that the apparently low levels of consumption in forests are due to production of wood. In fact, a range test of the proportion of foliage consumed in the different ecosystem types segregated only tropical grasslands, with a mean percentage consumption of 61.6%, from the temperate systems (mean, 4.7%–7.9%); the only tropical forest point for this variable had an intermediate level of consumption, 29.2%. The best fitted line describing the overall relationship of C and NFP suggested that the percentage of NFP consumed by herbivores rose from 2% at low NFP to 83% at high NFP. A 30-fold increase in primary productivity was associated with an increase in consumption by herbivores of three orders of magnitude.

Because both biomass and consumption were related to NAP by a regression slope greater than 1, while secondary productivity was linearly related with a larger dispersion of points, two conclusions are suggested: first, the efficiency with which consumption is converted into SP by the herbivores changes with NAP; second, that efficiency is different for different types of ecosystems and may explain part of the cross-ecosystem variation in SP that is not explained by NAP (see Fig. 7.1).

Consumption and SP were closely and positively related ($r^2 = .67$, df $= 23, p < .00001$) by

$$\log SP = 1.03 (\log C) - 1.42$$

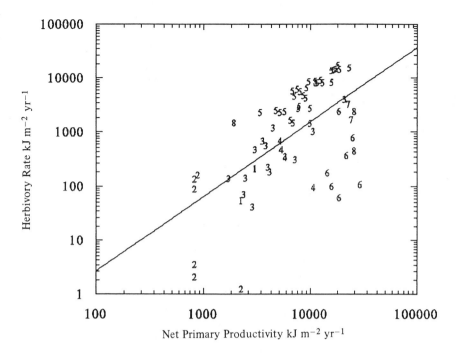

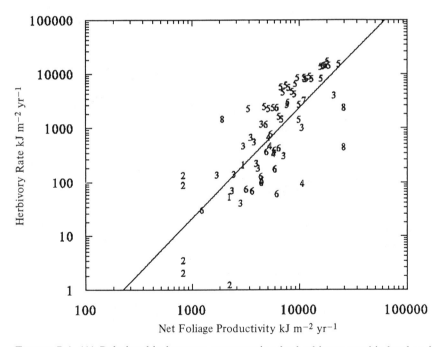

FIGURE 7.4. (A) Relationship between consumption by herbivore trophic level and net primary productivity. Ecosystem symbols as in Figs. 7.1 and 7.2. (B) As in Fig. 7.4A, but with wood production removed from forest ecosystems so only foliage production is represented. Ecosystem symbols as in Figs. 7.1 and 7.2.

with units as before (Fig. 7.5). However, an inspection of these data indicates that there are two significantly different families of points: those below the line are ecosystems dominated by vertebrates, and those above the line are dominated by invertebrates. The average efficiency of conversion of C into SP was 0.4 ± 0.06% in the former and 9.6 ± 1.8% in the latter. Although the number of points used to estimate these means was low, differences in efficiency have obvious physiological explanations. Thus, the type of dominant herbivores, such as invertebrates in temperate grasslands and salt marshes, and ungulates in tropical grasslands, explain much of the variation in SP unrelated to NAP (Figs. 7.2 and 7.5).

These data reveal several general patterns in the aboveground portions of terrestrial ecosystems:

1. Secondary production, herbivore biomass, and consumption are strong correlates of primary production.
2. Herbivore biomass and consumption vary exponentially with NAP, but SP varies linearly with primary production.
3. Low consumption levels in forests largely reflect the relative proportion of foliage production rather than intrinsic differences in foliage consumption.

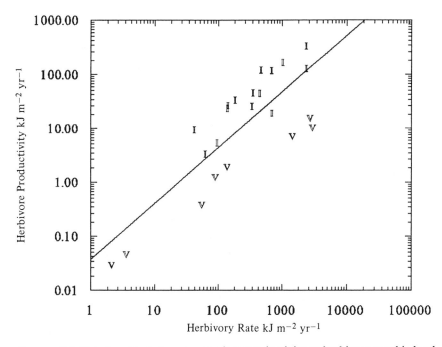

FIGURE 7.5. Relationship between secondary productivity at herbivore trophic level and consumption by that level. Dominant herbivores: I, invertebrates; V, vertebrates.

4. The proportion of vertebrates (homeotherms) and invertebrates (heterotherms) in different ecosystem types influences the efficiency of conversion of consumed food into new biomass and, therefore, explains part
   of the variation in secondary production not related to NAP.

Primary and Secondary Productivity in Terrestrial Ecosystems

The comparative ecosystem analysis reveals that although there are distinct
differences in energy flow patterns in different types of terrestrial ecosystems, these are not categorical differences but differences of degree in common patterns that transcend typology. Although the data assembled were
confined to the first two trophic levels, because of the limitations of the
ecological literature, there seems no reason to suppose that they do not
generally apply to higher trophic levels in the grazing food web and to the
detritus web as well. All major attributes of the herbivore trophic level —
energy amount (biomass), energy input (consumption), and energy flux
(production) — were positively related to net primary productivity.

We believe, however, that it would be incorrect to infer that energetic
properties of the herbivore level are merely a passive consequence of independently determined energy flow into the primary producers. Consumers
in a wide variety of ecosystems have major regulatory effects on ecosystem
structure and processes that are transmitted throughout the food web (Carpenter and Kitchell 1988; Huntly and Inouye 1988; McNaughton et al.
1988; Naiman et al. 1988; Pastor et al. 1988; Whicker and Detling 1988).
Consequently, it is more reasonable to regard net primary productivity as
an integrative and indicative variable that reflects processes throughout the
trophic web than to consider it an independent processes dragging other
trophic levels along with it.

Tropical and temperate forests fell below the fitted line relating C and
NAP. But when the relationship was reexamined with values of NAP restricted to foliage and eliminating the production sequestered in wood,
forests were comparable to other, quite different ecosystems. This is an
interesting result, because the foliage of tropical (Coley 1983a) and temperate (Feeny 1968; Edmunds and Alstad 1978) forests contains abundant and
diverse secondary compounds, while grasses are much less heavily defended
by those chemicals. Chemical defenses, therefore, are less important as
regulators of ecosystem processes than they are at the individual (Feeny
1976) and population (Coley et al. 1985) levels. That is, secondary chemicals may have an important role as regulators of feeding preferences and
modes of feeding, influencing the types of herbivores that dominate consumption in ecosystems with different types of primary producers. But
they seem unimportant factors in the regulation of overall energy flow in
ecosystems.

Tropical grasslands, particularly in African game parks, support significantly higher rates of consumption than have been measured in most other terrestrial ecosystems. Abundant ungulates in tropical nature reserves have been shown to be influential organizers of plant life form balance (McNaughton and Sabuni 1988), species composition (McNaughton 1983), growth form within species (McNaughton 1984), energy flow (McNaughton 1985), and nutrient cycling (McNaughton et al. 1988) in those ecosystems. This suggests that tropical grasslands may be more comparable to many aquatic grazing systems (Carpenter et al. 1985; Carpenter and Kitchell 1988) than to many other terrestrial ecosystems (McNaughton 1983, 1985). However, our literature search also suggests that there may have been a systematic exclusion of herbivory in terrestrial ecosystem studies. Measuring herbivory in grasslands dominated by large herbivores is more feasible technically than estimating the chronic herbivory from invertebrates in forest ecosystems.

We were surprised that we were able to find only two estimates of herbivory in temperate forests during herbivore "outbreak" years, particularly considering the economic importance attached to such outbreaks by the lumber industry. Given the short life cycles, cryptic nature, and often ephemeral occurrence of invertebrate herbivores, has their importance been grossly underestimated in terrestrial ecosystem studies? We were also surprised to find no measurements of herbivory in temperate grasslands during a grasshopper outbreak, although such eruptions are both common and of evident importance to ecosystem function when they do occur. Finally, although herbivores are acknowledged to have major effects in tropical forests (Coley 1983a,b; Brown and Ewel 1987), whole-system measurements are virtually nonexistent.

TYPE OF HERBIVORE

The dominant type of herbivore is an important characteristic of ecosystems that influences allocation of the energy consumed. Secondary production is higher and respiration costs lower per unit consumed for heterotherms, largely insects in the data presented here, than for homeotherms, largely ungulates in this data set. Temperate grasslands, old fields, and salt marshes have herbivore trophic levels dominated by insects, while tropical grasslands are most distinctly dominated by large ungulate herbivores in our data set. Consequently, consumption rates are higher but SP is lower per unit NAP for tropical grasslands compared with other ecosystems. Although no comprehensive measurement of secondary productivity in tropical forests was found, we suspect that tropical forests may have the highest secondary productivities of any terrestrial ecosystems because of the importance of insects as consumers (Coley 1983a,b). Nevertheless, undisturbed tropical forests also contain abundant avifaunas and primates, which would tend to deflate the herbivore–SP/NAP balance. Clearly, good inclusive data on herbivory in tropical forests would add considerably to our knowledge of the comparative properties of terrestrial ecosystems. Similarly, estimates of chronic herbivory in temperate ecosystems, and of acute herbivory during herbivore eruptions, would both add immensely to our understanding of ecosystem processes.

The tundra, during most years when rodent numbers are low, supports low levels of herbivore biomass, consumption, and secondary production. During peak rodent years, however, energy is transferred at rates typical of the general trend. Herbivore outbreaks are characteristic of deserts (White 1976) and temperate forests (Martinat 1987), as well as tundras (Batzli et al. 1981). All tend to fall below the trend line during non-outbreak years and near or above the line during years of high herbivore density. There are definite, but as yet obscure, differences between these pulse-organized ecosystems and others that sustain more even levels of consumption.

Our comparative ecosystem analysis suggests that a full understanding of the relationships between primary and secondary productivity would be reinforced by (a) more information on NAP and herbivore parameters in extreme, tropical forest and desert ecosystems; (b) expansion of data up the trophic web to higher trophic levels; (c) a much greater body of information on belowground processes and dynamics; and (d) greater consideration of the types of consumers dominating ecosystems with fundamentally different primary producers.

Conclusions

We were both surprised and gratified by the comparatively strong correlations among producer and herbivore trophic level properties across a vast range of ecosystem types and primary productivities. We believe that the results of this comparative ecosystem analysis have three principal implications for ecosystem science.

First, primary productivity appears to be an integrator and indicator of general ecosystem processes. It is likely that levels of net production also reflect belowground processes related to the decomposer food web, such as net mineralization and resultant plant nutrient availability. We consequently see no reason to suppose that process rates in detritus food webs would not generally reflect the process-rate relationships documented here. This suggests that NAP measurements, generally more feasible than either consumer- or decomposer-level measurements, may provide considerable information about overall ecosystem processes.

Second, as a result of the integrative nature of primary productivity, it may be possible for the large-scale application of ecology during the International Geosphere-Biosphere Program to gain substantial insights by satellite monitoring (Tucker et al. 1985) of the seasonal vegetation green wave (McNaughton 1979). This green wave is a likely indicator of processes occurring both below ground and above ground at other trophic levels.

Finally, we think it is likely that there are fundamental differences between ecosystems where herbivory is comparatively constant temporally and those that are temporally pulsed at time scales beyond interseasonal variation. The flow of primary production is usually directly into detritus food webs in such pulsed systems, but there are aperiodic or periodic flushes through herbivore webs, while the balances between these flows are more constant in other ecosystems. Such flow pulsation is not producer life form or NAP related, because it is as characteristic of low-productivity deserts and tundras as it is of high-productivity forests. More attention to the levels of chronic herbivory in forests, and to the magnitudes and frequencies of herbivore eruptions characteristic of a variety of ecosystems, would provide a much more reliable and thorough database than currently exists for a comparative analysis of the relationships between primary and secondary production in terrestrial ecosystems.

Acknowledgments. Preparation of this chapter was supported by NSF BSR-8817934. M. Oesterheld was partially supported by a fellowship from CONICET, Argentina, and D.A. Frank by a fellowship from the Syracuse University Graduate School.

REFERENCES

Batzli, G.O., R.G. White, and F.L. Bunnell. (1981). Herbivory: A strategy of tundra consumers. In: *Tundra Ecosystems: a Comparative Analysis: International Biological Program*, L.C. Bliss, O.W. Heal, and J.J. Moore, eds. Cambridge University Press, Cambridge, pp. 359–375.

Batzli, G.O., R.G. White, S.F. MacLean, F.A. Pitelka, and B.D. Collier. (1980). The herbivore-based trophic system. In: *An Arctic Ecosystem: The Coastal Tundra at Barrow, Alaska*, J. Brown, P.C. Miller, L.L. Tieszen, and F.L. Bunnell, eds. Dowden Hutchinson & Ross, Stroudsburg, pp. 335–410.

Bray, R.J. (1964). Primary consumption in three forest canopies. *Ecology* 45:165–167.

Brown, B.J. and J.J. Ewel. (1987). Herbivory in complex and simple tropical successional ecosystems. *Ecology* 68:108–116.

Bunderson, W.T. (1986). Primary production and energetic relationships of wild and domestic ungulate communities in African ecosystems. In: *Rangelands: A Resource under Siege*, P.J. Joss, P.W. Lynch, and O.B. Williams, eds. Australian Academy of Sciences, Canberra, pp. 397–401.

Burnridge, M.J. (1982). Effects of disease on animal productivity. In: *CRC Handbook of Agricultural Productivity*, M. Rechcigh, ed. CRC Press, Boca Raton, pp. 319–343.

Byerly, T.C. (1977). Ruminant livestock research and development. *Science* 195:450–456.

Cargill, S.M. and R.L. Jefferies. (1984). The effects of grazing by Lesser Snow Geese on the vegetation of a sub-arctic salt marsh. *J. Appl. Ecol.* 21:669–686.

Carpenter, S.R. and J.F. Kitchell. (1988). Consumer control of lake productivity. *BioScience* 38:764–769.

Carpenter, S.R., Kitchell, R.J., and J.R. Hodgson. (1985). Cascading trophic interactions and lake productivity. *BioScience* 35:634–639.

Caughley, G., G.C. Grigg, J. Caughley, and G.J.H. Hill. (1980). Does dingo predation control densities of kangaroos and emus? *Aust. J. Wildl.* 7:1–12.

Chew, R.M. and A.E. Chew. (1970). Energy relationships of the mammals of a desert shrub (*Larrea tridentata*) community. *Ecol. Monogr.* 40:1–21.

Coe, M.J., Cumming, D.H. and J. Phillipson. (1976). Biomass and production of large African herbivores in relation to rainfall and primary production. *Oecologia* 22:341–354.

Coley, P.D. (1983a). Herbivory and defensive characteristics of tree species in a lowland tropical forest. *Ecol. Monogr.* 53:209–233.

Coley, P.D. (1983b). Intraspecific variation in herbivory on twotropical tree species. *Ecology* 64:426–433.

Coley, P.D., Bryant, J.P., and C.S. Chapin III. (1985). Resource availability and plant antiherbivore defense. *Science* 230:895–899.

Coughenour, M.B., J.E. Ellis, D.M. Swift, D.L. Coppock, K. Galvin, J.T. McCabe, and T.C. Hart (1985). Energy extraction and use in a nomadic pastoral ecosystem. *Science* 230:619–625.

Coupland, R.T. and G.M. Van Dyne. (1979). System synthesis. In: *Grassland Ecosystems of the World; International Biological Programme*, R.T. Coupland, ed. Cambridge University Press, Cambridge, pp. 97–106.

DeAngelis, D.L., R.H. Gardner, and H.H. Shugart. (1981). Productivity of forest ecosystems studied during the IBP: the woodlands data set. In: *Dynamic Properties of Forest Ecosystems*, D.E. Reichle, ed. Cambridge University Press, New York, pp. 567–672.

East, R. (1984). Rainfall, soil nutrient status and biomass of large African savanna mammals. *Afr. J. Ecol.* 22:245–270.

Edmunds, G.F. and D.N. Alstad. (1978). Coevolution in insect herbivores and conifers. *Science* 199:941–945.

Erlinge, S. (1987). Predation and noncyclicity in a microtine population in southern Sweden. *Oikos* 50:347–352.

Feeny, P.P. (1968). Effects of oak leaf tannins on larval growth of the winter moth *Operophtera brumata*. *J. Insect Physiol.* 14:805–817.

Feeny, P.P. (1976). Plant apparancy and chemical defense. In: *Biochemical Advances Between Plants and Insects. Recent Advances in Phytochemistry*, J. Wallace and R.L. Mansell, eds. Plenum Press, New York, pp. 1–40.

Forbes, E.A. (1844). Report on the mollusca and radiata of the Aegean Sea. *Rep. Br. Assoc. Adv. Sci.* 13:130–193.

Forbes, S.A. (1887). The lake as microcosm. *Bull. Peoria Sci. Assoc.* 17:77–87.

French, N.R., R.K. Steinhorst, and D.M. Swift. (1979). Grassland biomass trophic pyramids. In: *Perspectives in Grassland Ecology; Ecological Studies*, N. French, ed. Springer-Verlag, New York, pp. 59–88.

Golley, F.B. (1968). Secondary productivity in terrestrial communities. *Am. Zool.* 8:53–59.

Golley, F.B., H.T. Odum, and R.F. Wilson. (1962). The structure and metabolism of a Puerto Rican red mangrove forest in May. *Ecology* 43:9–19.

Gosz, J.R., R.T. Holmes, G.E. Likens, and F.H. Bormann. (1978). The flow of energy in a forest ecosystem. *Sci. Am.* 238:93–102.

Grier, C.G. and R.S. Logan. (1977). Old-growth *Pseudotsuga menziesii* communities of a western Oregon watershed: biomass distribution and production budgets. *Ecol. Monogr.* 47:373–400.

Harris, W.F., P. Sollins, N.T. Edwards, B.E. Dinger, and H.H. Shugart. (1975). Analysis of carbon flow and productivity in a temperate deciduous forest ecosystem. In: *Productivity of World Ecosystems*, D.E. Reichle, J.F. Franklin, and D.W. Goodall, eds. National Academy of Sciences, Washington, D.C., pp. 116–122.

Haukioja, E. and S. Koponen. (1975). Birch herbivores and herbivory at Kevo. In: *Fennoscandian Tundra Ecosystems. Part 2. Animals and Systems Analysis*, F.E. Wielgolaski, ed. Springer-Verlag, Heidelberg, pp. 182–188.

Huntly, N. and R. Inouye. (1988). Pocket gophers in ecosystems: patterns and mechanisms. *BioScience* 38:786–793.

Leigh, E.G. and D.M. Windsor. (1982). Forest production and regulation of primary consumers on Barro Colorado Island. In: *The Ecology of a Tropical Forest: Seasonal Rhythms and Long-Term Changes*, E.G. Leigh, A.S. Rand, and D.M. Windsord, eds. Smithsonian Institution, Washington, D.C., pp. 111–122.

Lieth, H. (1975a). Primary productivity of the major vegetation units of the world. In: *Primary Productivity of the Biosphere*, H. Lieth and R.H. Whittaker, eds. Springer-Verlag, Berlin, pp. 201–215.

Lieth, H. (1975b). Modeling primary productivity of the world. In: *Primary Productivity of the Biosphere*, H. Lieth and R.H. Whittaker, eds. Springer-Verlag, Berlin, pp. 237–263.

Lieth, H. and E. Box. (1972). Evapotranspiration and primary productivity; C.W. Thornwaite Memorial Model. In: *Publications in Climatology*, Vol. 25, R.J. Mather, ed. C.W. Thornwaite Associates, New Jersey, pp. 37–46.

Martinat, P.J. (1987). The role of climatic variation and weather in forest insect outbreaks. In: *Insect Outbreaks*, P. Barbosa and J.C. Schultz, eds. Academic Press, New York, pp. 241–268.

McNaughton, S.J. (1979). Grassland-herbivore dynamics. In: *Serengeti. Dynamics of an Ecosystem*, A.R.E. Sinclair and N. Norton-Griffiths, eds. University of Chicago, Chicago, pp. 46–81.

McNaughton, S.J. (1983). Serengeti grassland ecology: the role of composite environmental factors and contingency in community organization. *Ecol. Monogr.* 53:291–320.

McNaughton, S.J. (1984). Grazing lawns: animals in herds, plant form, and coevolution. *Am. Nat.* 124:863–886.

McNaughton, S.J. (1985). Ecology of a grazing ecosystem: the Serengeti. *Ecol. Monogr.* 55:259–294.

McNaughton, S.J. and N.J. Georgiadis. (1986). Ecology of African grazing and browsing mammals. *Annu. Rev. Ecol. Syst.* 17:39–65.

McNaughton, S.J. and G.A. Sabuni. (1988). Large African mammals as regulators of vegetation structure. In: *Plant Form and Vegetation Structure*, M.J.A. Werger, P.J.M. van der Aart, H.J. During, and J.T.A. Verhoven, eds. SPB Publishing, The Hague, Netherlands, pp. 339–354.

McNaughton, S.J., R.W. Ruess, and S.W. Seagle. (1988). Large mammals and process dynamics in African ecosystems. *BioScience* 38:794–800.

McNaughton, S.J., M. Oesterheld, D.A. Frank, and K.J. Williams. (1989). Ecosystem-level patterns of primary productivity and herbivory in terrestrial habitats. *Nature* (London) 341:142–144.

Menhinick, E.F. (1967) Structure, stability, and energy flow in plants and arthropods in a *Sericea lespedeza* stand. *Ecol. Monogr.* 37:255–273.

Murphy, P.G. (1975). Net primary productivity in tropical terrestrial ecosystems. In: *Primary Productivity of the Biosphere*, H. Lieth and R.H. Whittaker, eds. Springer-Verlag, Berlin, pp. 217–231.

Naiman, R.J., C.A. Johnston, and J.C. Kelley. (1988). Alteration of North American streams by beaver. *BioScience* 38:753–762.

Nielson, O. (1978). Aboveground food resources and herbivory in a beech forest ecosystem. *Oikos* 31:273–279.

Odum, H.T. (1970). A Tropical Rain Forest; a Study of Irradiation and Ecology at El Verde, Puerto Rico. National Technical Information Center, Springfield, Virginia.

Ohmart, C.P., L.G. Stewart, and J.R. Thomas. (1983). Leaf consumption by insects in three Eucalyptus forest types in southeastern Australia and their role in short-term nutrient cycling. *Oecologia* 59:322–330.

Pastor, J., R.J. Naiman, B. Dewey, and P. McInnes. (1988). Moose, microbes, and the boreal forest. *BioScience* 38:770–777.

Pfeiffer, W.J. and R.G. Wiegert. (1981). Grazers on *Spartina* and their predators. In: *The Ecology of a Salt Marsh; Ecological Studies*, L.R. Pomeroy and R.G. Wiegert, eds. Springer-Verlag, New York, pp. 87–112.

Plowright, W. (1982). The effects of rinderpest and rinderpest control on wildlife in Africa. *Symp. Zool. Soc. Lond.* 50:1–28.

Robbins, C.T. (1983). *Wildlife Feeding and Nutrition*. Academic Press, New York.

Ross, J. (1982). Myxomatosis: The natural evolution of the disease. *Symp. Zool. Soc. Lond.* 50:77–95.

Schowalter, T.D., J.W. Webb, and D.A. Crossley, Jr. (1981). Community structure and nutrient content of canopy arthropods in clearcut and uncut forest ecosystems. *Ecology* 62:1010–1019.

Scott, J.A., N.R. French, and J.W. Leetham. (1979). Patterns of consumption in grasslands. In: *Perspectives in Grassland Ecology; Ecological Studies*, N. French, ed. Springer-Verlag, New York, pp. 89–106.

Sharp, D.D., H. Lieth, and D. Whigham. (1975). Assessment of regional productivity in North Carolina. In: *Primary Productivity of the Biosphere*, H. Lieth and R.H. Whittaker, eds. Springer-Verlag, Berlin, pp. 131–146.

Sinclair, A.R.E. (1975). The resource limitation of trophic levels in tropical grassland ecosystems. *J. Anim. Ecol.* 44:497–520.

Smalley, A.E. (1960). Energy flow of a salt marsh grasshopper population. *Ecology* 41:672–677.

Thomas, A.S. (1963). Further changes in vegetation since the advent of myxomatosis. *J. Ecol.* 51:151–183.

Tucker, C.J., J.R.G. Townshed, and T.E. Goff. (1985). African land-cover classification using satellite data. *Science* 227:369–375.

Van Hook, R.I. (1971). Energy and nutrient dynamics of spider and orthopteran populations in a grassland ecosystem. *Ecol. Monogr.* 41:1–26.

Van Dyne, M.G., M.F. Smith, R.L. Czaplewski, and R.G. Woodmansee. (1976).

Analysis and synthesis of grassland ecosystem dynamics. In: *Patterns of Primary Production in the Biosphere*, H. Leith, ed. Dowden Hutchison & Ross, Stroudsburg, pp. 199–204.

Whicker, A.D. and J.K. Detling. (1988). Ecological consequences of prairie dog disturbances. *BioScience* 38:778–785.

White, T.C.R. (1976). Weather, food and plagues of locusts. *Oecologia* 22:119–134.

Wiegert, R.G. and F.C. Evans. (1967). Investigations of secondary productivity in grasslands. In: *Secondary Productivity of Terrestrial Ecosystems*, K. Petrusewicz, ed. Krakow, Poland, pp. 499–518.

Wielgolaski, F.E. (1975). Functioning of Fennoscandian tundra ecosystems. In: *Fennoscandian Tundra Ecosystems*, F.E. Wielgolaski, ed. Springer-Verlag, Heidelberg, pp. 300–326.

8
Comparing Ecosystem Structures: The Chesapeake Bay and the Baltic Sea

ROBERT E. ULANOWICZ AND FREDRIK WULFF

Abstract

Much can be learned about the differences in how two distinct ecosystems are functioning by studying their respective quantitative networks of trophic exchanges. The feeding processes occurring in the ecosystems of the Baltic Sea and the Chesapeake Bay have been parsed into similar 17-node networks of carbon flows, and the use of network analysis to compare the structures reveals that the Baltic ecosystem functions less as a hypertrophic system than does the community in the Chesapeake: The diversity of trophic connections is higher in the Baltic community and the specificity of feeding (trophic articulation) is greater there as well. More pathways for recycle are evident in the Baltic system, and they involve a larger proportion of higher species. Trophic efficiencies are generally greater in the Baltic than they are in the Chesapeake. On a relative basis, about four times as much carbon fixed in primary productivity reaches the fishes of the Baltic than is transferred up the trophic web of the Chesapeake. This disparity is explained in part by a large indirect subsidy that Chesapeake fishes receive from inputs of allochthonous carbon that is processed for them by the benthic community of deposit feeders.

Introduction

The reasons for comparing ecosystems are both academic and practical. On the academic side, by juxtaposing different ecosystems, or configurations of the same ecosystem at different times, one hopes to discover why and how natural communities come to be structured as they are (e.g., Odum 1969). Motivating this activity is the widespread (but not universal) belief that ecosystems develop in a way that can be generalized and possibly quantified.

While academicians may be concerned with the process of ecosystems "becoming," the popular lay belief seems to be that ecosystems already exist in some ideal natural state that too often is degraded by the activities of humankind. Regardless of one's opinions about "ecogenesis," managers are continually being asked to judge how much a particular ecosystem as a whole has been damaged. To be able to render such judgments requires a background of copious ecosystem comparisons that, it is hoped, indicates some fundamental ecological principles.

How to compare ecosystems is no trivial question, as evidenced by the contents of this volume. At the core of the issue stands the matter of scale – which temporal and spatial scales are best for comparing systems?

At the smaller scales, investigating mechanisms behind the fluctuations of individual populations is almost sure to provide useful information about what ails a disturbed ecosystem. Some contend that the mechanistic approach is the only legitimate route to ecological knowledge (e.g., Lehman 1986). But others question the sufficiency of mechanistic descriptions. They contend that cybernetic and homeostatic processes inextricably involve entire groups of populations and function in a way that makes the prediction of overall behavior from a knowledge of the component mechanisms a highly problematic activity at best (e.g., Platt et al. 1981; Ulanowicz 1986).

As an alternative to reductionistic ecology, the investigator might identify one, or at most a few, measurable properties thought to characterize the status of a system. For example, Vollenweider (1969) employed the magnitude of nutrient loading and water residence time to classify the trophic status of lakes. Peters (1986) identified more than 30 other correlations, such as phosphorous burden versus chlorophyll level or fish yield versus nutrient concentration, that have been used to classify aquatic systems in an activity that he calls "empirical limnology." While such phenomenological endeavors afford useful comparisons at the whole-system level, they usually do not divulge exactly where in the system difficulties are arising. Furthermore, marine ecologists would find it hard to pursue an analogous "empirical marine ecology." It is far more difficult to perform experiments in the ocean than in lakes, and at this time one cannot easily identify a number of independent marine ecosystems that would compare with Peters' 48 Canadian Shield lakes.

Ideally, one seeks a comparison of at least two ecosystems that quantifies the differences in the overall structure and allows one to trace the origins of those differences.

Networks – A Bridge between Holism and Reductionism

We wish to argue that a compromise exists between the need for overwhelming amounts of data to pursue reductionism and the often unsatisfactory loss of information inherent in phenomenological holism, such as that advocated by Peters. Ecologists long have been fond of constructing whole-system budgets of material and energy transactions in ecosystems.

Their results are usually represented as a network of exchanges among the elements of the communities under study. Because these frameworks are cast in the material or energetic units common to any system, the task of system intercomparison is greatly facilitated. One could say metaphorically that such quantified networks resemble a skeletal structure of ecosystem processes.

There are investigators who will regard any representation of ecosystems in such simplistic terms as a brutish attempt to sidestep or ignore the wonderful intricacies at play in ecosystems (e.g., Engleberg and Boyarsky 1979). Virtually all naturalists (including these authors) will regret that elements such as mating rituals, territorial displays, or rates of flagellation by algae do not appear explicitly in the network of ecosystem exchanges. Our assumption is not that such processes are unimportant. Quite the opposite: Because they are important, their effects are broadly impressed upon the ecosystem, and, in particular, these nonmaterial agents play a significant (but not exclusive) role in organizing the framework of palpable exchanges. To continue the earlier analogy, the discovery of a skeleton of a prehistorical individual does not allow the anthropologist to perceive directly the physiology or environment of that distant ancestor. However, modern osteologists are able to infer a surprising amount about the life and times of the deceased through a meristic study of its remains.

It should be noted that one may relate the flow network to the "corpus" of an ecosystem without subscribing to the Clementsian notion of the ecosystem as a "superorganism." To claim that an ecosystem possesses structure and undergoes development is not to elevate its ontological status above that of an individual organism. However, the analogy with anatomy does cast the current status of ecosystems science in an enlightening perspective. Leonardo da Vinci sketched the structure of the human body long before the functions of many of the body parts were fully appreciated and certainly before most physiological mechanisms were discovered. Anatomical description was a prerequisite to the advances in physiology that were to follow. Ecology today needs a clearer picture of ecological structure if ever a full understanding of ecosystem functioning is to be achieved some distant tomorrow.

No sooner does one decide to use ecosystem flow networks to effect a "comparative anatomy" than one is immediately beset by two significant practical difficulties: (1) Considerable effort and resources are required to describe quantitatively the networks of two or more ecosystems. (2) The networks that result can be disturbingly complicated, sometimes so complex as to cause one to wonder whether any meaningful comparison is possible.

As regards the first obstacle, progressively more ecosystem networks are being described with moderate resolution (say 15 or more compartments). After enough quantified networks appeared, it was only a matter of time before the probability became high that two or more of them could serve as the basis of a "comparative anatomy." In this instance, Wulff had collected copious data on exchanges in the Baltic Sea ecosystem, while Baird and Ulanowicz (1989) had delimited a counterpart for the Chesapeake Bay. In late 1986, the authors became aware of each other's work and of the potential benefits that might derive from comparing their respective systems.

As for complexity, it lately has been the subject of much interest in physics and biology. Recent work in applied mathematics (described later) allows one to analyze complex networks in a systematic fashion, hereinafter referred to as "network analysis." In what follows, we wish to show how network analysis can be applied to the ecosystem networks of the Baltic Sea and the Chesapeake Bay to identify significant differences and similarities between the two communities.

The Ecosystems and Their Accompanying Networks

Reasonably detailed descriptions of the ecosystems of the Baltic Sea and the mesohaline region of the Chesapeake Bay can be found in Wulff and Ulanowicz (1989). Only a few pertinent observations are repeated here.

The Baltic Sea consists of three major basins, the Baltic proper, the Bothnian Sea, and the Bothnian Bay, which together span an area of 373,000 km^2. The Baltic proper connects over a shallow sill and through the Kattegat Strait to the North Sea (Fig. 8.1A). The average salinity at the surface of the Baltic is barely in the mesohaline range (about 7 ppt). The Baltic proper has a permanent halocline at 65 m, below which salinities range from 10 to 15 ppt. A seasonal thermocline forms during the summer at about 20 m. Salt-water replenishment enters sporadically via the Kattegat and is driven by meteorological events.

The Baltic proper has undergone considerable eutrophication during the last 50 years because of increased anthropogenic nutrient input (Larsson et al. 1985). Anoxic conditions in the deep waters are now common, and the deep sediments have turned anoxic, resulting in the extirpation of some associated benthic species. On the other hand, benthic biomass (Cederwall and Elmgren 1980) and fish catches (Ojaveer 1981) both have increased during the same period.

The Chesapeake Bay is a drowned river valley situated along the middle of the eastern North American coastline that covers a much smaller surface area of 12,500 km^2. The full range of salinities extends along its 290-km length, but attention here will be confined to the mesohaline zone (6–18 ppt) as shown in Fig. 8.1B. The bay has a relatively shallow average depth of only 9 m and exceeds 60 m at only a few isolated spots along its channel. During the summer months, a moderately strong combined halothermocline forms at about 8–10 meters depth.

The Chesapeake has always been slightly eutrophic, but during the past half century (since about 1940) nutrient loads have increased appreciably from domestic sewage and agricultural runoff. During the warmer months,

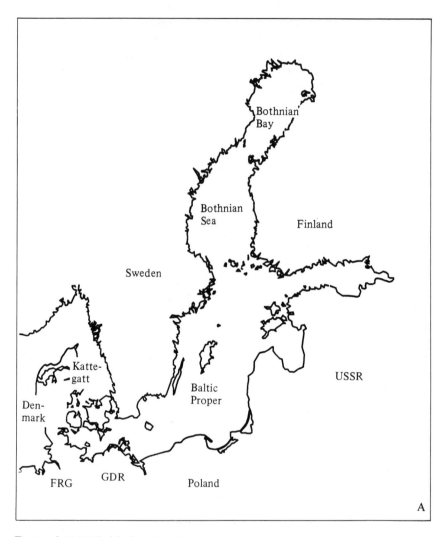

FIGURE 8.1. (A) Baltic Sea. (B) Chesapeake Bay.

anoxia is quite common below the thermocline. During the past 20 years, dinoflagellate blooms (mostly *Gymnodynium splendens*) have been occurring with increasing frequency and duration, and submerged aquatic grasses have fallen to a tiny fraction of their former stocks. Harvests of almost all commercial fish have declined.

One can safely conclude that both systems have been stressed during recent decades; however, it is not obvious which of the two communities is more heavily impacted. The rise in the Baltic fish catch and the corresponding decline in Chesapeake harvests indicate that the Chesapeake might be more hypertrophic. However, there is strong evidence that fish stocks in Chesapeake Bay have been overharvested (MDDNR 1985), and the remain-

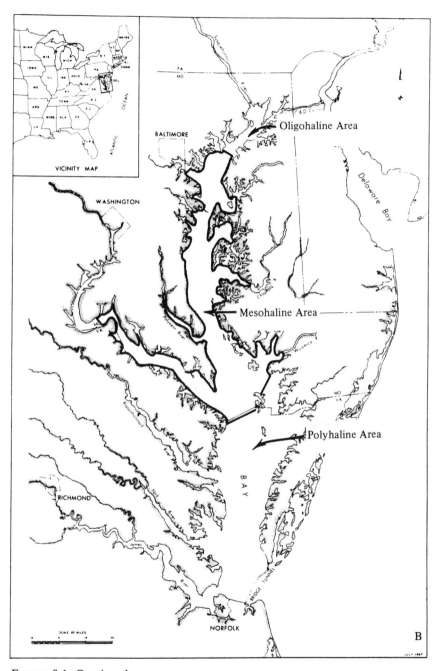

FIGURE 8.1. *Continued*

der of the ecosystem might not be as stressed as its Baltic counterpart. Conventional wisdom holds that the more oligohaline Baltic should be less resilient to nutrient stress. A systematic comparison of the two ecosystem networks should resolve which of the systems has sustained more overall damage, and network analysis should spotlight just where the functioning of each system has been impaired.

The compartments of the ecosystems, their annually averaged standing stocks of carbon (mg C m^{-2}) and the exchanges among them (mg C m^{-2} d^{-1}) are given in Figs. 8.2A and 8.2B. Those wishing to trace the derivation of any particular value in the schematics are referred to Wulff and Ulanowicz (1989) for Baltic values or Baird and Ulanowicz (1989) for Chesapeake magnitudes. The one-to-one correspondence between the compartments of the two networks is intentional. Several of the calculated network properties depend strongly on the choice of compartments, and it was thought that numerical comparisons would remain equivocal unless the systems were partitioned in exactly the same way. Originally, a 17-compartment flow model of the Baltic ecosystem had been created by Wulff, whereas 36 taxa were used by Baird and Ulanowicz (1989) to represent the biotic community in Chesapeake. It was difficult to settle on a common set of compartments into which both the original networks could be aggregated.

Although both networks represent spatial averages over the whole systems, the temporal frequencies of available data differed in the two cases. Networks characterizing the Chesapeake ecosystem during spring, summer, fall, and winter had been described (Baird and Ulanowicz 1989). Most of the Baltic flows were known only as annual averages, although the planktonic community could have been parsed on a finer scale. As data on all flows in both systems were available on an annual basis, this became the interval to be used for the comparison.

This search for common dimensions exemplifies the necessary tendency toward choosing "least common denominators" as the basis for comparison. Just as two fractions cannot be compared until their denominators have been changed to a common (and generally larger) reference value, the comparison of two ecosystems necessitates choosing the least resolved (i.e., largest) spatial, temporal, and taxonomic intervals for which data are available from both systems.

It should be noted in Figs. 8.2A and 8.2B that despite the common compartments, the biomasses, fluxes, and topologies of the networks differ in places. With regard to the differences in topology, during the past 20 years a considerable literature has evolved under the rubric of "food web theory," wherein regularities among trophic topologies of diverse ecosystems have been catalogued (see Cohen 1989 for overview). If one considers only those connections among the living elements in both systems (those numbered 1–12 in Figs. 8.2A and 8.2B), one may readily see how these two networks fit into the overall patterns that Cohen cites and how they stand in relation to each other.

Among the 12 living compartments in each system, there are 20 connections in the Baltic and 18 in the Chesapeake. Cohen cited how most food webs contain roughly twice as many links as the number of nodes. By such a yardstick both networks seem slightly underconnected, but not egregiously so. The convention among food web investigators is to classify taxa as top predators (those that are predators but never prey), intermediate taxa (those that act as both predator and prey), and basal compartments (those that are consumed but never consume other living materials). Cohen states that "on average, top species make up about one fourth of the total, intermediate species about one half, and basal species the remaining fourth." In the Chesapeake the percentages are 33%, 50%, and 17%, respectively, while in the Baltic the proportions are 33%, 58%, and 8%. Although the Baltic is somewhat sparse in top predators, the two food webs reasonably fit the general pattern.

Food web investigators also have observed that the proportions of trophic links between members of the three categories of compartments likewise exhibit reasonable regularity among all systems that have been catalogued. Thus, about 35% of the total linkages connect top predators to intermediate taxa; 30% link intermediates to other intermediates; 32%, intermediates to basals; and 8%, tops to basals. In the Chesapeake these ratios are 6 : 50 : 33 : 11, whereas in the Baltic they are 25 : 45 : 30 : 0. Both systems are more highly connected among the intermediates than is generally observed, and scarce in links from the intermediate to top consumers. The proportions of linkages from basal to intermediate species is virtually identical to Cohen's ensemble averages, but the Baltic appears to lack any basal-top connections. This latter irregularity results from the way meiofauna were treated in the two networks. In the Baltic, meiofauna are an intermediate species, serving as host to deposit feeders and benthic invertebrate carnivores. In the Chesapeake, most meiofauna are considered to be recycled to the sediment particulate organic matter (POC) and do not appreciably contribute to benthic predators, that is, it stands as a "top" species in that network.

These minor topological differences notwithstanding, the food web portions of both networks appear to be typical of the collection of webs that have been catalogued by food web theorists. The analysis of trophic topology does uncover a few discrepancies in the way the system was parsed, but any system-level differences that may exist are not made apparent. To compare the workings of the system as a whole, it becomes necessary to take explicit account of differences in flow magnitudes and the routes of transfers through the detrital pools.

Whole-System Comparisons

Table 8.1 lists the values of some properties of the Chesapeake and Baltic ecosystem networks. The total system throughput is the sum of all the flows occurring in each network. It measures the total activity of each system.

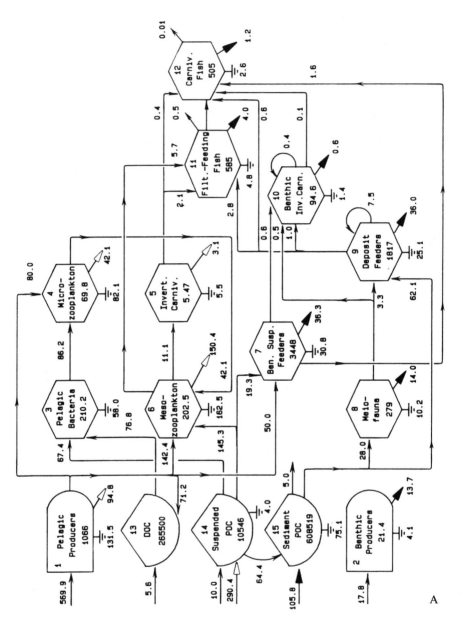

FIGURE 8.2. Diagrams of annually averaged carbon flows: (A) ecosystem of Baltic Proper; (B) Chesapeake Bay. Units of biomass (inside boxes), mg C m^{-2}; units of flows, mg C m^{-2} d^{-1}. For simplicity, particulate organic carbon (POC) flows between compartments are not connected: open arrows into space indicate flows to suspended POC; filled arrows into space, flows to sediment POC. Ground symbols indicate dissipative respirations.

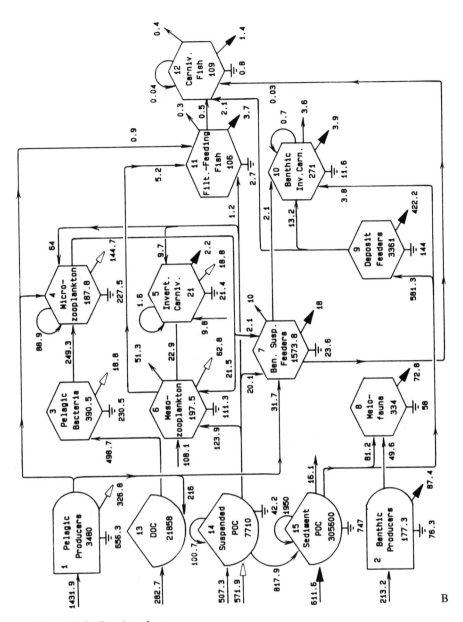

Figure 8.2. *Continued*

B

149

TABLE 8.1. Indices of total system behavior.

Index (units)[a]	Chesapeake	Baltic
Total system throughput (mgC m^{-2} d^{-1})	11,224	2,577
Flow diversity (bits)	2.94	3.10
Development capacity (mgC m^{-2} d^{-1} bits)	33,000	8,007
Relative ascendency (%of 3)	49.5	55.6
Overhead on inputs (% of 3)	2.6	0.9
Overhead on exports (% of 3)	0.4	0.0
Overhead on respirations (% of 3)	19.4	21.4
Redundancy (% of 3)	28.1	22.0
Finn cycling index (% of 1)	29.7	22.8

[a]See glossary for definitions.

This activity in the Chesapeake is about four times greater than that in the Baltic, because primary production in the former system exceeds that in the latter by almost the same ratio. Some of the elevated activity in the Chesapeake can be attributed to the higher annual temperatures at that latitude.

A second measure by which to compare whole systems is flow diversity. This was first suggested by MacArthur (1955) as a measure of the capacity of a system for homeostasis. Rutledge et al. (1976) later improved on MacArthur's index, and Ulanowicz (1986) scaled the improved flow diversity by the total system throughput to create a measure of the capacity of the system for development. The flow diversity is calculated using the Shannon–Weaver measure of complexity, which is logarithmic in nature. Hence, small differences in the value of this index can reflect much larger qualitative disparities. The Baltic network has a higher flow diversity (see Table 8.1), and this leads one to expect that this system has more remaining ability to cope with disturbance. This difference was unanticipated and constitutes the first clue that the Baltic might not be as strongly impacted as the Chesapeake.

Ulanowicz (1986) separated the development capacity into five components, each of which relates to some qualitative property of the overall structure. The component most relevant here is the relative ascendency, or the articulation of the network. In a system that is more highly articulated (or better organized), the effects of an event at any one compartment are propagated to only a small subset of other compartments. Ulanowicz (1980, 1986) suggested that ecosystems naturally develop in the direction of greater articulation. Articulation is presented in Table 8.1 as the percentage of development capacity that is expressed as organized flow. One sees that about 6% more of the capacity of the Baltic network appears as organized structure.

Odum (1969) proposed that mature ecosystems recycle a greater percentage of material and energy than do pioneer or disturbed communities. The last entry in Table 8.1 is the Finn cycling index, the fraction of the total system activity comprised by recycling. The Chesapeake community appears to recycle considerably more carbon than does the Baltic, which seems to contradict the emerging picture of the Baltic as the less disturbed system. However, Ulanowicz (1984) has remarked that perturbed systems often exhibit greater degrees of recycling. Presumably, augmented cycling in a disturbed system is its homeostatic response to retain in circulation resources that, before the perturbation, had been stored in the biomass of higher organisms. To determine if the Chesapeake is functioning in this fashion requires more details about the patterns of individual cycles in both ecosystems.

Structure of Cycling

A knowledge of the aggregate amount of cycling in a system is often insufficient information on which to judge how well an ecosystem is functioning. Ulanowicz (1983) constructed an algorithm that enumerates all the simple cycles present in a network and then removes those cycles from the underlying framework of dissipative flows. This program uncovered 22 routes for recycle in the Baltic network and 20 in that of the Chesapeake. More importantly, however, most of the recycling in the Chesapeake occurs over two short benthic cycles, where 424 mg C m^{-2} d^{-1} circulates between the deposit feeders and the sediment POC and 72.8 units circulates between the benthic meiofauna and the sediment detrital pool (Fig. 8.3). It is clear from Fig. 8.3 that a greater proportion of the recycled flow in the Baltic courses over longer cycles. For example, 42.1 units of flow traverse the circuit from suspended POC → pelagic bacteria → microzooplankton → mesozooplankton and back again.

That longer cycles are relatively more important to the Baltic system is shown in Fig. 8.4, where the proportions of recycle activity that occur over loops of various trophic lengths are plotted for the two systems. One concludes that the higher Finn cycling index for the Chesapeake networks was indeed misleading. Once the pattern and nature of the recycling was considered, it became apparent that the Baltic possesses a more developed apparatus for recycling.

One notices in Figs. 8.3A and 8.3B that two of the six Chesapeake compartments that engage in no cycling whatsoever are seen to facilitate the reuse of Baltic carbon (pelagic bacteria and planktivorous fish). The "microbial loop" appears to be more important to the dynamics of the plankton community in the Baltic than is the case in the Chesapeake.

Trophic Structure

From the analysis thus far, it appears that the Baltic remains less hypertrophic than the Chesapeake ecosystem. To test such a conclusion, we wish to elaborate the actual trophic structure of the two communities.

A

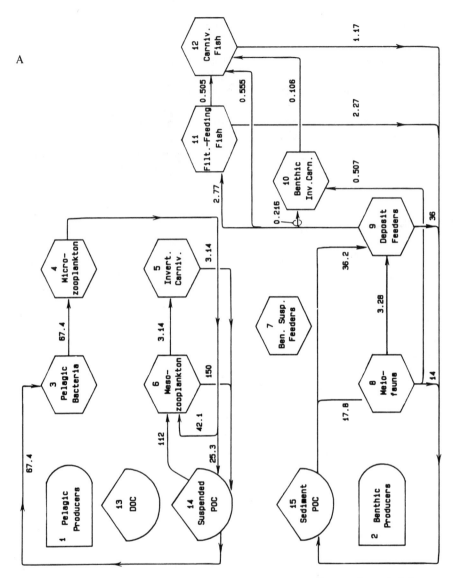

FIGURE 8.3. Networks for Baltic (A) and Chesapeake (B) show only cycled flows, that is, the amount of any compartmental throughput that returns to that compartment after traversing indicated loop.

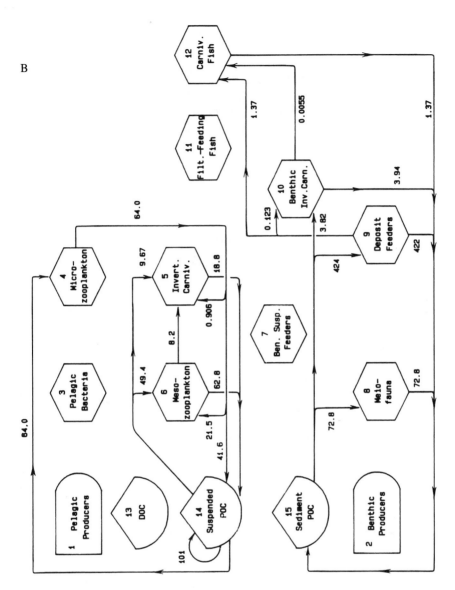

FIGURE 8.3. *Continued*

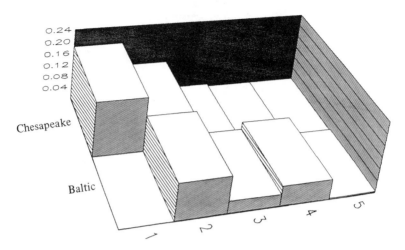

FIGURE 8.4. Proportions of total activity devoted to recycle (vertical axis) over cyclic pathways of various lengths (horizontal axis) in Chesapeake and Baltic ecosystems.

Depicting the trophic structure of a network, like other tools in network analysis, builds upon representing the exchanges in matrix form. That is, the magnitude of each exchange occurring in an n-compartment community can be entered into an n × n matrix in such a way that the row index designates the donor of the medium and the column location shows the recipient. It becomes useful to normalize each column of this exchange matrix by the total amount of all medium flowing into that recipient. The entries of each column then represent the fractions that each donor (prey) constitutes in the overall diet of that recipient (predator). The reason for normalizing in this fashion becomes apparent as soon as one multiplies this matrix of dietary coefficients by itself. Each element in the resulting product matrix represents the fraction of the predator's (column index) diet that is derived from each donor (row index) along all pathways exactly two trophic exchanges in length. Subsequent multiplication of the product matrix by the original matrix reveals the dietary contributions along all pathways of trophic length 3, etc.

The powers of the dietary matrix are useful in numerous ways, two of which are relevant to elaborating the trophic structure of the system. Levine (1980) showed how the amounts arriving at various trophic distances from primary sources can be used as weighting factors to estimate the average trophic level at which each heterotroph is feeding. In effect, each compartment is assigned a point along a trophic continuum. Levine's "trophic positions" for the 15 compartments (Fig. 8.5) show little disparity (< 0.3 trophic units) between the trophic positions of most compartments in the two communities, except for those of the planktivorous and carnivorous fishes. The planktivorous fish in the Baltic appear on the average to feed a full trophic position higher than their Chesapeake counterparts. This difference suggests examining the feeding behaviors of the filter-feeding fishes in both systems for more insight. In the Chesapeake, the dominant filter feeding fish species is the menhaden (*Brevoortia tyrannus*), which acts mostly as a herbivore on phytoplankton. In contrast, the Baltic counterparts consume mostly zooplankton (one trophic level higher) during the warmer months and switch to consuming benthic invertebrates during colder periods.

As an alternative to the trophic continuum, one may envision all transfers as occurring between a set of integral trophic levels. In reality, heterotrophic species often feed at several trophic levels. However, the power sequence of the dietary matrix discussed earlier indicates just how much each species feeds at each trophic level. Hence, this information can be used to apportion the activity of a given compartment among the imagined integral levels (Ulanowicz and Kemp 1979; Fig. 8.6). Only the living members of the ecosystem are apportioned in this manner. The attendant recycle through the nonliving detritus (which confounds the trophic apportionment scheme) is treated separately (Ulanowicz, in press). Virtually all ecosystem networks can be mapped into concatenations like the ones shown in Fig. 8.6. Therefore, one may speak of these chains as canonical trophic forms that should be particularly useful for comparing ecosystems widely disparate in structure.

The canonical trophic forms of the two systems reveal several obvious differences. The Baltic trophic chain possesses one more link than its Ches-

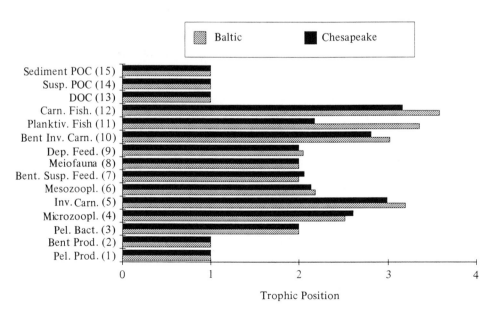

FIGURE 8.5. Average trophic position for each compartment in Chesapeake and Baltic networks.

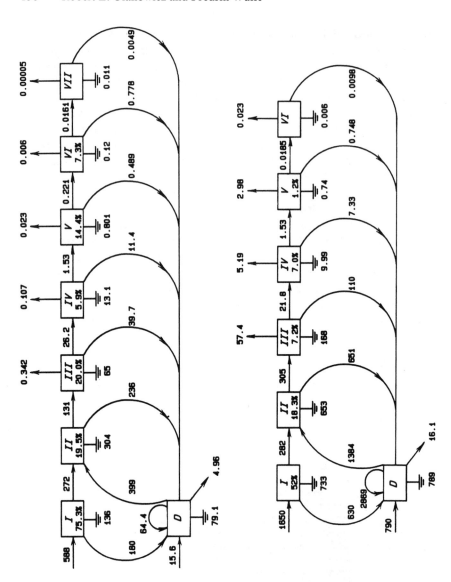

FIGURE 8.6. Canonical trophic aggregations of Baltic (A) and Chesapeake (B) ecosystem networks (see Figs. 8.2A and 8.2B). Boxes represent virtual trophic levels that are linear combinations of actual taxa. Percentages inside boxes refer to aggregate efficiencies of each virtual level.

apeake counterpart, probably because trophic efficiencies are higher in the Baltic. (The efficiency of a trophic level is calculated by dividing its output to the next level by its total input.) Furthermore, one sees that detritivory (the flows from D to II) are much larger in the Chesapeake system. Detritivory in the bay strongly predominates over herbivory (the flow from I to II) by a ratio of 4.9 to 1, whereas the corresponding quotient in the Baltic is 1.5 to 1.

Bilateral Indirect Relationships

Evidence that the trophic structure of the Baltic ecosystem appears less impacted than that of the Chesapeake concludes the macroscopic comparison of the two systems. However, an ecologist or ecosystem manager could still gain much from further comparison of the networks at finer resolution. For example, one might be charged with managing a particular species and wish to know the effects that all other compartments have on the given taxon over all trophic pathways, direct and indirect. The tool for quantifying indirect bilateral relationships is the power series of the dietary matrix discussed early. To be more specific, Leontief (1951) discovered that the infinite series generated by the powers of matrices normalized like the dietary matrix would always converge. Further, he showed that the limit of this series could be calculated through a few simple matrix operations on the dietary array. The end result is a "structure matrix" whose elements estimate the indirect trophic effects of any one compartment on any other element over all pathways of all lengths. Hannon (1973) introduced this econometric tool into ecology, and Szyrmer and Ulanowicz (1987) later modified the analysis to yield results with clearer ecological meanings.

Here we are interested in using "input-output" analysis to estimate how much of the various exogenous inputs eventually reach the "commercially useful products" that issue from the system. Once again, it is clear that the Baltic is more effective at producing valuable resources (Table 8.2). For example, one unit of Baltic phytoplankton production results in about four

TABLE 8.2. Percentage of input from three major sources that reaches the four commercially important compartments.

Source	Source	Benthic filter feeders	Planktivorous fish
Phytoplankton production	Baltic	11.9	1.7
	Chesapeake	2.9	0.4
Benthic production	Baltic	0	1.8
	Chesapeake	0	0
Allochthonous input	Baltic	8.9	0.8
	Chesapeake	2.0	0.2

times as much planktivorous fish as does its counterpart in the Chesapeake (and this despite the earlier observation that planktivorous fish in the Baltic feed at higher levels of the food chain).

Turning matters around, one might ask how much of the medium ingested by a particular heterotroph was once incorporated by various other components during its transit through the trophic web. In other words, one can calculate the "indirect diet" of each heterotroph. Elsewhere, Baird and Ulanowicz (1989) observed how two carnivorous fish populations in Chesapeake Bay exhibited apparently minor differences in their behaviors and direct diets; however, considerable niche separation arose once the ultimate sources of their diets were calculated. Pelagic primary production is obviously much more important to the carnivorous fish of the Baltic (95%) than it is to the same populations in Chesapeake Bay (42%) (Table 8.3). The Baltic fishes depend more heavily on the pelagic producers, mesozooplankton, and benthic suspension feeders, whereas benthic deposit feeders and detrital material are the primary resources for carnivorous fishes in the bay. There is a sharp disparity in the Chesapeake between the carnivorous fish, which depend on detrital food sources, and the plankton-feeding finfish, which rely mostly on the grazing food chain. In the Baltic this difference is much less pronounced. The contrast in system behavior of the two systems can be traced to the fact that primary and secondary production are both sufficiently high during the Maryland winter to support plankton feeding by fish in the bay. However, pelagic production becomes so

TABLE 8.3. Indirect diets (in % of total intake) of carnivorous and filter-feeding fishes in both systems.[a]

	Carnivorous fish		Planktivorous fish	
Compartment	Baltic	Chesapeake	Baltic	Chesapeake
Pelagic producers (1)	94.8	41.5	94.1	68.3
Benthic producers (2)	2.9	9.6	3.0	0
Pelagic bacteria (3)	8.0	13.2	11.7	9.6
Microzooplankton (4)	15.5	13.2	22.6	13.8
Invertebrate carnivores (5)	16.4	1.6	20.7	1.0
Mesozooplankton (6)	48.2	19.0	83.2	72.5
Benthic suspension feeders (7)	49.8	2.6	8.2	0
Meiofauna (8)	4.6	5.8	4.3	0
Deposit feeders (9)	24.5	79.8	26.3	0
Benthic invertebrate carnivores (10)	2.9	0.5	0.1	0
Planktivorous fish (11)	32.7	19.5	1.0	0
Carnivorous fish (12)	0.2	1.5	0.3	0
Dissolved Organic matter, DOC (13)	4.6	9.8	6.7	9.6
Suspended POC (14)	43.4	79.5	51.7	52.6
Sediment POC (15)	25.0	79.8	26.3	0

[a]Percents add to more than 100, because the same material visits several compartments along its way to the designated consumer. POC, particulate organic matter.

sparse in the wintertime Baltic that most zooplankton are forced to over-winter in diapause, thereby compelling the planktivorous fish to go to the bottom and utilize the benthic deposit feeders (Aneer 1980).

It appears that the Chesapeake ecosystem relies more on its benthic processes than does its Baltic counterpart. This is understandable, given the large difference in mean depths between the two systems. The predominant role of benthic metabolism in the Chesapeake becomes quite visible when one uses input-output analysis to calculate how much of the net primary production leaves the systems from each compartment (Fig. 8.7). It is immediately evident that the major egress of carbon from the Chesapeake ecosystem is via the respiration of benthic detritus, whereas the Baltic meso-zooplankton (predominantly mysids) respire most of the carbon from that system.

Finally, it is often informative to take explicit account of the antagonistic nature of predator–prey trophic interactions. That is, the immediate effect of feeding is positive to the predator population, but negative to the prey stocks. Ulanowicz and Puccia (1990) have outlined a variation on input-output techniques that permits simultaneous bookkeeping of both positive and negative trophic influences. The technique allows one to quantify indirect interactions such as competition for prey, indirect mutualism (e.g., the predator of a predator is beneficial to the original prey population), self-damping of heterotrophic populations, autocatalytic loops, etc.

As can be seen from Figs. 8.2A and 8.2B, the trophic topologies of the two networks are reasonably similar and only a few indirect influences bear mention. It was just remarked how the mesozooplankton play a strong role in the Baltic ecosystem, so it comes as no surprise that the indirect effects originating in this compartment are more obvious than those of its counterpart in Chesapeake Bay. In both systems, the mesozooplankton act as predators on the microzooplankton, which in turn consume pelagic bacteria. The method of Ulanowicz and Puccia (1990) reveals that the indirect mutualism inherent in this trophic concatenation is about eight times stronger in the Baltic than in the Chesapeake network, where the beneficial effect of mesozooplankton on bacteria is confounded by a subsidy of suspended POC to the microzooplankton. The same subsidy ameliorates the negative direct impact that mesozooplankton in the Chesapeake have upon the microzooplankton. Input-output analysis shows that the negative effect that mesozooplankton exert on benthic suspension feeders in the Baltic is about twice as strong as the same competitive relationship (mesozooplankton and benthic suspension feeders both consume suspended POC) in the Chesapeake.

It was remarked earlier that sediment POC played a much more significant role in the community metabolism of the Chesapeake. Further, the cycle analysis spotlighted how large amounts of sediment carbon are being recycled by the meiofauna and deposit feeders. So reflexive are these loops that sediment POC comes to exert a large autocatalytic effect on itself.

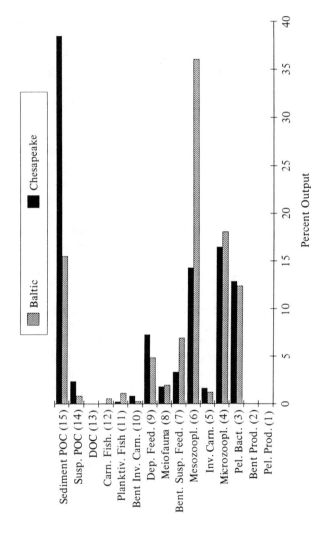

FIGURE 8.7. Fraction of net primary productivity that exits from the systems at each compartment.

Summary and Conclusions

In summary, the Baltic ecosystem appears to be less prone to hypertrophy than is the community in the Chesapeake. Evidence for this conclusion appears at all levels of comparison. The diversity of trophic processes is higher in the Baltic, and its species are trophically linked together in more narrow, or articulated, fashion. The amount of activity devoted to cycling is higher in the Chesapeake, but this fact is an indicator more of stress than of maturity. Routes of recycle are more numerous in the Baltic ecosystem, and individual cycles there are longer and contain a greater proportion of higher trophic level species. Overall trophic efficiencies are greater in the Baltic; one unit of primary production in that system creates four times as much fish as does a unit of productivity in the Chesapeake. The Baltic may be characterized as more pelagic than the Chesapeake. Carnivorous fish in the Sea ultimately depend on phytoplankton, mesozooplankton, and benthic suspension feeders for most of their sustenance, whereas Chesapeake fish rely indirectly on deposit feeders and particulate detritus for the better part of theirs.

Ideally, we would like to quantify the confidence level of each of the aforementioned conclusions. Unfortunately, explicitly stating the significance of most of the results of network analysis is not an easy task. For example, the statistical behavior of the Shannon–Weaver uncertainty or of the average mutual information used to compute the system articulation is recondite (Kullback 1978), and the tests are difficult to implement. Somewhat more accessible progress has been made toward evaluating the credibility of results issuing from input-output analysis by running sensitivity analyses on the constituent matrix manipulations (Bosserman 1981). Implementing these analyses involves a total effort comparable to that already expended up to this point. While such evaluations are in all ways desirable, a lack of time and resources has precluded their application to this study. Suffice it here to mention that in searching the various outputs for notable differences, we took care to confine our attention only to clear-cut disparities and never yielded to the temptation to employ marginal differences to paint any preconceived picture of how the systems compared one to another.

Perhaps the outcome of this comparison would have been better anticipated by everyone had the authors begun by contrasting the nutrient-loading characteristics of both systems. One notes, for example, that the ratio of catchment area to water surface area in the Chesapeake is sevenfold greater than it is in the Baltic. Furthermore, when one prepares a simple Vollenweider (1982) diagram of phosphorous loading versus the ratio of mean depth to water residence time (Fig. 8.8), the Chesapeake system plots in the region of excessively enriched systems, whereas the Baltic falls in the borderline region between undesirable and permissible zones. Presumably, several decades ago the Baltic was well inside the oligotrophic half of the

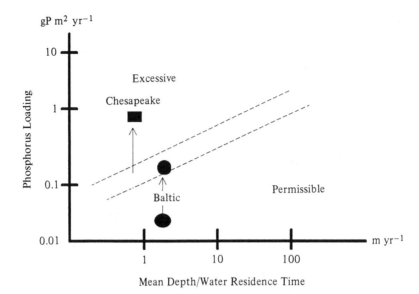

FIGURE 8.8. Diagram (Vollenweider 1982) of phosphorous loading plotted against ratio of mean depth to water residence time. Lower Baltic point represents estimated conditions in Baltic Sea earlier in the twentieth century.

plot and has since moved toward the zone characterizing degraded systems. Such preliminary shifts often are accompanied by increases in fish harvests (as was the case in the Baltic), whereas yields are likely to decrease with further degradation. We do not mean to imply that plotting a system on Vollenweider diagrams gives sufficient information to judge its trophic status. Useful as the Vollenweider technique may be, it nevertheless would be highly speculative to categorize a system on the basis of that analysis alone. The point of this essay has been to show that the quantified networks of two systems permit one to carry out an entire suite of comparisons. The breadth of the results issuing from the exercise allows not only the manager to draw some very defensible conclusions about the relative trophic status of the communities, but also affords the ecologist a bridge with which to connect analyses at the whole-system level to those concerning individual populations. Ecologists have long preached that reductionism and holism go hand in glove. The wherewithal to act on that notion is now available.

Acknowledgments. R.E. Ulanowicz was supported in part by LMER Grant #BSR-8814272 from the Division of Biotic Systems and Resources, the National Science Foundation. F. Wulff's work was underwritten by the Swedish National Science Foundation (NFR) and the Environmental Protection Board (SNV).

REFERENCES

Aneer, G. (1980). Estimates of feeding pressure on pelagic and benthic organisms by Baltic herring (*Clupea harentus* v: *membras L.*). *Ophelia* (Suppl.) 1:265-275.

Baird, D. and R.E. Ulanowicz. (1989). The seasonal dynamics of the Chesapeake Bay ecosystem. *Ecol. Monogr.* 59:329-364.

Bosserman, R.W. (1981). Sensitivity techniques for examination of input-output flow analyses. In: *Energy and Ecological Modelling*, W.J. Mitsch, R.W. Bosserman, and J.M. Klopatek, eds. Elsevier, New York, pp. 653-660.

Cederwall, H. and R. Elmgren. (1980). Biomass increase of benthic marcrofauna demonstrates eutrophication of the Baltic Sea. *Ophelia* (Suppl.) 1:287-304.

Cohen, J.E. (1989). Big fish, little fish: the search for patterns in predator-prey relationships. *The Sciences* 29:37-42.

Engelberg, J. and L.L. Boyarsky. (1979). The noncybernetic nature of ecosystems. *Am. Nat.* 114:317-324.

Finn, J.T. (1976). Measures of ecosystem structure and function derived from analysis of flows. *J. theor. Biol.* 56:363-380.

Hannon, B. (1973). The structure of ecosystems. *J. theor. Biol.* 41:535-546.

Kullback, S. 1978. *Information Theory and Statistics*. Peter Smith, Gloucester.

Larsson, U., R. Elmgren, and F. Wulff. (1985). Eutrophication and the Baltic Sea: causes and consequences. *Ambio* 14:9-14.

Lehman, J.T. (1986). The goal of understanding in ecology. *Limnol. Oceanogr.* 31: 1160-1166.

Leontief, W. (1951). *The Structure of the American Economy, 1919-1939*, 2d Ed. Oxford University Press, New York.

Levine, S. (1980). Several measures of trophic structure applicable to complex food webs. *J. theor. Biol.* 83:195-207.

MacArthur, R.H. (1955). Fluctuations of animal populations and a measure of community stability. *Ecology* 36:533-536.

MDDNR. (1985). First Annual Status Report on Striped Bass. Tidewater Administration, Maryland Department of Natural Resources, Annapolis, Maryland.

Odum, E.P. (1969). The strategy of ecosystem development. *Science* 164:262-270.

Ojaveer, E. (1981). Marine pelagic fishes. In: *The Baltic Sea*, A. Voipio, ed. Elsevier, Amsterdam, pp. 276-292.

Peters, R.H. (1986). The role of prediction in limnology. *Limnol. Oceanogr.* 31: 1143-1159.

Platt, T., K.H. Mann, and R.E. Ulanowicz. (1981). *Mathematical Models in Biological Oceanography*. UNESCO Monographs on Oceanography Methodology 7, UNESCO Press, Paris.

Rutledge, R.W., B.L. Basorre, and R.J. Mulholland. (1976). Ecological stability: an information theory viewpoint. *J. theor. Biol.* 57:355-371.

Szyrmer, J. and R.E. Ulanowicz. (1987). Total flows in ecosystems. *Ecol. Model.* 53:123-136.

Ulanowicz, R.E. (1980). An hypothesis on the development of natural communities. *J. theor. Biol.* 85:223-245.

Ulanowicz, R.E. (1983). Identifying the structure of cycling in ecosystems. *Math. Biosci.* 65:219-237.

Ulanowicz, R.E. (1984). Community measures of marine food networks and their possible applications. In: *Flows of Energy and Materials in Marine Ecosystems*, M.J.R. Fasham, ed. Plenum Press, New York, pp. 23–47.

Ulanowicz, R.E. (1986). *Growth and Development: Ecosystems Phenomenology.* Springer-Verlag, New York.

Ulanowicz, R.E. Ecosystem trophic foundations: Lindeman exonerata. In: *Complex Ecology: The Part-Whole Relation in Ecosystems*, B.C. Patten and S.E. Jorgensen, eds. Prentice-Hall, New York (in press).

Ulanowicz, R.E. and W.M. Kemp. (1979). Toward canonical trophic aggregations. *Am. Nat.* 114:871–883.

Ulanowicz, R.E. and C.J. Puccia. (1990). Mixed trophic interactions in ecosystems. *Coenoses* 5:7–16.

Vollenweider, R.A. (1969). Moeglichkeiten und Grenzen Elementarer Modelle der Stoffbilanz von Seen. *Arch. Hydrobiol.* 66:1–36.

Vollenweider, R.A., ed. (1982). *Eutrophication of Waters: Monitoring, Assessment and Control.* OECD, Paris. 154 pp.

Wulff, F. and R.E. Ulanowicz. (1989). A comparative anatomy of the Baltic Sea and Chesapeake Bay ecosystems. In: *Flow Analysis of Marine Ecosystems*, F. Wulff, K.H. Mann and J.G. Field, eds. Springer-Verlag, Berlin, pp. 232–256.

Glossary

To assist the reader in following the text, we attempt to clarify the words and terms listed below that either are newly coined, used by analogy or otherwise unusual:

Anatomy: The structural characteristics of a living or a once- living system. Used here in reference to ecosystems to mean the configuration of trophic relationships among the compartments of an ecosystem.

Articulation: Degree of specificity or definition in the effect of a given event. In a highly articulated trophic network, each prey has only one, or very few, predators, and vice versa.

Ascendency: A property of a network of flows describing both the aggregate intensity of process activities and the level of definition (see *articulation*) or specificity with which these processes occur. Mathematically, it is the product of the total system throughput (see following) and the average mutual information inherent in the network (Rutledge et al. 1976; Ulanowicz 1986).

Autocatalysis: A circumstance wherein the activity of a network element augments itself, usually via one or more intermediaries. If the activity of element A catalyzes the activity of B, B catalyzes C, and C in turn catalyzes A, then the cycle or loop A-B-C-A is said to be autocatalytic in all its members. Autocatalysis inherent in networks of ecosystem interactions (not limited to feeding relations) is assumed to be the principal agent driving the system toward configurations with ever-higher network ascendencies.

Canonical form: The most elementary or simplest of a group of equivalent mathematical expressions, used here in reference to flow networks. If a network can be mathematically transformed into another aggregated representation, then the results of repeated application of the transform will converge to the simplest possible representation, or canonical form.

Complexity: The degree to which a system or network is difficult to analyze or separate, used here mostly in reference to the multiplicity and overlap of the various pathways through the system. Assumed to be equivalent to the flow diversity (see following).

Cycle: A pathway in which the beginning and terminal compartments are the same. The object of the analysis described here is to decompose all cycles into simple ones, that is, those in which no compartment appears more than once.

Development: Any change in network topology that contributes toward an increase in *ascendency* (see earlier). The *development capacity* is a measure of system activity and complexity and is equal to the product of the *total system throughput* times the *flow diversity*. The development capacity is always greater than or equal to the network ascendency and thereby appears as an upper bound on the growth and development of the system.

Finn Cycling Index: That fraction of the total system activity (*total system throughput*) devoted to the recycling of medium (see Finn 1976).

Flow Diversity: A measure of the complexity of flow pathways inherent in a network. This property is taken to be the result of applying the Shannon–Weaver index to the fractions of the total activity represented by the individual intercompartmental transfers and exogenous connections (i.e., inputs, exports, and respirations).

Gain: The factor by which an input to a compartment is amplified in the output. The gain of a *cycle* is the amplification a signal undergoes in one circuit around the cycle and usually is a small number, less than 1.

Hypertrophy: Ecosystems that are well nourished are called eutrophic. Those that for either natural or anthropogenic reasons are adversely overenriched are termed hypertrophic.

Information: That which causes one to change a probability assignment. It is equivalent to the decrease in uncertainty such reassignment engenders. The information inherent in the network structure is manifested by the constraints the connection topology places upon the calculated flow diversity. (See "average mutual information" in Rutledge et al. 1976.)

Input-Output Analysis: The quantitative study of the origins or fates of a medium after more than a single transfer. The study of indirect material causality (see Hannon 1973).

Organization: A flow network is said to be more highly organized the more articulated it becomes.

Overhead: That fraction of the *development capacity* encumbered from appearing as coherent activity (*system ascendency*). Generally, in any flow network there are four categories of overhead: (1) that caused by uncertainties in the inputs, (2) multiplicities of exports and (3) respirations, and (4) parallelism among internal pathways (see *redundancy*).

Reductionism: The search for explanation at smaller scales. In its strictest applications it becomes the belief that all causes behind any observed event lie at smaller temporal and spatial scales.

Redundancy: The degree to which pathways in a network are duplicated so that the connection between any two arbitrary compartments cannot be severed by elimination of a single intervening link. One of four components of the network *overhead* (see earlier).

Reflexivity: The degree to which an event is turned back on itself as its own cause.

Relative Ascendency: The fraction of the *development capacity* that appears as structured activity. The quotient of the *ascendency* by the *development capacity*. A relative measure of a system's organization.

Taxon: One of a group of elements that collectively define a formal system. A node in a flow network, not necessarily related to a Linnean taxonomic unit.

Throughput: The amount of flow of medium through an entity. That flowing through a system node or compartment is called *compartmental throughput*. The aggregate of all compartmental throughput in a system is called the *total system throughput*, which gauges the total activity of the system.

Trophic Level: (1) The average number of times an arbitrary particle is ingested by system *taxa* on its way to being incorporated by a given component (see Levine 1980). (2) An abstract element in a virtual chain of trophic transfers representing the totality of feeding relations in a system (see Ulanowicz, in press).

Part III Stress and Disturbance

9
Comparative Responses of Aquatic Ecosystems to Toxic Chemical Stress

ROBERT W. HOWARTH

Abstract

I reviewed a variety of studies on the effects of toxic chemical stress on aquatic ecosystems to search for common patterns of response or sensitivity. Information was obtained from whole-ecosystem experiments, mesocosm experiments, and well-conducted studies of toxic discharges (such as accidental oil spills) into natural ecosystems. Whole-ecosystem experiments have provided the most powerful tool for determining the ecological effects of toxic substances in lakes, but there have been no experiments at this scale in marine ecosystems. A variety of whole-ecosystem and mesocosm experiments in both freshwaters and seawater have found that single-species tests conducted in the laboratory can grossly underestimate the potential for environmental damage from toxic substances.

My review suggests a similarity in response across different types of aquatic ecosystems to a variety of toxic substances. Toxic chemical stress can affect both the structure and function of aquatic ecosystems. However, structural changes are generally more predictable and may occur earlier or at lower levels of stress than do changes in ecosystem processes. I found several examples of structural changes occurring with no noticeable change in rates of ecosystem processes. However, I found no examples of ecosystem processes being affected without an accompanying change in structure. When ecosystem processes are affected by toxic chemical stress, rates can either increase or decrease. The most predictable structural change appears to be the loss of sensitive species. Changes in the structure of aquatic ecosystems can be of importance even when not accompanied by changes in ecosystem processes such as primary production. For instance, fish populations can be lost as a result of changes in food web structure.

Ecosystems dominated by opportunistic species appear more resistant to stress than those dominated by more specialized organisms. Ecosystems with greater diversity are probably less resistant to stress. The relative sensitivity of an ecosystem (or a mesocosm designed to mimic a natural ecosystem) needs to be considered when extrapolating results on the effects of toxic chemical stress, as well as in designing and interpreting experiments or monitoring programs in these systems. Identification of general patterns in the sensitivity and resistance of ecosystems to toxic chemical stress can aid in environmental management by helping to identify which ecosystems require special protection or particularly close monitoring.

I hypothesize that ecosystems with more open element cycles may be more resistant to toxic chemical stress than are more closed ecosystems. If true, then eutrophication has some effects in common with toxic chemical stress. As systems are stressed by eutrophication or by toxic substances, they become more resistant to further change. However, many desirable characteristics of the original ecosystems may already have been lost. If the hypothesis proves true, it should be useful in designing experiments to test the ecological effects of toxic substances. It may also prove extremely useful in siting decisions and other aspects of environmental management.

Introduction

Human activity is toxifying the earth by mobilizing and redistributing trace metals (Nriagu and Pacyna 1988) and petroleum hydrocarbons (National Academy of Sciences 1985), by speeding the cycling of eutrophying and acid-producing ions (Schindler 1977; Galloway et al. 1984; Nixon et al. 1986), and by synthesizing thousands of new organic compounds, many of which are potentially harmful in the environment (Farrington 1989). Such altered and accelerated element cycles, along with inputs of synthetic substances, have had major effects on aquatic ecosystems, and the scientific community is still struggling to develop adequate strategies to control or limit ecological damage. To date, the management strategies for toxic chemical stresses in natural ecosystems have had little if any basis in ecosystem science (Schindler 1987).

The major approach for managing the release of toxic chemicals into the environment has been to predict effects from short-term bioassay data collected in laboratory studies of single species (Schindler 1987; Howarth 1989). Unfortunately, short-term bioassays ignore virtually all the ecological and biogeochemical interactions that occur in the real world. Further, these tests most frequently involve species which are easily cultured and relatively insensitive to the effects of toxic substances (Schindler 1987; Howarth 1989). As management tools, they are horribly permissive. For instance, when used to predict the effects of oil pollution in marine waters, such assays as the LD_{50} approach are found to overestimate the minimum oil concentration capable of causing ecological harm by some 100 fold or more (Table 9.1), thereby greatly underestimating potential effects in natural ecosystems (Howarth 1989). Short-term bioassays also underestimate the damage to lakes caused by acid precipitation (Schindler et al. 1985) and can mislead scientists concerning the causes of eutrophication (Schindler

TABLE 9.1. Concentrations of dissolved oil hydrocarbons estimated to have deleterious effects on organisms or ecosystems.

	Micrograms/liter	Source
LD_{50} value	1,000–3,000	U.S. Dept. of Interior 1985
Inhibition of sexual reproduction in macroalgae	0.2	Steele 1977
Disruption of snail feeding	1–3	Jacobson and Boylan 1973
Decreased viability of larval fish	2–10	Vandermeulen and Capuzzo 1983
Alteration in zooplankton composition (mesocosms)	5–15	Davies et al. 1980
Alteration in benthic community composition (mesocosms)	29	Gray 1987
Alteration in phytoplankton community composition (mesocosms)	< 40	Lee et al. 1978

1977; Smith 1984; Hecky and Kilham 1988; Howarth 1988). The value of short-term, single-species toxicological data as a management tool can be greatly increased through more sophisticated approaches to determining population sensitivities, such as using life history tables (Gentile et al. 1982; Schindler 1987). However, such approaches are not as yet in common use in environmental management, and even with such approaches, it is difficult to predict ecological effects of toxic substances in natural systems. Single-species tests are simply unable to predict indirect effects or changes in processes which may occur in whole ecosystems.

If we are to better manage and control toxic chemical stresses in aquatic ecosystems, we need a framework for understanding their effects at the ecosystem level. Several workers have proposed such frameworks for ecosystems in general (Margalef 1975; Woodwell 1983; Odum 1985; Rapport et al. 1985) or for aquatic ecosystems in particular (Schindler 1987). However, the applicability of these frameworks for aquatic ecosystems has not been fully explored, and many of the predicted responses to stress may not occur generally. For example, Schindler (1987) noted that many of the functional responses to stress predicted by Odum (1985) did not occur in the whole-lake acidification experiments at the Experimental Lakes Area (ELA). Schindler (1987) suggested that structural changes (such as changes in food webs and alteration in dominant species) are the major responses of aquatic ecosystems to stress, and that functional changes (such as changes in primary production or nutrient cycling) may not occur because of biotic compensation.

In this chapter, I further explore the responses of aquatic ecosystems to stress, with particular reference to general patterns in either functional or structural changes. I also suggest a construct for comparing the relative sensitivities (or resistances) of ecosystems to toxic chemical stress. I am

defining "stress" in the sense of Odum (1985) as "a detrimental or disor-
ganizing influence." Thus, we can think of chemical stress as a forcing
function related to toxic chemical exposure (Levine 1989).

There is some suggestion that terrestrial ecosystems differ significantly
from aquatic ecosystems in their responses to stress (Schindler 1987), or at
least in our ability to measure those responses. The temporal scale of
response is undoubtedly quite different because of the much longer life
span of the dominant primary producers in terrestrial ecosystems. Also,
ecosystem-level responses to toxic chemical stress have received less study
in terrestrial ecosystems than in aquatic ecosystems (Weinstein and Birk
1989). Consequently, while I make some comparisons with terrestrial ecosystems, I focus on the responses of aquatic ecosystems.

The Comparative Approach

The organizers of the Cary Conference asked that I use a comparative
approach and suggested that I address such questions as: "Is there a common response to a given stress or disturbance in various ecosystems?"; "Is
there a common response to various stresses or disturbances in a single type
of ecosystem?"; and "Is there a predictable sequence of responses to stress
or disturbance?" I in fact attempt to address these and similar questions,
and such questions clearly call for a comparative approach to ecosystem
analysis. However, I am severely hampered in this attempt by a relative
scarcity of data.

The most notable successes in comparative ecology involve the analysis
of large sets of data of easily measured variables, such as phosphorus,
nitrogen, and chlorophyll concentrations. Thus, we know that phosphorus
concentrations are reasonable predictors of chlorophyll concentrations in
lakes (Dillon and Rigler 1974; Schindler 1978), that cyanophytes are only
common in lakes with a molar N : P ratio less than 64 : 1 (Smith 1983), and
that percent abundance of cyanophytes in freshwater lakes can be reasonably predicted from the phosphorus concentration (Trimbee and Prepas
1987).

Data become more scarce when we start addressing questions involving
ecosystem processes using a comparative approach. For example, in 1988
some colleagues and I published a review of the biogeochemical controls
on nitrogen fixation that was based on a comparative approach (Howarth
et al. 1988). There we were dealing with a rate process that is more difficult
to quantify and is much less frequently measured than are standing stock
nutrient concentrations and phytoplankton biomasses. The more limited
data set resulted in more limited conclusions. In fact, we were forced to
infer information on nitrogen fixation rates from cyanophyte abundance
data to reach many of our conclusions. Similarly, Marino et al. (1990) used
data on cyanophyte abundance in a comparative study of 13 saline lakes to

infer information on nitrogen fixation in those lakes; it would have proven much more difficult to obtain the direct nitrogen fixation data in such a comparative study.

As we move to questions of ecosystem response to toxic chemical stress, the data set becomes even more circumscribed. The best data come from controlled experiments at the whole-ecosystem level or from carefully observed assessments of environmental impact at the ecosystem scale, as discussed previously. This leaves us with a handful of studies at most (Schindler 1987). By cautiously including observations from mesocosm experiments, I believe we can usefully extend the database, but it is still severely limited. Therefore, we move increasingly toward speculation when we apply the comparative approach to ecosystem analysis to the question of response of ecosystems to toxic stress.

The difficulty of comparing the responses of ecosystems to toxic chemical stress is compounded by scaling problems. One-time additions of a toxic substance may well have different effects than do continuous additions or pulsed additions. Also, point-source additions may behave differently than nonpoint additions (Auerbach 1981). Likewise, the depth of an aquatic ecosystem and its surface-to-volume ratio can affect the distribution and fate of toxic substances and therefore the doses seen by the biota. Ideally, we should use the concentrations or activities of substances to which the biota are exposed as the measure of the level of stress. This is, of course, precisely what Schindler et al. (1985) did for their acidification study, using pH as the forcing function. However, for other toxic stresses, such information is seldom available. When available, concentrations can vary greatly over time, leaving a question of how to integrate average exposure or dose over time. Thus, at present, we must use qualitative estimates of exposure to toxic chemical stress.

Sources of Information: Whole-Ecosystem Experiments, Observations, and Mesocosm Experiments

Information on the ecological effects of toxic stress can be obtained from well-designed whole-ecosystem studies, from ecosystem-level observations of stress response with appropriate reference systems, and from careful extrapolation of mesocosm experiments. Studies with single species can provide valuable information on mechanisms underlying ecological effects and on the comparative toxicity of different substances; by themselves, however, single-species bioassay tests have proven unable to predict ecological effects of toxic stress (Howarth 1989).

Whole-ecosystem experiments are a tremendously powerful tool and perhaps provide the most reliable information on effects (Likens 1985; Schindler 1987). Unfortunately, there are only a handful of such experiments that address the effects of toxic chemical stress. Of these, many are in stream ecosystems, and a few (at most) are whole-lake studies. In marine ecosystems, there are no experiments at the scale of the type performed at ELA.

Why are there so few whole-ecosystem experiments on toxic chemical stress? I suggest a major reason is a lack of commitment on the part of funding agencies. Laboratory-based experiments on toxicity to individual species appear less expensive to agencies, and until they accept that information obtained from such experiments is inadequate and that they need information at the ecosystem scale, they will resist funding ecosystem-scale experiments. The complete lack of whole-ecosystem marine experiments also reflects the difficulty of finding "replicate" systems amenable to experimental addition of toxic substances along crowded, developed coastlines. However, the possibility of finding appropriate sites for marine whole-ecosystem experiments may exist in such places as Hudson's Bay (D. Schindler, personal communication) or the coast of Norway (R. Wright, personal communication).

An alternative to experimental additions of stress-producing substances to whole ecosystems is to make ecosystem-scale observations in systems receiving actual pollution discharges. Such studies would seem particularly valuable in marine ecosystems, where it is so difficult to set aside experimental reserves. Unfortunately, such observations are rare, and when made, are often flawed (Howarth 1989). For instance, three major studies examined the effects of chronic oil pollution associated with oil development in the Gulf of Mexico (GURC 1974; Bedinger et al. 1980; Middleditch 1982). In all three studies, actual concentrations of oil hydrocarbons were never measured, and so actual exposure is not known. It seems likely, in fact, that the "control" sites in all three studies may have been as contaminated with oil as were the study sites near oil rigs (Bender et al. 1979; Sanders and Jones 1981; Carney 1987), and none of the studies adequately dealt with the spatial and temporal variability caused by being near the mouth of the Mississippi River (Carney 1987). On the other hand, studies of changes associated with benthic faunal structure with increasing distance away from an oil platform in the North Sea, which included data from before the start of oil development, have yielded valuable information (Addy et al. 1978; Kingston 1987).

Why not make ecosystem-scale observations of the effects of new developments part of our environmental management strategy? Current environmental impact assessments have little or no feedback to reality. Likely environmental consequences are predicted before an action or development, but adequate follow-up studies to determine the accuracy of the predictions are surprisingly rare (Rosenberg et al. 1981; Hecky et al. 1984; Schindler 1987). Adequate follow-up studies would require long-term observation of the potentially affected site, both before and after the event of interest, in comparison to one or more unaffected reference sites (Hecky et al. 1984; Likens 1985; Schindler 1987). That is, the event would be treated as a controlled ecosystem experiment. This sort of follow-up, inherent in the concept of adaptive management (Walters 1986), is rare.

While not as desirable as well-planned studies of pollution-generating developments, we can also obtain information from studies of environmental accidents such as oil spills. Very little has been learned following most oil spills, but careful study of a few spills has added markedly to our understanding of oil pollution (Teal and Howarth 1984; National Academy of Sciences 1985; Howarth 1989). These spills have in common that they occurred in ecologically well-studied and well-characterized environments, usually near high-quality marine laboratories, so that long-term pre- and postspill observations of oiled and reference sites were possible (Teal and Howarth 1984). In one case, a spill occurred during a nearby, ongoing study of factors controlling primary production and sedimentation (Elmgren and Frithsen 1982).

Mesocosm experiments represent yet another way to obtain information on the effects of toxic chemical stress (Menzel 1977; Grice and Reeve 1982; Oviatt et al. 1982; Gearing 1989). Schindler (1987) correctly noted that mesocosm experiments should not be considered "ecosystem scale" experiments, because the size of the mesocosm limits the possible range of ecological responses. For example, fish communities in mesocosms, if any fish occur at all, are necessarily much simpler than those in larger, more open ecosystems. Thus, the nature of top-down controls or cascading effects (Carpenter et al. 1985) would be expected to be different in mesocosms than in the larger ecosystems (Schindler 1987). Experimental ponds, while bigger than many mesocosms, can suffer from this same problem.

Mesocosms can also suffer from a temporal scaling problem, and limnocorrals often seem to deviate from the natural system within a few weeks (Gearing 1989). Thus, Schindler (1987) states that mesocosms "simulate some features of real ecosystems, if experiments are not of long duration." On the other hand, some of the Marine Ecosystem Research Laboratory (MERL) experiments have been run for more than 2 years, with the control tanks continuing to closely resemble Narragansett Bay in terms of both ecosystem structure and function, except for the lack of a fish community (Gearing 1989). It would seem that larger mesocosms mimic the real world longer, and that flow-through mesocosms (such as the MERL tanks, which receive regular inputs of Narragansett Bay water) are probably better than closed limnocorrals. Continuous inputs of new water minimize one of the typical problems of limnocorrals, artificial prevention of nutrient replenishment from the watershed (Schindler 1987).

Despite the limitations of mesocosms, I believe we can learn a great deal about ecosystem responses to toxic chemical stress through careful extrapolation of data from mesocosm experiments. There are marked similarities in effects observed in an oil-addition mesocosm experiment at MERL at the University of Rhode Island (Oviatt et al. 1982) and those seen following a particularly well-studied oil spill in the Baltic Sea (Elmgren and

Frithsen 1982) and other oil spills (Teal and Howarth 1984). Also, the effects of cadmium additions to mesocosms at ELA were similar to those observed in whole-lake acidification experiments (Schindler 1987). In summary, I believe that we should give the greatest credence to information from whole-ecosystem experiments and from ecosystem-scale observations of natural systems subject to toxic stress agents, when suitable reference systems exist. But when used with an awareness of their limitations, mesocosm experiments also are extremely useful.

Structural versus Functional Responses to Stress

Both structural and functional responses can result from toxic stress in aquatic ecosystems. But does one sort of change precede the other, occurring earlier or at lower doses of toxic chemicals? Is one sort of change a better early warning signal of further damage to come?

A whole-lake acidification experiment at ELA demonstrated structural changes, in the nature of loss of sensitive species, without any marked change in ecosystem processes such as primary production and decomposition (Schindler et al. 1985; Schindler 1987). Noting that cadmium additions also affected structure but not ecosystem processes, Schindler (1987) concluded that measures of ecosystem function are "relatively poor indicators of early stress." He went on to state that he had "not seen convincing evidence" that "any perturbation, whether natural or anthropogenic, has been sufficiently extreme to cause long-term decreases" in rates of ecosystem processes such as production, decomposition, and nutrient cycling.

On the other hand, Odum (1985) predicted that stress should cause a variety of changes in ecosystem processes such as production, respiration, and nutrient cycling rates. And in response to Schindler (1987), Levine (1989) stated that her recent search of the literature uncovered roughly 100 articles on functional responses to toxic stress, with most reporting significant decreases in process rates, but some reporting increases. Many of the studies mentioned by Levine (1989) appear to be extremely short term, or to have been performed with small microcosms where the validity of extrapolation to real ecosystems is even more problematical than with mesocosms. In my review of ecosystem-scale and mesocosm-scale experiments and observations, I did indeed find examples in which functional processes were depressed by toxic chemical stress (Table 9.2). However, in most of these cases, the depression was either quite short lived or the dose of the toxic chemical was fairly high. I found approximately equal numbers of studies where processes were stimulated or where they were relatively unaffected (Table 9.2). Following are brief summaries of some observed functional responses to toxic stresses, identified by type of ecosystem and type of stress.

TABLE 9.2. Summary of structural and functional responses of various aquatic ecosystems to several types of toxic stress. [a]

Stress agent	Ecosystem type	Structural change?	Functional change?	Reference
Acid	Lake (experiment)	Yes	No	Schindler et al. 1985
Acid	Lakes (survey)	Yes	No	Dillon et al. 1979
Acid	Lakes (experiment)	Yes	+ / No / −	Perry & Troelstrup 1988
Acid	Stream (experiment)	Yes	−	Hall & Likens 1981; Hall et al. 1980, 1985
Oil	CEPEX (marine)	Yes	−	Lee et al. 1978
Oil	MERL (marine)	Yes	+	Elmgren et al. 1980
Oil	Marine spill	Yes	+	Johansson 1980
Oil	Marine spill	Yes	+ / −	Cabioch et al. 1981
Oil	Marine spill	Yes	No	Winfrey et al. 1982
Cu	Lake (experiment)	Yes	(−)	McKnight 1981
Cd	Lake (experiment)	Yes	No	Schindler 1987
Cu	CEPEX (marine)	Yes	(−)	Thomas and Siebert 1977; Thomas et al. 1977
Heavy metals	Stream survey	Yes	+	Crossey and LaPoint 1988

[a] Under functional change, "no" means no functional response was seen; "+" means an increase occurred in the rate of some ecosystem process such as primary production; "−" means a decrease occurred in some rate process; a symbol in parentheses indicates that a change was observed, but the alteration was of very short duration. Where several symbols are given, more than one process was affected (some increasing, and some decreasing).

Functional Responses

ACIDIFICATION OF LAKES

In the ELA acidification experiment, primary production increased, but this was largely because of natural variation (Schindler et al. 1985; Schindler 1987). Acidification did not change decomposition rates (Kelly et al. 1984; Schindler et al. 1985, 1986), although rates of sulfate reduction and denitrification were increased by the additions of sulfuric and nitric acids, respectively (Rudd et al. 1986). A survey of Ontario lakes acidified by actual acid precipitation found no decrease in primary production (Dillon et al. 1979). Preliminary results from a whole-lake acidification experiment at Little Rock Lake reported increased rates of decomposition for birch leaves, decreased rates for leatherleaf leaves, and no effect for oak and aspen leaves (Perry and Troelstrup 1988). Preliminary results of the Little Rock Lake experiment also suggest no substantial effect on rates of primary production (T. Frost, personal communication). Ecosystem processes in lakes apparently are not very sensitive to acidification.

HEAVY METALS IN LAKES AND MARINE ECOSYSTEMS

Addition of copper to a reservoir resulted in decreased primary production, but for only 10 days (McKnight 1981). Similarly, cadmium additions to experimental ponds depressed primary production for 10 days, but rates then recovered to control levels (Kettle and DeNoyelles 1986). Cadmium additions to mesocosms at ELA had no major effects on ecosystem processes (Schindler 1987). Copper additions to marine mesocosms [the Controlled Ecosystem Pollution Experiment (CEPEX) bags] depressed primary production for 20 days, but rates then recovered (Thomas et al. 1977).

OIL IN MARINE ECOSYSTEMS

Primary production in marine ecosystems can be either stimulated or inhibited by oil pollution (Teal and Howarth 1984; National Academy of Sciences 1985; Howarth 1989). Following an oil spill in the Baltic Sea, primary production rates increased, apparently because of a decrease in zooplankton grazing pressure (Johansson 1980). This same effect of increased primary production following decreased zooplankton grazing was also observed in an oil-addition experiment to the MERL mesocosms (Elmgren et al. 1980; Oviatt et al. 1982). In the huge spill of the *Amoco Cadiz* off the Brittany coast, primary production was decreased in the most heavily polluted areas but was stimulated further away from the wreck (Cabioch et al. 1981); in this case, the increase in primary production was attributed to a release of nutrients from animals killed by the spill (Cabioch et al. 1981). In yet another spill, no conclusive effect on primary production was found

(Lannergren 1978). In a mesocosm experiment using large limnocorrals in an estuary in British Columbia (the CEPEX bags), primary production was suppressed by low concentrations of oil; the suppression was believed to result from hydrocarbon-degrading bacteria blooming and competing with the phytoplankton for nutrients (Lee et al. 1978).

Ecosystem processes other than primary production have seldom been measured following oil spills or in oil-addition experiments in mesocosms. Two exceptions are the measurement of microbial activity in heavily oiled sediments following a spill in Brittany and the measurement of sediment fluxes in the MERL experiment. No effects of the oil spill in Brittany on rates of sulfate reduction or rates of methane production could be determined (Winfrey et al. 1982). On the other hand, the oil addition in the MERL experiment resulted in decreased rates of sediment oxygen uptake and nutrient exchange (Oviatt et al. 1982).

ACIDIFICATION OF STREAMS

Acidification of a stream resulted in mobilization of cations, significant alteration of nitrogen cycling, and decreased periphyton biomass, perhaps suggesting lower primary production (Hall and Likens 1981; Hall et al. 1980, 1985).

HEAVY METALS IN STREAMS

Field observations of a stream receiving a point source of mine wastes contaminated with heavy metals showed that the metal pollution resulted in a significant increase in whole-system respiration and significant decreases in the P:R ratio (Crossey and La Point 1988). There was also a suggestion of an increase in gross primary production, but rates were highly variable, and any change was not of sufficient magnitude to be clearly associated with the treatment. Although effects on ecosystem processes were observed, Crossey and La Point (1988) concluded that community structure changes were less expensive to monitor and were at least as sensitive a measure of change.

OTHER TOXIC SUBSTANCES IN STREAMS AND LAKES

Organic pesticide additions to a stream resulted in no major disruptions of nutrient cycles (Eaton et al. 1986). Chlorine additions at concentrations of 250 μg/l to experimental stream mesocosms resulted in decreased decomposition of the macrophyte *Potamogeton crispus*; no effects were observed at lower doses of chlorine (Perry and Troelstrup 1988). Atrazine additions to experimental ponds depressed primary production at concentrations of 100 μg/l or more but not at concentrations of 20 μg/l (Larsen et al. 1986).

This brief review of functional responses in mesocosms and whole ecosystems to toxic chemical stress is certainly incomplete, but should be representative. In some cases, changes in ecosystem processes such as primary production result from stress. However, rates can be either increased or decreased as a result of toxic chemical stress, and the changes frequently seem short lived. Further, it is important to note that I found no examples in which functional changes were seen while structure remained unchanged (see Table 9.2). On the other hand, there are many examples of structure being altered without significant change in ecosystem processes. For example, major changes in phytoplankton and zooplankton community composition occurred in the acidification experiment at ELA, even though no major changes in ecosystem processes were observed (Schindler et al. 1985). Similarly, while copper additions to a reservoir and cadmium additions to marine mesocosms resulted in only short-term decreases in primary production, major changes in phytoplankton species composition persisted for much longer following both of these perturbations (Thomas and Seibert 1977; McKnight 1981). Thus, structural changes are indeed more readily apparent and more sensitive indicators of the response of aquatic ecosystems to toxic chemical stress.

The Nature of Structural Changes

Ford (1989) recently reviewed the effects of toxic stress agents on the structure of aquatic ecosystems, synthesizing information from an impressive array of literature. She concluded that losses of sensitive species are the earliest structural indicators of change; species richness declines under greater stress. These decreases in species richness may or may not be accompanied by decreases in biomass and abundances, depending on the specifics of the ecosystem and the stress agent (Ford 1989).

Interestingly, the specific stress agent does not seem to matter greatly, and similar structural responses are frequently seen from a variety of stresses (Rapport et al. 1985; Ford 1989). For instance, very similar changes in the structure of marine benthic communities occur, whether the stress is from oil pollution, organic loading, or other toxic pollution (Pearson and Rosenberg 1978; Boesch and Rosenberg 1981; Gray 1989). Sensitive species such as ampeliscid amphipods and ostracods are rapidly replaced by pollution-tolerant, opportunistic species such as *Capitella* spp. (Pearson and Rosenberg 1978; Boesch and Rosenberg 1981; Teal and Howarth 1984; Howarth 1989). Of course, if opportunistic species dominate a community before a pollution event, and sensitive species are absent, change is less likely.

Changes in phytoplankton species composition frequently result from toxic chemical stress, whether in lakes or marine ecosystems. Often, the same sorts of changes occur, regardless of the stress. For example, Schindler (1987) reported similar changes from eutrophication and from acidification in ELA experiments. However, in other cases, different stresses can result in different changes, as seen by different responses to arsenic and to silver additions to mesocosms in Chesapeake Bay (Sanders and Cibik 1988).

Several authors have suggested that stress tends to favor domination by smaller organisms (Margalef 1975; Woodwell 1983; Odum 1985; Rapport et al. 1985; Gray 1989). In many cases in aquatic ecosystems, this indeed seems to be true. When oil pollution has caused changes in phytoplankton composition, smaller forms often have been favored (Parsons et al. 1976; Lee and Takahashi 1977). Metal additions to both freshwater and marine ecosystems and mesocosms also frequently cause a shift to smaller phytoplankton forms (Thomas and Siebert 1977; McKnight 1981). This was also observed following arsenic additions to estuarine mesocosms (Sanders and Cibik 1988). On the other hand, an addition of silver to the same mesocosms resulted in larger phytoplankton species, not smaller (Sanders and Cibik 1988), and lake acidification resulted in domination by larger phytoplankton species (Schindler et al. 1985; Schindler 1987). I conclude that stress causes shifts toward opportunistic species, but such species may or may not be smaller than the forms they replace.

Significance of Structural Changes

Structural changes are more sensitive indicators of stress in aquatic ecosystems than are functional changes. But are such structural changes merely early warning signals of functional alterations to come if stress persists or worsens? Or should we be concerned about the structural changes themselves? The significance of structural changes has received surprisingly little study, but some examples illustrate reason for concern.

In the ELA whole-lake acidification experiments, food webs were altered and fish recruitment failed without any major change in ecosystem processes (Schindler et al. 1985). In the CEPEX marine mesocosm experiments, structural changes occurred, but we do not know if fish production or recruitment is affected by such changes; mesocosms are simply too small, and the CEPEX experiments too short in duration, to examine the response of fish. However, Greve and Parsons (1977) argued that changes in phytoplankton species composition of the sort observed in the CEPEX experiments may well alter food webs to the detriment of fish. Their hypothesis is that when large diatoms dominate, food webs lead to large zooplankton, favoring young fish. On the other hand, when large diatoms are replaced by smaller flagellates, food webs lead to smaller zooplankton, which would favor gelatinous zooplankton instead of fish as predators.

An adequate test of the Greve and Parson (1977) hypothesis requires an experiment at the whole-system scale in a marine ecosystem, and of course this has not yet been performed. However, supportive evidence has come from other mesocosm experiments. Sanders and Cibik (1988) found that arsenic additions to estuarine mesocosms caused a shift to smaller phytoplankton forms, and the zooplankton also shifted towards species less suitable for a food chain leading to harvestable fish. The zooplankton that dominated before the toxic stress, the copepods *Acartia tonsa* and *Eurytemora affinis*, were found to have a much lower fecundity and survival in a laboratory study when fed the phytoplankton assemblage resulting in the mesocosms from the arsenic addition (Sanders 1986).

Another common effect of toxic chemical stress in marine ecosystems is the loss of ampeliscid amphipods from the benthic community (Elmgren and Frithsen 1982; Oviatt et al. 1982; Teal and Howarth 1984). Because these are a prime source of food for many species of demersal fish, their loss could have drastic effects on the food web (Howarth 1989).

A Closer Look at a Functional Response to Stress — Controls on Primary Production in Aquatic Ecosystems

I concluded previously that structural responses are better than functional responses as indicators of stress in aquatic ecosystems because they occur more predictably and earlier, or at lower levels of toxic chemical stress. However, functional responses also can occur (see Table 9.2). Is there any generality in the nature of the functional responses to stress? An examination of changes in primary production is enlightening.

We often see very little change in primary production, even though toxic chemical stress causes large changes in community structure. Examples include lake acidification (Schindler et al. 1985; Dillon et al. 1979) and additions of toxic metals to lakes (McKnight 1981; Kettle and DeNoyelles 1986; Schindler 1987) and to the marine CEPEX mesocosms (Thomas and Siebert 1977; Thomas et al. 1977). Such results perhaps should not be surprising because of the great functional redundancy among primary producer species. In contrast to terrestrial ecosystems, new populations can grow in very quickly to replace those which are damaged by pollution. Primary production in these systems presumably is controlled "from below" by the availability of nutrients, and nutrient availability is not greatly affected by these particular toxic stresses. "Top-down," cascading trophic effects of the sort hypothesized by Carpenter et al. (1985) do not seem to have a major influence on primary production in these particular studies. Similarly, oil additions to the CEPEX mesocosms decreased primary production, but this effect seemed to result from decreased nutrient availability as hydrocarbon-degrading bacteria consumed nutrients (Lee et al. 1978); again, primary production appeared to be controlled principally "from below."

On the other hand, toxic chemical stress sometimes increases primary production, as seen following two oil spills and in oil-addition experiments in the MERL mesocosms (Johansson 1980; Elmgren et al. 1980; Cabioch et al. 1981; Oviatt et al. 1982). And in at least two of these cases, the increase in primary production is attributed to reduced zooplankton grazing (Elmgren et al. 1980; Johansson 1980; Oviatt et al. 1982). In the third case, many animals were killed and faunal community composition was greatly changed; the authors of this study believe an increased nutrient availability from the death of the animals caused the increased primary production (Cabioch et al. 1981), but an alteration in grazing is as likely. Thus, in at least some of these ecosystems, toxic chemical stress seems to alter primary production by altering "top-down" controls.

Do differences in these studies suggest different responses to different stress agents? Or are the differences we observe a result of differences in the basic functioning of different types of aquatic ecosystems? As noted, vastly different types of toxic stress often seem to have very similar effects on ecosystem or community structure. Thus, I suspect that the differences we observe in controls on primary production may be a result of differences in the functioning of different types of aquatic ecosystems.

Toward a General Theory

Better environmental management will require a better theory or model relating ecosystem response to toxic chemical stress. Aside from the intrinsic theoretical interest in developing such a model, I see at least three reasons for pursuing a better theory:

1. Theory can aid in designing experiments and in interpreting experimental data. On what basis is an experimental system chosen? Does the experimental system represent other systems at risk from toxic chemical stress, or is it more or less sensitive than these ecosystems?
2. Theory can improve extrapolation of data from studies to other systems. Ecosystem experiments are necessary, but are both expensive and rare. We need to be able to extrapolate data from these experiments and from mesocosm experiments as far as possible, yet with care. Such extrapolation requires a model or framework.
3. Theory may aid in actual management decisions, such as siting of activities releasing toxic chemicals. Are some types of ecosystems more sensitive to stress than others, and therefore in need of greater protection? Probably, and we need a better framework for identifying this sensitivity.

Odum (1985) proposed a general framework for understanding the response of ecosystems to stress. But as noted by Schindler (1987), many of the responses predicted by Odum may not be generally applicable. Perhaps the most useful generalization for the response of aquatic ecosystems stressed by toxic substances concerns structural changes. Sensitive species are lost and are replaced by more opportunistic species. Food webs are altered, perhaps shortened (Odum 1985; Schindler 1987), and the structural complexity of the ecosystem is reduced. At present, our best measure of such change is merely the disappearance of the sensitive species; various measures of diversity are much less powerful (Schindler 1987; Ford 1989).

Ecosystems that experience little stress, whether from prior toxic chemical stress, physical stress, or biotic disturbances such as invasions, will tend to be dominated by sensitive species. Apparently, species that are limited to a narrow range of environmental conditions, that are otherwise highly specialized in their niches, or that are easily affected by physical disturbances, are also those which are most sensitive to toxic chemical stress. Ecosystems dominated by these sensitive species are the ecosystems most sensitive to changes from toxic chemical stress.

This simple conclusion may have surprising power, because the relative sensitivities of dominant species are fairly well known for many aquatic ecosystems (see Patrick 1949, 1968, 1972; Pearson and Rosenberg 1978; Boesch and Rosenberg 1981; Sanders and Jones 1981; Teal and Howarth 1984; Økland and Økland 1986; Ford 1989). As an example, we might expect areas in the North Sea or on Georges Bank to be more sensitive to the effects of oil pollution than are areas near the Louisiana coast in the Gulf of Mexico. The benthic fauna of the former includes sensitive species such as ampeliscid amphipods, whereas the benthic fauna in the in the Gulf of Mexico is dominated by more opportunistic species such as *Mulina* (Sanders and Jones 1981; Howarth 1987). In fact, oil production clearly has affected the benthic fauna near rigs in the North Sea (Addy et al. 1978; Kingston 1987), while this is not as obviously so in the Gulf of Mexico (Howarth 1989). As another example, we would expect coral reefs to be much more sensitive to toxic chemical stress than are salt marshes.

This simple construct may also aid in interpreting and extrapolating data. For instance, very little information exists concerning the minimum concentration of oil necessary to cause changes in the structure of marine phytoplankton communities. In experiments with the CEPEX mesocosms, very low level oil additions (initial concentrations of 20 and 40 μg/l, decreasing to 0 within about 2 weeks) resulted in large phytoplankton changes (Parsons et al. 1976; Lee et al. 1978). However, following an oil spill in the Baltic Sea, oil concentrations of 50–60 μg/l (Kineman and Clark 1980) caused no changes in phytoplankton species composition (Johansson 1980). And in experiments in the MERL mesocosms, dissolved oil concentrations of 90 and 190 μg/l caused observable but fairly small changes in phytoplankton composition (Vargo et al. 1982). In all these cases, the oils were refined fuel oils, and it is doubtful that the nature of the dissolved oil varied greatly (Teal and Howarth 1984). Thus, the differences in response among these observations probably reflect differences in sensitivity in the ecosystems or mesocosms. Indeed, for the oil spill in the Baltic Sea, microflagellates comprised 75% to 90% of the phytoplankton composition before the spill (Johansson 1980), and these plankton may be relatively insensitive to oil pollution (Teal and Howarth 1984). Many of the species present in the feed water from Narragansett Bay for the MERL experiments also are ones thought to be oil tolerant (Vargo et al. 1982). When predicting the likely sensitivity of other marine ecosystems to oil pollution, we should consider the dominant phytoplankton.

Similarly, we should be mistrustful of extrapolating toxicity data from experimental ponds, which tend to be dominated by opportunistic species, to oligotrophic and mesotrophic lakes, which probably have more sensitive species. Thus, while Larsen et al. (1986) found no effect of atrazine additions to ponds at concentrations of 20 μg/l, it may be unwise to assume that this would be the case in a larger, more oligotrophic lake.

Note that I am defining sensitivity in terms of the concentration or activity of a toxic substance. That is, I am arguing that we can compare the responses of systems to a given concentration of oil or cadmium, or compare the response of lakes to acidification at a given pH. Thus, I am not considering the buffering capacity of a lake or the ability of a system to tie up pollutants in unavailable forms.

Openness of the Ecosystem as a Determinant of Sensitivity

Although the construct that I suggested is useful, it is a fairly qualitative approach and suffers from lack of relationship to any other aspect of the ecosystem. Sensitive species may be absent from an ecosystem either because of some basic characteristic of the system or because these species have previously been displaced by toxic stress. Is there some characteristic of an ecosystem that in the absence of previous stress allows us to predict sensitivity?

In the spirit of speculation encouraged by the organizers of the Cary Conference, I hypothesize that we may be able to predict the sensitivity of an aquatic ecosystem to a toxic chemical stress based on the relative openness of the ecosystem. More open systems may be less sensitive and more resistant to toxic chemical stress (Table 9.3). The openness of an ecosystem can be defined in several manners, but I suggest the most useful measure is a "recycling ratio," or the relative rate of uptake of a critical element in net primary production (per unit area) compared to the overall input of that

TABLE 9.3. Some characteristics of open and closed ecosystems.

Open systems	Closed systems
More generalists	More specialized species
Stress-tolerant species	Stress-sensitive species
Ecosystem is resistant[a]	Ecosystem is sensitive

[a]DeAngelis (1980): open systems are resilient.

element to the ecosystem. The critical element should be one which is limiting primary production, or colimiting production in association with other equally limiting elements (Howarth 1988). This recycling ratio is similar to the "index of recycling" defined by DeAngelis (1980), but is based on inputs to the ecosystem instead of losses.

The recycling ratio is measurable for many aquatic ecosystems and so offers a quantifiable definition of openness. To some extent, an ecosystem is more or less open depending on events in adjacent, "upstream" ecosystems and on the physics and hydrology of the system. A lake with a watershed dominated by relatively leaky terrestrial ecosystems, such as agriculture, will be more open than a comparable lake whose watershed is dominated by less nutrient-leaky forests. And a lake with a longer residence time of water will tend to be less open than an otherwise comparable lake with a shorter water residence time. A salt marsh is so open because of the tides regularly carrying nutrients in and out. However, the biota can also partially regulate the relative openness or "closed-ness" of an ecosystem by controlling the coupling of element recycling back to primary producer organisms. Coral reefs are relatively closed ecosystems in part because of tight recycling of elements between heterotrophs and primary producers within the structure of the reef (D'Elia 1988).

At least qualitatively, ecosystem openness is correlated with the presence or absence of sensitive species and therefore with ecosystem sensitivity to stress. Relatively closed ecosystems (for instance, the Sargasso Sea, with a recycling ratio of 75–400; Fig. 9.1) have considerable structural complexity, more specialized species, and many sensitive species. Relatively open ecosystems (a salt marsh, with a recycling ratio less than 1; Fig. 9.1) have much less structural complexity and a relative lack of sensitive species and are dominated by more opportunistic species. Presumably, the Sargasso Sea is much more sensitive to toxic chemical stress than is a salt marsh, given comparable doses or exposures.

During eutrophication, ecosystems become more open (see Fig. 9.1). They also become structurally simpler, more dominated by opportunistic species, and therefore, presumably, more resistant to toxic chemical stress. Thus, there are many similarities between eutrophication and the response of an aquatic ecosystem to toxic stress. An ecosystem that has been disturbed either by increased nutrient inputs or by toxic chemical stress is probably more resistant to further stress from toxic chemicals.

My construct of using the openness of an ecosystem to determine sensitivity to stress has some similarities to and is partially inspired by the framework of Odum (1969, 1985). But note that while I am arguing that more closed ecosystems are less resistant to stress, Odum (1985) maintained the opposite, that is, that more "mature," closed ecosystems are more resistant to stress because of greater functional complexity.

Based on mathematical models of ecosystem element cycles, DeAngelis (1980) proposed that more open ecosystems are more resilient, that is that

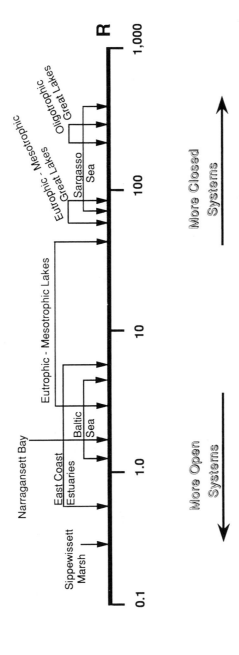

FIGURE 9.1. Recycling ratio continuum for aquatic ecosystems, in which R = uptake of limiting or critical element into net primary production divided by annual input of that element to the ecosystem. See text for discussion.

they recover more rapidly following a disturbance. A similar argument was proposed by Pomeroy (1970). If indeed so, then more open systems are both more resistant to stress and more resilient following stress. Boesch and Rosenberg (1981) also argued that the marine benthic communities that are more resistant are also most resilient.

Why might relatively closed ecosystems have more specialized species? Presumably, there is more opportunity for biotic adjustment of element cycles, and tighter coupling among element cycles, in more closed ecosystems. Conversely, there is more of a forcing function from a single element when the ecosystem is more open. This suggests that there is more opportunity to develop niche specialization (in the sense of Hutchinson 1959) in more closed ecosystems. That is, low nutrient status, with more reliance on recycling of elements, may be a driving evolutionary force leading to greater specialization and diversity. It is not clear to me why more specialized species should necessarily be more sensitive to toxic chemical stress. However, that this is indeed the case, and that more diverse communities are therefore more sensitive to stress, has previously been noted (Sanders 1968; Goodman 1975; Pearson and Rosenberg 1978; Boesch and Rosenberg 1981).

Even in the absence of toxic chemical stress, it is clear that the relative openness of an ecosystem is not the only factor controlling its structure. Natural disturbances can also be important, as Fisher and Grimm (Chapter 10, this volume) discusses for stream ecosystems. However, for many plankton-based aquatic ecosystems, I suspect that natural disturbances of this sort may be less important determinants of structure than is the openness of element cycles. On the other hand, disturbance may be the more critical factor controlling the structure of terrestrial ecosystems. In any event, many terrestrial ecosystems have rather low recycling ratios compared to many relatively closed aquatic ecosystems. Data from O'Neill and Reichle (1980) suggest a range of nitrogen recycling ratios of 2.6 to 50 for various forests around the world studied as part of the International Biological Program (IBP). Forests at the lower end of this range have recycling ratios which are comparable to those found in fairly open, eutrophic estuaries (Fig. 9.1). That the recycling ratio for a forest can be so low is in part a result of the high proportion of carbon compared to other elements in the structural biomass of a forest. Plankton-based aquatic ecosystems have much less structural material and so produce less organic carbon per unit of nitrogen or phosphorus assimilated.

Closing Thoughts

Whole-ecosystem experiments, mesocosm experiments, and observations of actual toxic discharges into natural waters have all provided valuable information on the effects of toxic stress. Of these approaches, whole-ecosystem experiments have probably proved the most powerful tools for understanding toxic stress in lakes. Unfortunately, no experiments at this

TABLE 9.4. Concluding thoughts and hypotheses on the response of aquatic ecosystems to toxic stresses.

Structural changes are more predictable and may occur earlier or at lower levels of stress than do changes in ecosystem processes.

When functional changes occur, rates can either increase or decrease.

The earliest indicator of structural change is loss of sensitive species.

Loss of sensitive species occurs predictably regardless of the agent of toxic chemical stress.

The significance of structural change requires more study; particularly in marine ecosystems where whole-ecosystem studies are lacking.

Ecosystems dominated by generalists/opportunists are most resistant to stress.

More open ecosystems (defined by a "recycling ratio") may be more resistant to stress.

---

scale have been conducted in marine ecosystems. Although mesocosm experiments often provide insights that are similar to those gained from whole-ecosystem experiments, one can never be sure that the results from a mesocosm experiment will scale up to larger systems. It has proven difficult to obtain information on the effects of toxic stress on fish populations from mesocosm experiments simply because of the small spatial and short temporal scale of the experiments.

That single-species tests conducted in the laboratory can severely underestimate the effects of toxic chemical stress has been demonstrated in many mesocosm and whole-ecosystem experiments. Therefore, if we are to better protect nature, we need more whole-system experiments, and we need to conduct these in sensitive ecosystems. Such "experiments" could include careful before-and-after observations of the effects of new industrial development and discharges, with parallel observation of appropriate reference sites. We also need more mesocosm and whole-ecosystem studies in which a variety of doses are examined (such as the acidification experiments at ELA; Schindler et al. 1985), thereby allowing a determination of the minimum harmful dose. Future ecosystem-scale experiments and environmental management strategies would benefit from an improved theory relating ecosystem response to toxic chemical stress.

Acknowledgments. Tom Frost, Nelson Hairston, Jr., Roxanne Marino, Karen McGlathery, David Rucknick, and an anonymous reviewer provided valuable comments on earlier versions of this manuscript. Diane Sherman and Roxanne Marino assisted with preparation of the manuscript and figures. This is a contribution from Cornell's Biogeochemical Applied Research Facility.

REFERENCES

Addy, J.M., D. Levell, and J.P. Hartely. (1978). Biological monitoring of sediments in Ekofisk oilfield. In: *Proceedings of the Conference Assessment of Ecological Impacts of Oil Spills.* American Institute of Biological Sciences, Washington, D.C., pp. 519–539.

Auerbach, S.I. (1981). Ecosystem response to stress: a review and critique. In: *Stress Effects on Natural Ecosystems*, G.W. Barrett and R. Rosenberg, eds. Wiley and Sons, New York, pp. 29–41.

Bedinger, C.A., R.E. Childers, J.W. Cooper, K.T. Kimball, and A. Kwok. (1980). *Central Gulf Platform Study. Vol. 1. Pollutant Fate and Effects Studies. Part 1. Background, Program Organization and Study Plan*. Southwest Research Institute, Houston, Texas.

Bender, M.E., D.J. Reish, and C.H. Ward. (1979). Independent appraisal. Reexamination of the offshore ecology investigation. *Rice Univ. Stud.* 65:35–118.

Boesch, D.F. and R. Rosenberg. (1981). Response to stress in marine benthic communities. In: *Stress Effects on Natural Ecosystems*, G.W. Barrett and R. Rosenberg, eds. Wiley and Sons, New York, pp. 29–41.

Cabioch, L., J.C. Dauvin, F. Gentil, C. Retiere, and V. Rivain. (1981). Perturbations induites dans la composition et le fonctionnement des peuplements benthiques sublittoraus sous l'effet des hydrocarbures de l'Amoco Cadiz. In: *Amoco Cadiz, Fates and Effects of the Oil Spill*, CNEXO, Paris, pp. 513–526.

Carney, R.S. (1987). A review of study designs for the detection of long-term environmental effects of offshore petroleum activities. In: *Long-Term Effects of Offshore Oil and Gas Development: An Assessment and Research Strategy*, D.F. Boesch and N. Rabalais, eds. Elsevier, New York, pp. 651–696.

Carpenter, S.R., J.F. Kitchell, and J.R. Hodgson. (1985). Cascading trophic interactions and lake productivity. *BioScience* 35:634–639.

Crossey, M.J. and T.W. LaPoint. (1988). A comparison of periphyton community structural and functional responses to heavy metals. *Hydrobiologia* 162:109–121.

Davies, J.M., I.E. Baird, L.C. Massie, S.J. Hay, and A.P. Ward. (1980). Some effects of oil-derived hydrocarbons in a pelagic food web from observations in an enclosed ecosystem and a consideration of their implications for monitoring. *Rapp. P.V. Reun. Cons. Int. Explor. Mer.* 179:201–211.

DeAngelis, D.L. (1980). Energy flow, nutrient cycling, and ecosystem resilience. *Ecology* 61:764–771.

D'Elia, C.F. (1988). The cycling of essential elements in coral reefs. In: *Concepts of Ecosystem Ecology*, L.R. Pomeroy and J.J. Alberts, eds. Springer-Verlag, New York, pp. 195–230.

Dillon, P.J. and F.H. Rigler. (1974). The phosphorus chlorophyll relationship in lakes. *Limnol. Oceanogr.* 19:767–773.

Dillon, P.J, N.D. Yan, W.A. Scheider, and N. Conroy. (1979). Acidic lakes in Ontario, Canada: characterization, extent and responses to base and nutrient addition. *Ergeb. Limnol.* 13:317–336.

Eaton, J.A.J., R. Hermanutz, R. Kiefer, L. Mueller, R. Andersen, R. Erickson, B. Nordling, J. Rogers, and H. Pritchard. (1986). Biological effects of continuous and intermittent dosing of outdoor experimental streams with chloropyrifos. In: *Aquatic Toxicology and Hazard Assessment: Eighth Symposium*, R.C. Bahner and D.J. Hansen, eds. ASTM STP 891, American Society for Testing and Materials, Philadelphia.

Elmgren, R. and J.B. Frithsen. (1982). The use of experimental ecosystems for evaluating the environmental impacts of pollutants: a comparison of an oil spill in the Baltic Sea and two long-term, low-level oil addition experiments in mesocosms. In: *Marine Mesocosms*, G.D. Grice and M.R. Reeve, eds. Springer-Verlag, New York, pp. 153–165.

Elmgren, R., G.A. Vargo, J.F. Grassle, J.P. Grassle, D.R. Heinle, G. Langois, and S.L. Vargo. (1980). Trophic interactions in experimental marine ecosystems perturbed by oil. In: *Microcosms in Ecological Research*, J.P. Giesy, ed. U.S. Dept. of Energy, Washington, D.C., pp. 779–800.

Farrington, J.W. (1989). Bioaccumulation of hydrophobic organic pollutant compounds. In: *Ecotoxicology: Problems and Approaches*, S.A. Levin, M.A. Harwell, J.R. Kelly, and K.D. Kimball, eds. Springer-Verlag, New York, pp. 279–308.

Ford, J. (1989). The effects of chemical stress on aquatic species composition and community structure. In: *Ecotoxicology: Problems and Approaches*, S.A. Levin, M.A. Harwell, J.R. Kelly, and K.D. Kimball, eds. Springer-Verlag, New York, pp. 99–129.

Galloway, J.N., G.E. Likens, and M.E. Hawley. (1984). Acid precipitation: natural versus anthropogenic components. *Science* 226:829–831.

Gearing, J.N. (1989). The role of aquatic microcosms in ecosyxicologic research as illustrated by large marine systems. In: *Ecotoxicology: Problems and Approaches*, S.A. Levin, M.A. Harwell, J.R. Kelly, and K.D. Kimball, eds. Springer-Verlag, New York, pp. 411–448.

Gentile, J.H., S.M. Gentile, N.G. Hairston, Jr., and B.K. Sullivan. (1982). The use of life-tables for evaluating the chronic toxicity of pollutants to *Mysidopsis bahia*. *Hydrobiologia* 93:179–187.

Goodman, D. (1975). The theory of diversity-stability relationships in ecology. *Q. Rev. Biol.* 50:237–266.

Gray, J.S. (1987). Oil pollution studies of the Solbergstrand mesocosms. *Phil. Trans. R. Soc. Lond. B* 316:641–654.

Gray, J.S. (1989). Effects of environmental stress on species rich assemblages. *Biol. J. Linn. Soc.* 37:19–32.

Greve, W. and T.R. Parsons. (1977). Photosynthesis and fish production: hypothetical effects of climatic change and pollution. *Helgol. Wiss. Meeresunters.* 30: 666–672.

Grice, G.D. and M.R. Reeve. (1982). Introduction and description of experimental ecosystems. In: *Marine Mesocosms: Biological and Chemical Research in Experimental Ecosystems*, G.R. Grice and M.R. Reeve, eds. Springer-Verlag, New York, pp. 1–9.

GURC. (1974). Final Project Planning Council Concensus Report. Gulf Universities Research Consortium Report No. 138.

Hall, R.J. and G.E. Likens. (1981). Chemical flux in an acid-stressed stream. *Nature* (London) 292: 329–331.

Hall, R.J., C.T. Driscoll, G.E. Likens, and J.M. Pratt. (1985). Physical chemical and biological consequences of episodic aluminum additions to a stream ecosystem. *Limnol. Oceanogr.* 30:212–220.

Hall, R.J., G.E. Likens, S.B. Fiance, and G.R. Hendrey. (1980). Experimental acidification af a stream in the Hubbard Brook Experimental Forest, New Hampshire. *Ecology* 61:976–989.

Hecky, P.E. and P. Kilham. (1988). Nutrient limitation of phytoplankton in freshwater and marine environments: a review of recent evidence on the effects of enrichment. *Limnol. Oceanogr.* 33:796–822.

Hecky, R.E., R.W. Newbury, R.A. Bodaly, K. Patalas, and D.M. Rosenburg. (1984). Environmental impact prediction and assessment: the Southern Indian Lake experience. *Can. J. Fish. Aquat. Sci.* 41:730–732.

Howarth, R.W. (1987). Potential impacts of petroleum on the biotic resources of Georges Bank. In: *An Atlas of Georges Bank*, R. Backus, ed. M.I.T. Press, Cambridge, Massachusetts.

Howarth, R.W. (1988). Nutrient limitation of net primary production in marine ecosystems. *Annu. Rev. Ecol. Syst.* 19:89–110.

Howarth, R.W. (1989). Determining the effects of oil pollution in marine ecosystems. In: *Ecotoxicology: Problems and Approaches*, S.A. Levin, M.A. Harwell, J.R. Kelly, and K.D. Kimball, eds. Springer-Verlag, New York, pp. 69–87.

Howarth, R.W., R. Marino, and J.J. Cole (1988). Nitrogen fixation in freshwater, estuarine, and marine ecosystems. 2. Biogeochemical controls. *Limnol. Oceanogr.* 33:688–701.

Hutchinson, G. E. (1959). Homage to Santa Rosalia, or why are there so many kinds of animals? *Am. Nat.* 93:145–159.

Jacobson, S.M. and D.B. Boylan. (1973). Effect of seawater soluble fraction of kerosene on chemotaxis in a marine snail, *Nassarius obsoletus. Nature* (London) 241:213–215.

Johansson, S. (1980). Impact of oil on the pelagic ecosystem. In: *The Tsesis Oil Spill*, J.J. Kineman, R. Elmgren, and S. Hansson, eds. NOAA, U.S. Dept. of Commerce, Washington, D.C., pp. 61–80.

Kelly, C.A, J.W.M. Rudd, A. Furutani, and D.W. Schindler. (1984). Effects of lake acidification on rates of organic matter decomposition in sediments. *Limnol. Oceanogr.* 29:687–694.

Kettle, W.D. and F. DeNoyelles, Jr. (1986). Effects of cadmium stress on the plankton communities of experimental ponds. *J. Freshwater Ecol.* 3:433–444.

Kineman, J. and R.C. Clark, Jr. (1980). NOAA acute phase experiments on pelagic and surface oil. In: *The Tsesis Oil Spill*, J.J. Kineman, R. Elmgren, and S. Hansson, eds. NOAA, U.S. Dept. of Commerce, Washington, D.C., pp. 83–94.

Kingston, P.F. (1987). Field effects of platform discharges on benthic macrofauna. *Phil. Trans. R. Soc. Lond. B* 316:545–565.

Lannergren, C. (1978). Net- and nanoplankton: effects of an oil spill in the North Sea. *Bot. Mar.* 21:353–356.

Larsen, D.P., F. deNoyelles, Jr., F. Stay, and T. Shiroyama. (1986). Comparisons of single-species, microcosm and experimental pond responses to atrazine exposure. *Environ. Toxicol. Chem.* 5:179–190.

Lee, R.F., and M. Takahashi. (1977). The fate and effect of petroleum in controlled ecosystem enclosuers. *Rapp. P.-V. Reun. Cons. Int. Explor. Mer* 171:150–156.

Lee, R.F., M. Takahashi, J.R. Beers. (1978). Short term effects of oil on plankton in controlled ecosystems. In: *Proceedings of Conference on Assessment of Ecological Impacts of Oil Spills*. American Institute of Biological Sciences, Washington, D.C., pp. 635–650.

Levine, S.N. (1989). Theoretical and methodological reasons for variability in the responses of aquatic ecosystem processed to chemical stress. In: *Ecotoxicology: Problems and Approaches*, S.A. Levin, M.A. Harwell, J.R. Kelly, and K.D. Kimball, eds. Springer-Verlag, New York, pp. 145–174.

Likens, G.E. (1985). An experimental approach for the study of ecosystems. *J. Ecol.* 73:381–396.

Margalef, R. (1975). External factors and ecosystem stability. *Schweiz. Z. Hydrol.* 37:102–117.

Marino, R., R.W. Howarth, J. Shamess, and E. Prepas. (1990). Molybdenum and sulfate as controls on the abundance of nitrogen-fixing cyanobacteria in saline lakes in Alberta. *Limnol. Oceanogr.* 35:245–259.

McKnight, D. (1981). Chemical and biological processes controlling the response of a freshwater ecosystem to copper stress: a field study of the $CuSO_4$ treatment of Mill Pond Reservoir, Burlington, Mass. *Limnol. Oceanogr.* 26:518–531.

Menzel, D.W. (1977). Summary of experimental results: controlled ecosystem pollution experiment. *Bull. Mar. Sci.* 27:142–145.

Middleditch, B.S. (1982). *Environmental Effects of Offshore Oil Production.* The Buccaneer Gas and Oil Field Study. Plenum Press, New York.

National Academy of Sciences. (1985). *Oil in the Sea: Inputs, Fates, and Effects.* Washington, D.C.

Nixon, S.W., C. Oviatt, J. Frithsin, and B. Sullivan. (1986). Nutrients and the productivity of estuarine and coastal marine ecosystems. *J. Limnol. Soc. S. Afr.* 12:43–71.

Nriagu, J.O. and J.M. Pacyna. (1988). Quantitative assessment of worldwide contamination of air, water and soils by trace metals. *Nature* (London) 333:134–139.

Økland, J. and K.A. Økland. (1986). The effects of acid deposition on benthic animals in lakes and streams. *Experientia* (Basel) 42:471–486.

Odum, E.P. (1969). The strategy of ecosystem development. *Science* 164:262–270.

Odum, E.P. (1985). Trends expected in stressed ecosystems. *BioScience* 35:419–422.

O'Neill, R.V. and D.E. Reichle. (1980). Dimensions of ecosystem theory. In: *Forests: Fresh Perspectives from Ecosystem Analysis*, R.H. Waring, ed. Oregon State University Press, Corvallis, pp. 11–26.

Oviatt, C.A., J. Frithsen, J. Gearing, and P. Gearing. (1982). Low chronic additions of No. 2 fuel oil: chemical behavior, biological impact and recovery in a simulated estuarine environment. *Mar. Ecol. Prog. Ser.* 9:121–136.

Patrick, R. (1949). A proposed biological measure of stream conditions based on a survey of the Conestoga Basin, Lancaster County, Pennsylvania. *Proc. Acad. Nat. Sci. Phila.* 101:277–341.

Patrick, R. (1968). The structure of diatom communities in similar ecological conditions. *Am. Nat.* 102:173–183.

Patrick, R. (1972). Aquatic communities as indices of pollution. In: *Indicators of Environmental Quality*, W.A. Thomas, ed. Plenum Press, New York, pp. 93–100.

Parsons, T.R., W.K.W. Li, and R. Waters. (1976). Some preliminary observations on the enhancement of phytoplankton growth by low levels of mineral hydrocarbons. *Hydrobiology* 51:85–89.

Pearson, T.H., and R. Rosenberg. (1978). Macrobenthic succession in relation to organic enrichment and pollution of the marine environment. *Oceanogr. Mar. Biol. Annu. Rev.* 163:229–311.

Perry, J.A. and N.H. Troelstrup. (1988). Whole ecosystem manipulation: a productive avenue for test system research? *Environ. Toxicol. Chem.* 7:941–951.

Pomeroy, L R. (1970). The strategy of mineral cycling. *Annu. Rev. Ecol. Syst.* 1:171–190.

Rapport, D.J., H.A. Regier, and T.C. Hutchinson. (1985). Ecosystem behavior under stress. *Am. Nat.* 125:617–640.

Rosenberg, D.M., V.H. Resh, S.S. Balling, M.A. Barnaby, J.N. Collins, D.V. Durbin, T.S. Flynn, D.D. Hart, G.A. Lamberti, E.P. Elravy, J.R. Wood, T.E.

Blank, D.M. Schultz, D.L. Marrin, and D.G. Price. (1981). Recent trends in environmental impact assessment. *Can. J. Fish. Aquat. Sci.* 38:591–624.

Rudd, J.W.M., C.A. Kelly, V. St. Louis, R.H. Hesslein, A. Furutani, and M. Holoka. (1986). Microbial consumption of nitric and sulfuric acids in acidified north temperate lakes. *Limnol. Oceanogr.* 31:1267–1280.

Sanders, H.L. (1968). Marine benthic diversity: a comparative study. *Am. Nat.* 102:243–282.

Sanders, H.L. and C. Jones. (1981). Oil, science, and public policy. In: *Coast Alert: Scientists Speak Out*, T.C. Jackson and D. Reische, eds. Friends of the Earth Publishers, San Francisco.

Sanders, J.G. (1986). Direct and indirect effects of arsenic on the survival and fecundity of estuarine zooplankton. *Can. J. Fish. Aquat. Sci.* 43:694–699.

Sanders, J.G. and S.J. Cibik. (1988). Response of Chesapeake Bay phytoplankton communities to low levels of toxic substances. *Mar. Pollut. Bull.* 19:439–444.

Schindler, D.W. (1977). Evolution of phosphorus limitation in lakes. *Science* 195:260–262.

Schindler, D.W. (1978). Factors regulating phytoplankton production and standing crop in the worlds freshwaters. *Limnol. Oceanogr.* 23:478–486.

Schindler, D.W. (1987). Determining ecosystem responses to anthropogenic stress. *Can. J. Fish. Aquat. Sci.* 44(Suppl. 1):6–25.

Schindler, D.W., M.A. Turner, P. Stainton, and G. Linsey. (1986). Natural sources of acid nertralizing capacity in low alkalinity lakes of the Precambrian Shield. *Science* 235:844–847.

Schindler, D.W., K.H. Mills, D.F. Mailey, D.L. Findlay, J.A. Shearer, I.J. Davies, M.A. Turner, G.A. Linsey and D.R. Cruikshank. (1985). Long-term ecosystem stress: the effects of years of experimental acidification on a small lake. *Science* 228:1395–1401.

Smith, S.V. (1984). Phosphorus versus nitrogen limitation in the marine environment. *Limnol. Oceanogr.* 29:1149–1160.

Smith, V.H. (1983). Low nitrogen to phosphorus ratios favor dominance by blue-green algae in lake phytoplankton. *Science* 221:669–671.

Steele, R.L. (1977). Effects of certain petroleum products on reproduction and growth of zygotes and juveniles stages of the alga *Fucus edentatus* de la Pyl (Phaeophyceae: Fucales). In: *Fate and Effect of Petroleum Hydrocarbons in Marine Ecosystems and Organisms*, D. Wolfe, ed. Pergamon, New York, pp. 115–128.

Teal, J.M. and R.W. Howarth. (1984). Oil spill studies: a review of ecological effects. *Environ. Manage.* 8:27–44.

Thomas, W.H. and K.L.R. Seibert. (1977). Effects of copper on the dominance and the diversity of algae: controlled ecosystem pollution experiment. *Bull. Mar. Sci.* 27:23–33.

Thomas, W.H., O. Holm-Hansen, D.L.R. Seibert, F. Azman, R. Hodson, and M. Takahashi. (1977). Effects of copper on phytoplankton standing crop and productivity: controlled ecosystem pollution experiment. *Bull. Mar. Sci.* 27:23–33.

Trimbee, A.M. and E.E. Prepas. (1987). Evaluation of total phosphorus as a predictor of the relative biomass of blue-green algae with emphasis in Alberta lakes. *Can. J. Fish. Aquat. Sci.* 44:1337–1342.

U.S. Department of Interior. (1985). Final Environmental Statement, North Aleutian Basin Lease Sale #92. Minerals Management Service, Dept. of Interior, Anchorage, Alaska.

Vandermeulen, J.H. and J.M. Capuzzo. (1983). Understanding sublethal pollutant effects in the marine environment. In: *Ocean Waste Management Policy and Strategies*, M.A. Champ and M. Trainor, eds. Center for Academic Publications, Melbourne, Florida.

Vargo, G.A., M. Hutchins, and G. Almquist. (1982). The effect of low, chronic levels of no. 2 fuel oil on natural phytoplankton assemblages in microcosms: 1. Species composition and seasonal succession. *Mar. Environ. Res.* 6:245–264.

Walters, C. (1986). *Adaptive Management of Renewable Resources*. Macmillan, New York.

Weinstein, D.A. and E. Birk. (1989). The effects of chemicals on the structure of terrestrial ecosystems: mechanisms and patterns of change. In: *Ecotoxicology: Problems and Approaches*, S.A. Levin, M.A. Harwell, J.R. Kelly, and K.D. Kimball, eds. Springer-Verlag, New York, pp. 181–203.

Winfrey, M.R., E. Beck, P. Boehm, and D.M. Ward. (1982). Impact of crude oil on sulphate reduction and methane production in sediments impacted by the Amico Cadiz oil spill. *Mar. Environ. Res.* 7:175–194.

Woodwell, G.M. (1983). The blue planet: of wholes and parts and man. In: *Disturbance and Ecosystems*, H.A. Mooney and M. Godron, eds. Springer-Verlag, Berlin, pp. 2–10.

10
Streams and Disturbance: Are Cross-Ecosystem Comparisons Useful?

STUART G. FISHER AND NANCY B. GRIMM

Abstract

Disturbance is a pervasive influence in all ecosystems, but its effects vary widely. This is true both when quite different ecosystems are compared and when a suite of similar ecosystems, such as streams, is considered. Comparative approaches to ecosystem studies can provide insight into a complex process such as disturbance, but the power to resolve mechanistic questions diminishes as variance among the ecosystems that are compared increases. As a result, the usefulness of the comparative approach in ecosystem studies will depend on the goals of the investigation. Broad principles of ecosystem science may be revealed by comparison of very different ecosystems, but an understanding of underlying mechanisms shaping ecosystem structure and functioning still requires detailed studies of one or a few similar ecosystems. This is especially true when the process of interest is itself complex. Thus, a process such as disturbance may be best studied by comparing similar systems, whereas principles governing simpler attributes, such as decomposition rate or nutrient ratios, may be better elucidated by examining a wide range of ecosystem types.

Introduction

Change is a universal attribute of earthly environments and is immutably intertwined with processes responsible for the origin and evolution of life on this planet. Disturbance is change that is especially rapid and which results in substantial mortality of organisms, alterations in the distribution of abiotic resources, and consequent disruption of whole ecosystems (Sousa 1984, Pickett and White 1985, Risser 1987). But disturbances vary as widely as the ecosystems they influence and, to date, no conceptual synthesis of broad application has been generated in this field.

Is the comparative approach useful in gaining an understanding of such a complex process as disturbance in ecosystems? What can be gained by comparing similar, as opposed to disparate, ecosystems in this analysis? To answer these questions, we focus on running water ecosystems and the extent to which studies of single streams, similar streams, very different streams, and juxtapositions of streams and other ecosystems might contribute insight into disturbance as a factor shaping ecosystem structure and functioning. We will use this exercise with streams and disturbance as a model for assessing the utility of the comparative approach.

Our approach is predicated on the contention of Shugart and O'Neill (1979) that ecosystem scientists must be willing to generate hypotheses about whole ecosystems—to seek ecosystem properties that apply to a range of disparate ecosystem types. Disturbance is one such property; a few examples of ecosystem-level hypotheses related to disturbance are (1) ecosystems stable to one type of disturbance are stable to others; (2) stability of an ecosystem is a function of the size spectrum of its organisms—those with smaller organisms are more resilient and less resistant; and (3) resistance and resilience are inversely related (Webster et al. 1975). Our analysis will evaluate the ability of comparison to generate and test these kinds of hypotheses rather than simply to cluster ecosystems based on empirical descriptors such as chlorophyll, total biomass, or primary production.

Stream Ecology: A Brief Overview

Stream ecosystem research during the past 30 years has underscored the extreme openness of streams to material exchange with adjacent ecosystems and a pronounced linkage between up- and downstream segments. The trophic structure of streams in wooded areas depends almost entirely on imported organic matter derived ultimately from tree leaves (Cummins 1974). A variable fraction of this input is processed locally, and the remainder is exported to downstream reaches or, in some cases, to lateral floodplains (Mulholland 1981). A substantial literature exists on leaf litter inputs and processing (Kaushik and Hynes 1971; Petersen and Cummins 1974; Webster and Benfield 1986), organic matter budgets (Fisher and Likens 1973), and responses to organic matter enrichment (Cummins et al. 1972).

Studies of inorganic nutrient dynamics in stream ecosystems lagged slightly behind this early emphasis on energy (Meyer and Likens 1979; Elwood et al. 1981). Although nutrient budget studies abound in the field, perhaps the most significant advance in nutrient cycling provided by stream ecology is the concept of nutrient spiraling (Newbold et al. 1981). This concept is a simple but elegant one that incorporates both in situ cycling and downstream transport as competing forces in a comprehensive ecosystem model of nutrient dynamics. These studies of energy and materials have generated a view of streams as open, spatially distributed ecosystems that receive and process large amounts of organic matter and appreciably slow the export of nutrients to the world's oceans.

While the structure and functioning of stream ecosystems are now probably as well known as in any ecosystem type, there are two areas in which progress in stream ecology has been somewhat retarded. The first of these, the topic of this third Cary Conference, is use of a comparative approach. The second is incorporation of concepts of disturbance and recovery in stream studies. Difficulties faced by stream ecologists wishing to employ a comparative method are similar to those faced by workers with any ecosystem type. Ecosystems are enormously complex, and it is difficult to study several simultaneously, especially with replication. But beyond this, streams drain virtually every terrestrial ecosystem type known and, given the strong influence of the land on streams, are exceedingly diverse. Many early paradigms, developed largely in small streams draining forested regions, have now been tested in streams of hot and cold deserts (Cushing and Wolf 1982; Fisher et al. 1982; Fisher 1986), chaparral (Hart and Howmiller 1975; Cooper et al. 1986), tropical forest (Stout 1980; Covich 1988), grasslands (Gurtz et al. 1988), and arctic tundra (Peterson et al. 1986). Comparisons among these diverse streams have been unplanned and are largely anecdotal.

The River Continuum Concept (RCC; Vannote et al. 1980), which so greatly influenced the course of stream research for more than a decade, has a built-in comparative component in that it hypothesizes patterns, for example, of energy utilization and organic matter storage, based on stream size (e.g., order) in a single drainage. The RCC was tested and supported, with some important qualifications, in several North American streams that were widely distributed, albeit in temperate forests. Vitousek and Matson (Chapter 14, this volume) proposed that studies of gradients provide several advantages over a comparison of discrete units. The river continuum gradient (in organic matter supply, temperature, depth, etc.) provides this advantage within a drainage; however, between drainages the gradient axes may change substantially. While gradients may exist in streams everywhere, they may involve such different variables that, in fact, "New Zealand streams really are different," as Winterbourn and co-workers queried in their challenge of the universality of RCC patterns developed in North America (Winterbourn et al. 1981).

RCC efforts point up another problem with comparative studies of running waters, namely the lack of replication available for large rivers. A minor contributor to this situation is the limited number of large rivers and the unique ways humans have altered them. Aside from this, however, large rivers transect elevational, latitudinal, and geological gradients that are simply unique on the planet. While one may argue that the inherent value of the Mississippi or the Nile or the Ob or the Amazon warrants their careful study as ecosystems, it would be more difficult to argue that their collective study would produce a very robust conceptual construct. Even if it did, to what other river ecosystems might it be applied? There is growing evidence that even small regional rivers transect unique geologic structures which influence water chemistry (Lay and Ward 1987). The size of river to which this would apply would be some function of the ratio of river length to diameter of the geologic or biological patches across which it flows.

The uniqueness of larger rivers may limit the extent to which the RCC can be extended downstream with generality, but there are millions of river miles in smaller orders in North America alone that should provide a wealth of opportunity for comparative analysis. Comparisons in stream ecology have been of two types: those involving limited attributes of very similar streams (specific comparisons) and anecdotal comparisons of broad features of very different streams (general comparisons). Examples of specific comparisons include (1) Tate and Gurtz's (1986) study of leaf decomposition in permanent and intermittent stream reaches; (2) Lay and Ward's (1987) study of the effect of substratum chemistry on algae in two streams; and (3) Silsbee and Larson's (1983) comparison of streams of logged and unlogged watersheds in the Smoky Mountains of North Carolina. Such studies often involve planned experiments conducted by a single investigator or group of investigators. At the other extreme are attempts of stream ecologists to place their broad findings in context by comparison with accounts of different streams published by other investigators at different times.

Whichever the case, specific or general, differences in streams have been attributed to a host of factors (Table 10.1) with varying success. Although some of these factors are related to disturbance, until recently disturbance per se has received surprisingly little attention as a factor responsible for variation in a broad array of streams (see Resh et al. 1988).

TABLE 10.1. Factors to which differences in streams have been attributed in selected studies.

Factor	Reference(s)
Flow regime	Gray 1981; Rae 1987; McElravy et al. 1989
Permanence	Hill and Gardner 1987; Delucchi 1988
Substratum type and mobility	Gurtz and Wallace 1984; Soluk 1985
Temperature	Vannote and Sweeney 1980; Ward 1985
Stream order (size)	Vannote et al. 1980; Minshall et al. 1983; Naiman et al. 1987
Slope, power	Vannote et al. 1980
Water chemistry	Minshall and Minshall 1978
Substratum chemistry	Lay and Ward 1987
Light	Murphy et al. 1981; Towns 1981
Terrestrial vegetation	Hawkins et al. 1983
Organic matter supply	McArthur and Marzolf 1986; Gurtz et al. 1988
Catchment history	Webster and Waide 1982; Golladay et al. 1987
Catchment land use	Dance and Hynes 1980
Biotic interactions	Hart 1983, 1985; McAuliffe 1984; Power et al. 1985

Disturbance in Streams

Disturbance has probably received less attention as an organizer of streams than of other ecosystems, but there is ample evidence that a wide range of disturbance types and intensities characterizes an equally wide range of streams (Table 10.2). These disturbances may be autogenic (generated by processes within the ecosystem) or allogenic (imposed from outside). Generally, autogenic disturbance, which is mostly biological, occurs at a smaller spatial and temporal scale than allogenic disturbance, which is usually abiotic. Temporal and spatial scales of both types of disturbance are positively correlated (Table 10.2; see also Urban et al. 1987). Furthermore, a given ecosystem may be subject to a host of superimposed disturbance events, both allogenic and autogenic, and these may interact. So, for example, flash floods may have different effects, depending on how long it has been since a major fire occurred in the watershed. The impact of a fire in the watershed may be a function of time since last volcanic activity. Clearly streams of Mt. St. Helens today respond differently to both floods and fires than they did before the most recent set of eruptions. However, studies of the interaction of disturbance events are virtually unknown in the ecological literature (Vogl 1980).

TABLE 10.2. Spatial and temporal scales of disturbances affecting streams and rivers.

Type of disturbance	Spatial scale[a] (m)	Temporal Scale (yr)	Autogenic or allogenic?
Rolling sand grain	0.001	0.001	Autogenic
Invertebrate grazing	0.01	0.001	Autogenic
Fish grazing scars	0.01	0.01	Autogenic
Fish nesting	0.1	1–10	Autogenic
Crayfish burrows	1–10	1	Autogenic
Fish spawning	1–100	1	Autogenic
Algal mat sloughing	0.1–1	0.1–1	Autogenic
Muskrat, platypus, desmon activity	1–100	.01–1	Autogenic
Terrestrial animals: cattle, mammoth, human	0.1–100	0.01	Allogenic
Beaver	10–100	1–10	Autogenic
Hippopotamus activity	100–1000	1	Autogenic
Stream drying	1–10,000	0.1–10	Allogenic
Toxic chemical spills	1,000–10,000	0.1–100	Allogenic
Flash floods	1,000–10,000	0.01–1	Allogenic
Fire, hurricane, deforestation	100–10,000	1–100	Allogenic
Volcanic activity	1,000–100,000	>100	Allogenic
Regional floods	10,000–10^6	1–100	Allogenic

[a]Spatial scale is the linear extent of a disturbance, ° temporal scale is the recurrence interval of the event.

In addition to disturbance, other variables influence stream structure and functioning. Some of these change through time periodically or gradually (e.g., temperature, light, photoperiod); others do not (latitude, bedrock geology). Given this host of influences on streams, can we sort out effects of disturbance from those of other independent variables? And can we do this using a comparative approach? That is, can we gain insight about factors responsible for variation among ecosystems, particularly the relative role of disturbance, by studying a suite of ecosystems? How different should ecosystems in the study be? If we are interested in causes of color in fruit, should we compare apples with oranges or with other varieties of apples?

This question is akin to the principle of the experimental control, although in this case, the approach is an inductive one. If we are interested in comparing ecosystems to reveal the role of disturbance, we would be best advised to hold all factors except disturbance as constant as possible, that is, to select a set of ecosystems that is very uniform, despite a sacrifice in generality of the resulting analysis. Resh et al. (1988), for example, recommended that streams selected for comparative studies of disturbance be similar in hydrologic regime and geomorphology. The alternative to selection of like systems is to compare diverse ones in the interest of generality and accept a high degree of uncertainty that observed differences are attributable to the variable of interest, that is, disturbance.

Examples of these two extreme approaches are investigated in the sections that follow. First, we briefly explore the consequences of comparing such different ecosystems as streams and forests; we then consider comparisons among similar streams. We will show that both approaches have merit, depending on the goal of the study.

Comparisons between Streams and Other Ecosystems

There are clear advantages to be gained in comparison of disparate ecosystems. Chief among these is the generation of broad concepts, the generality of which may not be apparent before the comparison. Steele (1985) reaped this benefit in an insightful comparison of periodic phenomena in oceanic and terrestrial ecosystems. At first blush, forests and streams appear to have little in common, yet their comparison may lead us to fruitful insights. The dominant organisms of forests are large and slow growing and dominate the physical structure of that ecosystem (Table 10.3). Streams, on the other hand, are structured abiotically, for example, by substrate and slope, but also ironically by the nature of adjacent riparian forests. The shapes of these two systems are quite different and have consequences for both the extent and spatial orientation of within- and between-system exchanges. Thus, transport in forests is largely vertical and involves water and nutrient movements via transpiration or percolation through soils, biological trans-

TABLE 10.3. Comparison of features of forest and stream ecosystems.

Attribute	Forests	Streams
Life span of plants (yr)	1–1000	0.01–3.0
Size of plants	Large	Small
Nutrient cycling	≈ Closed	Open
Edge-to-area ratio	Low	High
Transport linkages	Vertical	Longitudinal, unidirectional, bilateral
Ecosystem physical structure	Biotic	Abiotic

port of photosynthate between roots and aerial organs, and gravity-driven litter- , branch- , and bolefall.

The surface-to-volume ratio of streams is very high, exchanges across their boundaries are pronounced, and many linkages are predominantly unidirectional because of the overriding influence of gravity-driven water currents. While this sort of cross-ecosystem comparison may seem elementary, examination of the familiar in unfamiliar ways may provide useful conceptual or methodological perspectives. For example, a model of nutrient retention during succession developed for forests (Vitousek and Reiners 1975) and extended to streams by incorporating unidirectional flow and export (Grimm and Fisher 1986) was supported by an analysis of nitrogen budgets for a desert stream (Grimm 1987). More generally, one might ask, are spatially distributed patterns of soils (in a vertical plane) and rivers (horizontally) really quite similar? What do pedogenesis and the River Continuum Concept have in common? Both result in part from differential transport of water and its dissolved and suspended load. Might the methods and models of the soil scientist be adaptable to running waters, and vice versa?

If the goal of ecosystem science is to discover principles that govern all ecosystems, and it should be, we may be limited to a broad-brush picture. For example, if we were to seek common themes associated with disturbance in different ecosystems, we might begin with the sort of comparison illustrated in Table 10.4. Although some of the attributes listed may be controversial or even wrong, comparisons of ecosystem phenomena at this level can produce a host of interesting hypotheses about ecosystems that are eminently testable.

For example, hurricanes in forests and spates in streams both involve rapid movement of the ecosystem medium, but the effect of each is quite different, probably because of the competency of the medium to transport the particles it encounters. Thus, while spates export organic matter and nutrients, high winds have virtually no immediate effect on the pool sizes of either. Drying in streams is much more like wind action in forests in that major material pools are retained and nutrient mobilization occurs slowly following disturbance through enhanced decomposition. Fire rapidly oxidizes organic matter and instantly mobilizes nutrients. These patterns might

TABLE 10.4. Characteristics of contrasting allogenic disturbances affecting forests and streams.

| Characteristic | Forest | | | Stream | |
	Fire	Wind	Flood	Drying
Shape	Variable	Variable	Elongate	Beaded
Mode of action	Rapid oxidation	Biotic disruption and removal	Biotic disruption	Loss of medium
Onset	Rapid	Variable	Variable	Slow
Refuges	Above or below ground	Coves, wind shadow	Aerial and hyporheic	None
Dampening by biota	Very small	Large	Small	Small
Resilience	Regrowth and recolonization	Regrowth	Regrowth and recolonization	Recolonization
Effect on:				
Aboveground organic mater	Oxidation	No change	Exported	Enhanced decomposition?
Nutrients	Rapid mobilization	Slow mobilization	Exported	Slow mobilization
Boundaries	None	None	Expansion	Contraction
Plants	High mortality	Moderate mortality	High mortality and export	Very high mortality
Animals	Emigration or high mortality	Low mortality	Variable	Emigration or high mortality

then lead us to compare not grasslands and rivers, but fire and flood; not forests and streams, but cyclone and drought. Hypotheses might be more properly organized around the cause (the disturbance) than the ecosystem affected by that cause. Examples of such hypotheses are (1) a disturbance which mobilizes nutrients and removes organic matter (by export or oxidation) is followed by autotrophic succession (recovery); or (2) resilience is independent of mortality but sensitive to changes in available nutrients.

Testing these hypotheses, like their generation in the first place, will involve examination of an array of very different ecosystems. Results of the tests will be widely applicable, but not very revealing of the mechanisms underlying ecosystem response. Response mechanisms undoubtedly vary among ecosystems and are best elucidated by in-depth studies of individual systems or of a subset of ecosystems that are mechanistically more similar. On the other hand, such mechanistic details may acquire greater meaning in a context provided by comparative studies.

Comparisons among Streams

While streams differ from each other, they are certainly more similar as a group than streams are to forests. Because of this, comparisons among streams may yield a more specific result; that is, differences may be attributable to one or a few causes because so many other factors are controlled.

When examining disturbance in streams, we will adopt the following language: disturbance is a punctuated killing that alters the ratio of organisms to resources. It is not predictable with certainty, as is the turning of the seasons, for example. Stability describes ecosystem response to disturbance and involves resistance, the capacity to avoid change, and resilience, the ability to recover rapidly after disturbance. We will organize our discussion of disturbance in streams around resistance and resilience concepts.

RESISTANCE IN STREAMS

Streams are subjected to a variety of agents of disturbance; however, floods, a feature central to running waters, have probably been most thoroughly documented (Table 10.5). Resistance of one or more ecosystem components has been quantified in several studies. Often that component is biotic, usually a restricted part of a community. Seldom are functional characteristics, such as productivity or nutrient cycling, examined in this context, and in only a few cases have detrital or abiotic components been considered. Resistance can be described as the percent change caused by a disturbance. Resistance is highly variable (Table 10.5), and such variance is determined by both the nature of the disturbance (e.g., its magnitude or timing) and the component studied (life history stage or growth form).

Size is no sure protection against disturbance. Although Gray (1981)

TABLE 10.5. Resistance, as percent reduction in standing crop, of stream biota of different types to variety of disturbances.

Disturbance	Component	Reduction (%)	Determinants	Reference
Flood	Suspended bacteria	0		Goulder 1986
Volcano	Microorganisms	100		Ward et al. 1983
Flood	Algae	68–96	Growth form, taxon, life history stage	Power and Stewart 1987
Floods	Algae	0–100	Disturbance magnitude, taxon	Grimm and Fisher 1989
Volcano	Algae	100		Rushforth et al. 1986
Desiccation	Algae	–	Habitat harshness	Peterson 1987
Floods	Benthic invertebrates	0–99	Disturbance magnitude	Grimm and Fisher 1989
Flood	Benthic invertebrates	58–97	Taxon	Hoopes 1974
Annual flood	Benthic invertebrates	>95		Siegfried and Knight 1977
Siltation	Chironomidae	>90		Gray and Ward 1982
Flood	Benthic invertebrates	94		Molles 1985
Habitat loss	Snails	35–99	Taxon	Lodge and Kelly 1985
Flood	Exotic fishes	100	Evolutionary history	Collins et al. 1981
Annual flood	CPOM, 10st-order stream	≈2	Geomorphology	Bilby and Likens 1980
Annual flood	CPOM, 4th-order stream	10–80	Geomorphology, season	Fisher 1977
Debris torrent	CPOM	≈100	Geomorphology	Lienkaemper and Swanson 1987
Floods	Sand volume	+ to 30	Disturbance magnitude	Fisher et al. (unpublished)

Determinants given are factors believed to cause variation in resistance.
[a]Coarse particulate organi matter.

found small immature insects to be much less resistant to flash floods than larger ones, microorganisms, which are smaller yet, may be unaffected (Goulder 1986). In some cases, resistance may be negative; that is, some component increases across an event. This is unusual and often involves immigration or transport of, for example, sand or detritus. Given that certain disturbances in streams (e.g., flash floods) usually accentuate transport, negative resistance is not surprising. In truth, however, the difference between before and after states may be the result of several turnovers of the component of interest during the disturbance event. Again, this is more likely if the component is abiotic and not subject to mortality.

Resistance may also be a function of structural heterogeneity of the more inclusive stream ecosystem. While midchannel communities may be in some peril when spates occur, those of floodplain or main-channel pools, backwaters, or the hyporheic zone may experience much lower mortality. Many stream invertebrates are insects that have a flying adult stage. This aerial stage is immune to flooding, as long as flood waters recede rapidly (Gray and Fisher 1981). The picture that emerges is that streams which are structurally complex are not uniformly affected by disturbance and the event is "incorporated" (sensu Urban et al. 1987). Detection of incorporation requires that the ecosystem be viewed at a scale which includes a representative array of highly disturbed and less disturbed patches. Incorporation may not occur in streams with simplified structure (canyon rivers, irrigation canals, channelized streams).

Resistance not only varies among components and scales examined but may exhibit a variable threshold response. In Sycamore Creek, Arizona, algal resistance to flash flooding (as change in chlorophyll *a* standing crop) shows a distinct threshold response when the effect of many floods is examined (Fig. 10.1). Over a wide range of floods, resistance varies nearly fourfold, presumably because peak discharge does not tell the whole story. Rate of rise and fall of the hydrograph, suspended particulates in transport, and even time of day might reduce some of this substantial variance. We do know that substrate size does not confer resistance to the algal community of Sycamore Creek; however, different algal taxa respond differently to floods. For example, tightly attached diatoms are more resistant than bluegreen mats, which are only weakly associated with their substrata (Grimm and Fisher 1989).

Resistance of invertebrates varies among taxa and life history stage (Table 10.6). In Sycamore Creek, mayflies and chironomids show high mortality during floods, whereas adult beetles and hemipterans are scarcely affected. Larval beetles, on the other hand, exhibit very high mortality; thus, several beetle species time their life cycles to avoid the larval stage when probability of flooding is highest (Gray 1981). Those taxa with low resistance nonetheless persist in Sycamore Creek through high resilience. In this case, resilience takes the form of vagility, rapid development, and reproduction throughout the year. Several distinct life history tactics there-

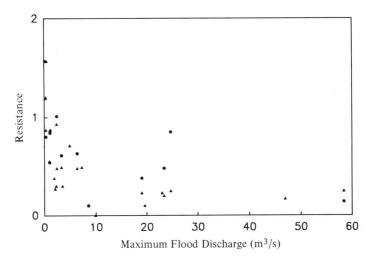

FIGURE 10.1. Algal resistance to flash flooding in Sycamore Creek, Arizona, as a function of disturbance magnitude (flash flood peak discharge). Resistance values range from 0 (complete elimination) to 1.57 (no change) and are calculated as arcsin-square root of $(\Delta C/100)$, where ΔC is percentage change in chlorophyll a standing crop across 27 disturbances (Grimm and Fisher 1989).

fore exist involving trade-offs between resistance and resilience, either of which, taken alone, may be highly variable among invertebrates.

In summary, resistance varies among streams and among components within streams. Variation is determined in part by disturbance type, intensity, and timing; the physical structure of the ecosystem and the scale under consideration; and the taxonomic groups and life history stages of the biota. These generalizations derive from analysis of results from a variety of studies of aquatic ecosystems, most of which did not themselves involve

TABLE 10.6. Life history tactics exhibited by aquatic fauna of Sycamore Creek, Arizona.[a]

1. Small size, rapid development, high flood mortality, aerial recolonization, very abundant [*examples*: Ephemeroptera, Chironomidae (Diptera)]
2. Medium size; slow development; seasonal reproduction (outside monsoon season); high flood mortality as larvae, low as adults; abundant (*example*: Coleoptera)
3. Medium size; slow development; continuous reproduction; high flood mortality; rare (*example*: Trichoptera)
4. Large size; very slow development; continuous reproduction; low flood mortality; abundant (*examples*: fish, Belostomatidae)
5. Large size; very slow development; seasonal reproduction (monsoon season); resting stage; mortality unknown, very rare (*example*: Tabanidae)

[a]Representatives of the fauna fit into all five categories listed, but only three of these (1, 2, 4) are highly successful in Sycamore Creek, which is subject to catastrophic flash floods during July–August "monsoon" season.

comparison (see Table 10.5). As a result, high variance might be expected as we seek general pattern from this diverse list. However, even when we examine resistance at similar sites or at different times in a single stream (Sycamore Creek), substantial variance persists. We can explain some of this variance, but enough remains to render comparison with other streams problematic, especially because few other streams are known well enough to estimate variance. Again the problem of control arises. How do we compare an array of variable streams with respect to a few key measures of disturbance when those streams vary in so many other ways? Whatever the solution, some compromise in the specificity of the analysis is inevitable.

RESILIENCE IN STREAMS

Resilience has been measured in several streams, again usually for a specific taxonomic group. Time of recovery is roughly related to organism size (Table 10.7). This holds across a wide range of stream types and several agents of disturbance. Undoubtedly the principle at work here is that it takes longer to produce a large organism than a small one, which in turn raises several questions about the currency in which resilience is measured (Table 10.8). Simple time to predisturbance state is often used, but this factor varies widely among components. If time is measured in generations of the population of interest, this variance will decrease substantially. What can be said, for example, about a chironomid species that requires 30 days but three generations to reach preflood levels in comparison to a stonefly that requires 2 years but just one generation? Obviously generation time may intimate something about fecundity, immigration, and perhaps survivorship that is more meaningful biologically than calendar time, but these are still population attributes. For the entire ecosystem, recovery might

TABLE 10.7. Resilience, as recovery time to predisturbance conditions, of stream biota of different types to a variety of disturbances.

Disturbance	Organisms	Recovery time (days)	Reference
Flood	Benthic bacteria	10–12	Cooper 1983
Volcano	Bacteria	≥ 14	Ward et al. 1983
Flood	Lotic algae	21	Power and Stewart 1987
Flood	Lotic algae	21–30	Fisher et al. 1982
Floods	Lotic algae	20–78	Grimm and Fisher 1989
Hailstorm	Thermal Cyanobacteria	152	Brock and Brock 1969
Flood	Benthic invertebrates	30–40	Fisher et al. 1982
Floods	Benthic invertebrates	20–40	Grimm and Fisher 1989
Siltation	Chironomidae	≤ 21	Gray and Ward 1982
Flood	Benthic invertebrates	≤ 120	Hoopes 1974
Flood	Benthic invertebrates	365	Molles 1985
Flood	Fishes	300	Matthews 1986
Drought	Fishes	365	Ross et al. 1985

TABLE 10.8. Some measures of resilience
for ecosystems.

Time to predisturbance state
    absolute time
    number of generations
    percent of life span of dominant species
    percent of return interval of disturbance
Initial rate of increase
    number/time
    biomass/time

---

better be reckoned as time standardized to the life span of the dominant species. In a stream this might be *Cladophora* or some species of macrophyte, and in a forest, an oak. Such a comparison may indeed reduce the difference between ecosystems, but accentuates differences between species within an ecosystem.

Resilience measured in interdisturbance intervals would be useful in comparing streams with different flood frequencies, although we are aware of no cases in which this has been done. An organism such as the chironomid, which can recover to preflood levels in half an average interflood interval (60 days in Sycamore Creek) would be more resilient than a belostomatid that requires two interflood periods. We might expect the latter to show greater resistance than the former to disturbance events that it is almost certain to encounter. This is true in Sycamore Creek. Finally, some measure of initial velocity (= slope) may be useful for two reasons. First, it emphasizes productivity when measured in biomass units, and second, it avoids the problems inherent in recognizing a "recovered" state, especially when trajectories are not classically asymptotic (Grimm and Fisher 1989).

A comparative analysis of the consequences of selecting among these currencies is beyond the scope of this paper and probably exceeds the data available for streams. In the analysis that follows, units are absolute time. Rather than gather data on recovery and seek a pattern, we could turn the problem around and ask to what extent disturbance and subsequent recovery are important in shaping what we see in this ecosystem or that. That is, are ecosystem structure and functioning influenced by disturbance in a spectrum of streams? Baker (1977) provided a basis for such a comparison with a map showing tendency of U.S. streams to flash flood. His scheme showed New England and midwestern streams to be very stable and those of central Texas and Arizona to be flashy. Fisher and Grimm (1988) compared temporal patterns of chlorophyll *a* in Sycamore Creek, a flashy stream in Arizona, with Fort River, a far more stable stream in Massachusetts (Table 10.9). By multiple regression analysis, nearly 68% of variation in chlorophyll *a* of Sycamore Creek, but none of Fort River, was explained by flood-related variables (Table 10.10). Days after flood and magnitude of the last flood are the best predictors of current state of Sycamore Creek,

TABLE 10.9. Comparison of flood disturbance features of Sycamore Creek, Arizona, and Fort River, Massachusetts.[a]

Variable	Sycamore Creek	Fort River
Catchment area (km^2)	424	105
Annual maximum discharge (m^3/sec)	40.9	22.4
Annual mean discharge (m^3/sec)	0.62	1.6
Ratio maximum: mean discharge	67	14
Floods per year	5.5	30
Mean interflood period (days)	67	12

[a]Floods are defined as events producing a 25% increase in discharge and exceeding 0.25 m^3/sec. Sycamore Creek is a flash flood-prone stream in the Sonoran Desert. Fort River drains a temperate, deciduous forest in New England, USA.

indicating that disturbance is indeed an important variable in that stream. The only significant variable in the Fort River analysis was daily insolation, which explained only a small percentage of the variation in chlorophyll *a*. Biggs (1988) did a similar analysis in an attempt to attribute variance in 66 different New Zealand foothills streams sampled once (Table 10.11). In that study, a similar percentage of variance was explained by the series of independent variables used, and again variables related to disturbance (discharge at sampling and time since flooding) figured prominently. The efficacies of these models to explain variance are similar, despite the fact that one data set is derived from a single stream sampled many times and the other from many streams sampled once.

The same multiple regression analysis performed on Sycamore Creek invertebrates was much less successful. Only 26% of variance in invertebrate numbers could be explained and only discharge and time since disturbance were significant (Fisher and Grimm 1988). Invertebrate data were

TABLE 10.10. Stepwise multiple regression analysis of algal standing crop (as chlorophyll *a*) in Sycamore Creek, Arizona and Fort River, Massachusetts.[a]

Predictor variable	Sycamore Creek Partial correlation	r^2	Fort River Partial correlation	r^2
Days since flood	0.68	0.31	—	—
Discharge	−0.50	0.64	—	—
Nitrate N	0.17	0.66	—	—
Discharge last flood	−0.33	0.68	—	—
Water temperature	0.13	0.70	—	—
Day length	0.12	0.71	—	—
Light	—	—	0.29	0.08

[a]Data are log transformed. Sycamore Creek, $n = 352$; Fort River, $n = 69$.

TABLE 10.11. Multiple regression analysis for periphyton biomass (ash-free dry mass) in 66 New Zealand rivers.[a]

Predictor variable	Percent of variance explained
River discharge	49.9
Conductivity	15.6
Days since flood	12.2
Water temperature	1.1
Total r^2 (%)	78.8

[a]Data from Biggs (1988).

not available for Fort River for comparison, but Skinner and Arnold (1988) found no relationship between flooding and invertebrates in a Pennsylvania stream, an area with a low propensity toward scouring floods.

Residual variance in invertebrates in Sycamore Creek was tentatively attributed to food quality by Grimm and Fisher (1989). The aquatic insects of that stream are largely collector-gatherers that feed heavily on fine particulate detritus, decreasing nitrogen in that material by 50% with each gut passage. Insects recovered rapidly after flash flooding in a monotonic fashion except when nitrate availability was low, in which case dramatic crashes in population densities occurred. Grimm and Fisher (1989) hypothesized that crashes were caused by high C : N in detritus caused by N limitation of bacterial conditioning. Late in successional sequences, invertebrate densities were correlated with nitrogen availability (Fig. 10.2). Whatever the mechanism, the point is that resilience of invertebrates was indirectly related to availability of an inorganic nutrient. Decline in invertebrate numbers under conditions of low nitrogen may be proximally explained by

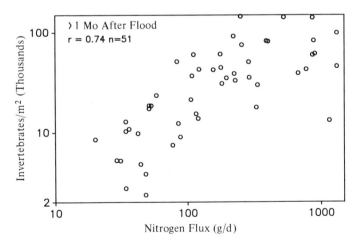

FIGURE 10.2. Correlation of invertebrate density (samples collected >1 month after spate) with inorganic nitrogen flux [(nitrate- + ammonium N) × discharge]; $r = 0.74$; $n = 51$. Reprinted from Grimm and Fisher (1989).●

competition for a limited resource—a biotic interaction. That point of intense biotic interaction is not reached when nitrogen supply is ample and population density is better predicted by time since disturbance.

Inorganic nitrogen availability is not an important predictor of algal standing crop (as chlorophyll *a*) in Sycamore Creek, despite the fact that primary productivity is limited by nitrogen (Grimm and Fisher 1986). This paradox is explained by shifting algal community structure. The typical postflood recovery pattern illustrated in Fig. 10.3 involves an ascension of nitrogen-fixing Cyanobacteria late in succession when and if nitrogen concentration declines (Grimm and Fisher 1989). Cyanobacteria do not appear in abundance when nitrogen concentrations are high. So while nitrogen may be a good predictor of chlorophyte abundance in Sycamore Creek, it is not associated with chlorophyll, which can come in a variety of packages.

In this one stream, resilience of several important components after disturbance is clearly influenced by nitrogen availability, although in diverse ways. When nitrogen is lowest, biotic interactions assume greatest importance; when nitrogen is ample, disturbance is more significant. When is nitrogen limiting? It is limiting late in succession when stream discharge is low and available nitrogen has been incorporated in biomass and is rapidly recycled by both algae and microorganisms. If floods are frequent, succession is truncated and the point at which nitrogen becomes limiting is not reached. (This pattern is discussed later.) Nitrogen in stream water is highly variable within a year and between years (Grimm and Fisher 1986), and we might reasonably expect the resilience component of stability to be so as well.

This detailed consideration of resilience of a single stream ecosystem is not a digression, but an attempt to make a point about ecosystem compari-

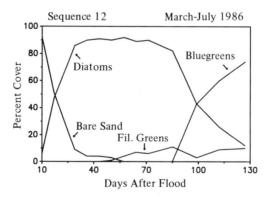

FIGURE 10.3. Changes in percentage cover of three algal types after flash flood in Sycamore Creek, Arizona, March–July, 1986, illustrating common successional pattern in algal patch structure. Late-stage increase in Cyanobacteria occurs when nitrogen availability is low.

sons. Within any given ecosystem, a whole-system property such as stability is complex, and its resolution requires a substantial research effort in each system that is to be compared. Even then, it is a considerable challenge to identify one or a few summary statistics that capture the essence of the property (stability) but are discrete enough to permit a meaningful comparative analysis.

In sum, while resilience has been measured in a variety of streams, comparisons are difficult because there is no agreement on the currency to be used to describe it. Further, ecosystem components vary in resilience. Even within a taxonomic group, some representatives may be resilient, while others show no resilience but are highly resistant. Natural selection, working on life history attributes, may generate multiple solutions to a given problem. Finally, variation in response of a single stream may be as high as variation within a set of similar streams. To detect significant differences in the analysis of variance may require a large number of samples.

Disturbance and Control of Stream Ecosystem Functioning

The multiple regression analyses just described are used to explain variation in, for example, benthic algal standing crop (Biggs 1988; Fisher and Grimm 1988). This approach is correlative, only suggestive of cause-and-effect relationships, and can be only as complete as the list of independent variables included. Although we can explain 70% to 80% of variance, a substantial fraction remains unexplained. The advantage of the multiple regression approach is that it asks the right question: what variables are important in shaping ecosystem-level attributes? Answers to this question for an array of streams would provide a good start in developing a rigorous comparative analysis of running water ecosystems.

Ecologists have debated the relative importance of biotic and abiotic factors in the control of community structure and ecosystem functioning for years (e.g., Andrewartha and Birch 1954; Cody and Diamond 1975; Connell 1978). In stream ecology too, the issue has centered on disturbance versus biotic interactions such as competition and predation (Power et al. 1988; Resh et al. 1988). The argument is that disturbance, if severe and frequent enough, prevents populations from becoming so large that resources become limiting and, for example, competitive exclusion occurs (McAuliffe 1984; Minshall and Petersen 1985). For our purposes here, it would be useful to know in what sorts of streams disturbance are paramount and where biotic interactions control the system. A categorization of this type would be useful in understanding streams and in guiding research, but it is probably a false dichotomy. Both controls may operate simultaneously or sequentially in a given stream (e.g., Hemphill and Cooper 1983; Power et al. 1985), and the challenge is to determine their relative importance and the conditions under which each predominates.

We propose a simple model for stream ecosystems subject to both flooding and drying disturbances (Fig. 10.4). In desert and arid land streams where flooding is severe, ecosystem-state variables such as productivity are changed by floods and recover thereafter, usually rapidly. If the period between floods is sufficiently long, biotic interactions may become important as the organism-to-resource ratio rises. In Sycamore Creek, one form of this is nutrient limitation, which may affect a host of attributes such as algal community composition, insect secondary production, and so forth.

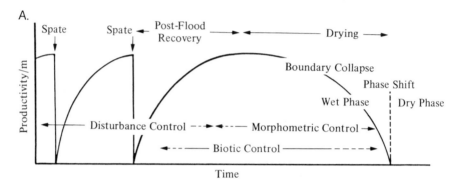

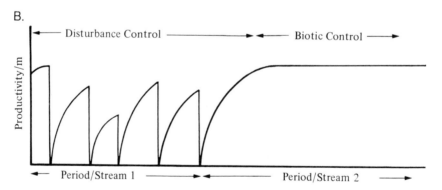

FIGURE 10.4. Conceptual model of temporal shift in control of ecosystem functioning in a stream influenced by flash flooding (spates). Dependent variable is productivity per linear meter, thus decline in drying period is caused by system shrinkage. (A) Biotic interactions intensify with time after initial flood disturbance, but in late stages, these interactions may be disrupted by morphometric constraints. Controlling agent (or agents in times of overlap) is defined as best predictor of system state. (B) In streams where floods are frequent (stream 1), disturbance-related variables are influential in shaping ecosystem structure and functioning. Where flow is more stable (stream 2), biotic interactions control ecosystem. This difference may also apply to a single stream that experiences periodic shifts in disturbance probability, e.g., a monsoonal weather pattern.

Without additional flooding, the stream begins to shrink (Fig. 10.4A). Boundaries collapse; some biotic interactions intensify, while others are disrupted. This morphometric control, which is analogous to late stages of lake eutrophication (Likens 1972), eventually leads to a phase shift as surface, then hyporheic, water is lost. While many organisms do not survive this transition, others may, and organic matter processing and nutrient cycling persist through the dry period until water is restored by another flood.

This scenario involves, in sequence, control by disturbance, biotic interactions, morphometric constraint, and a phase shift. At any one time any of these factors may be dominant, but overall control of the ecosystem depends on the regime of disturbance by either flood or drought. If floods are closely spaced, time since disturbance will explain much variance in system state (Fig. 10.4B). If floods occur rarely, biotic interactions will predominate and some resource factor may be the best predictor. Morphometric stresses will ensue if floods are rarer still. Morphometric and drying constraints on stream structure and functioning are accentuated in arid regions but are by no means restricted to them. Even in well-wetted areas, dry streambeds persist during much of the year in first-order channels above the point at which flow becomes permanent.

Comparisons among streams have been useful in generating the concepts illustrated in this model, but a second level of comparative analysis, where the fit of each stream to this model is known, might be very productive. This would require rather intensive study of several streams selected because they collectively cover the range of conditions incorporated in the model. Only with a fairly thorough study of each stream would the exercise be useful. Disturbance is complex and cannot be reduced to one or a few variables, but we are convinced that a reasonable effort should be rewarded with a natural geography of stream ecosystem structure and functioning.

Summary: Cross-Ecosystem Comparison as an Optimization Problem

We began this chapter with a question: Are cross-ecosystem comparisons useful? The answer depends on what is being compared and what one wants the comparison to generate. We saw, for example, that comparisons of disparate ecosystems (e.g., forests and streams) were useful in generating new insights and identifying gross patterns but did not produce a very detailed or mechanistic understanding of the processes under examination. Mechanistic understanding increased slightly as we compared ecosystems that were more similar; however, even simple indices of disturbance showed great variation among streams and within a single stream. High variance will make it difficult to identify significant differences among streams or will require large sample sizes.

Part of the problem is that disturbance is just one of many causes of system state. It is difficult to control for the others, especially because some of them are yet unknown. Hilborn and Stearns (1982) have pointed out that hypotheses can be made simple in their assumptions or in their statements about causes, but not both. Because the causes of system state are complex, assumptions have to be simple. The more similar ecosystems are, the fewer assumptions must be made in connecting causes to a hypothetical effect. We proposed a framework for examining the multiple causes impinging upon stream ecosystems in space and time (Fig. 10.4), but all causes are yet unknown and streams are highly diverse. Hypotheses about disturbance in streams might best be tested in similar streams first before encompassing the breadth of conditions exhibited by running waters globally.

Comparative ecology can be viewed as an optimization problem (Fig. 10.5). The process of gaining information is largely inductive where the generality of hypotheses resulting from the study of one or a few similar

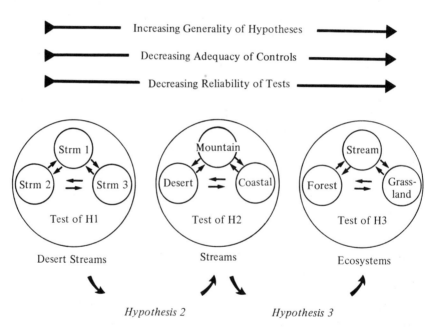

FIGURE 10.5. Inductive reasoning may be applied to comparative ecosystem science at different levels. The generality of conclusions from studies at one level may be tested at the next highest (more diverse) level. As hypotheses are tested by examining their fit to more and more diverse groups, the reliability of the test diminishes because controls become inadequate. Hypotheses that are adjusted to fit a diverse set of systems sacrifice mechanistic detail. The challenge is to apply the comparative method at the most inclusive level that will provide the mechanistic detail required by the study. This will vary according to goals of the research, the extent to which the ecosystem types compared are known, and the taste of the individual investigator.

ecosystems (e.g., desert streams) is tested in the larger arena of all streams. Broader hypotheses at that stage are tested by examining a set of diverse ecosystem types. At each stage, generality increases or the reliability of the test decreases; this results from the decreasing adequacy of controls, or the number of assumptions necessary in the comparison. Because hypotheses tested inductively are so easily falsified, satisfyingly detailed mechanistic explanatory hypotheses about ecosystems are seldom general. Hypotheses that apply to a wide array of ecosystems are often overly simplistic and contribute little to our understanding of a given ecosystem type.

In conclusion, although we would encourage comparative work at all levels, we believe that regional studies would be most profitable at this stage of the science in advancing our understanding of disturbance in stream ecosystems.

REFERENCES

Andrewartha, H.G. and L.C. Birch. (1954). *The Distribution and Abundance of Animals*. University of Chicago Press, Chicago.

Baker, V.R. (1977). Stream-channel response to floods, with examples from central Texas. *Geol. Soc. Am. Bull.* 88:1057–1071.

Biggs, B.J.F. (1988). Algal proliferations in New Zealand's shallow stony foothills-fed rivers: toward a predictive model. *Int. Ver. Theor. Angew. Limnol. Verh.* 23:1405–1411.

Bilby, R.E. and G.E. Likens. 1980. Importance of organic debris dams in the structure and function of stream ecosystems. *Ecology* 61:1107–1113.

Brock, T.D. and M.L. Brock. (1969). Recovery of a hot spring community from a catastrophe. *J. Phycol.* 5:75–77.

Cody, M.L. and J.M. Diamond, eds. (1975). *Ecology and Evolution of Communities*. Harvard University Press, Cambridge.

Collins, J.P., C. Young, J. Howell, and W.L. Minckley. (1981). Impact of flooding in a Sonoran Desert stream, including elimination of an endangered fish population (*Poeciliopsis o. occidentalis*, Poeciliidae). *Southwest. Nat.* 26:415–423.

Connell, J.H. (1978). Diversity in tropical rain forests and coral reefs. *Science* 199:1302–1310.

Cooper, A.B. (1983). Effect of storm events on benthic nitrifying activity. *Appl. Environ. Microbiol.* 46:957–960.

Cooper, S.D., T.L. Dudley, and N. Hemphill. (1986). The biology of chaparral streams in southern California. In: *Proceedings of the Chaparral Ecosystem Research Conference*, J. DeVries, ed. Report No. 62, Water Resources Center, Davis, California, pp. 139–152.

Covich, A.P. (1988). Geographical and historical comparisons of neotropical streams: biotic diversity and detrital processing in highly variable habitats. *J. North Am. Benthol. Soc.* 7:361–386.

Cummins, K.W. (1974). Structure and function of stream ecosystems. *BioScience* 24:631–641.

Cummins, K.W., J.J. Klug, R.G. Wetzel, R.C. Petersen, K.F. Suberkropp, B.A. Manny, J.C. Wuycheck, and F.O. Howard. (1972). Organic enrichment with leaf leachate in experimental lotic ecosystems. *BioScience* 22:719–722.

Cushing, C.E. and E.G. Wolf. (1982). Organic energy budget of Rattlesnake Springs, Washington. *Am. Midl. Nat.* 107:404–407.

Dance, K.W. and H.B.N. Hynes. (1980). Some effects of agricultural land use on stream insect communities. *Environ. Pollut.* 22:19–28.

Delucchi, C.M. (1988). Comparison of community structure among streams with different temporal flow regimes. *Can. J. Zool.* 66:579–586.

Elwood, J.W., J.D. Newbold, A.F. Trimble, and R.W. Stark. (1981). The limiting role of phosphorus in a woodland stream ecosystem: effects of P enrichment on leaf decomposition and primary producers. *Ecology* 62:146–158.

Fisher, S.G. (1977). Organic matter processing by a stream-segment ecosystem: Fort River, Massachusetts, USA. *Int. Rev. Gesamten Hydrobiol.* 62:701–727.

Fisher, S.G. (1986). Structure and dynamics of desert streams. In: *Pattern and Process in Desert Ecosystems*, W.G. Whitford, ed. University of New Mexico Press, Albuquerque, pp. 119–139.

Fisher, S.G. and N.B. Grimm. (1988). Disturbance as a determinant of structure in a Sonoran Desert stream ecosystem. *Int. Ver. Theor. Angew. Limnol. Verh.* 23: 1183–1189.

Fisher, S.G. and G.E. Likens. (1973). Energy flow in Bear Brook, New Hampshire: an integrative approach to stream ecosystem metabolism. *Ecol. Monogr.* 43:421–439.

Fisher, S.G., L.J. Gray, N.B. Grimm, and D.E. Busch. (1982). Temporal succession in a desert stream ecosystem following flash flooding. *Ecol. Monogr.* 52:93–110.

Golladay, S.W., J.R. Webster, and E.F. Benfield. (1987). Changes in stream morphology and storm transport of seston following watershed disturbance. *J. North Am. Benthol. Soc.* 6:1–11.

Goulder, R. (1986). Seasonal variation in the abundance and heterotrophic activity of suspended bacteria in two lowland rivers. *Freshwater Biol.* 16:21–37.

Gray, L.J. (1981). Species composition and life histories of aquatic insects in a lowland Sonoran Desert stream. *Am. Midl. Nat.* 106:229–242.

Gray, L.J. and S.G. Fisher. (1981). Postflood recolonization pathways of macroinvertebrates in lowland Sonoran Desert stream. *Am. Midl. Nat.* 106:249–257.

Gray, L.J. and J.V. Ward. (1982). Effects of sediment releases from a reservoir on stream macroinvertebrates. *Hydrobiologia* 96:177–184.

Grimm, N.B. (1987). Nitrogen dynamics during succession in a desert stream. *Ecology* 68:1157–1170.

Grimm, N.B. and S.G. Fisher. (1986). Nitrogen limitation in a Sonoran Desert stream. *J. North Am. Benthol. Soc.* 5:2–15.

Grimm, N.B. and S.G. Fisher. (1989). Stability of periphyton and macroinvertebrates to disturbance by flash floods in a desert stream. *J. North Am. Benthol. Soc.* 8:293–307.

Gurtz, M.E., G.R. Marzolf, K.T. Killingbeck, D.L. Smith, and J.V. McArthur. (1988). Hydrologic and riparian influences on the import and storage of coarse particulate organic matter in a prairie stream. *Can. J. Fish. Aquat. Sci.* 45:655–665.

Gurtz, M.E. and J.B. Wallace. (1984). Substrate-mediated response of stream invertebrates to disturbance. *Ecology* 65:1556–1569.

Hart, D.D. (1983). The importance of competitive interactions within stream populations and communities. In: *Stream Ecology: Application and Testing of General*

Ecological Theory, J.R. Barnes and G.W. Minshall, eds. Plenum Press, New York, pp. 99–136.

Hart, D.D. (1985). Grazing insects mediate algal interactions in a stream benthic community. *Oikos* 44:40–46.

Hart, S.D. and R.P. Howmiller. (1975). Studies on the decomosition of allochthonous detritus in two southern California streams. *Int. Ver. Theor. Angew. Limnol. Verh.* 19:1665–1974.

Hawkins, C.P., M.L. Murphy, N.H. Anderson, and M.A. Wilzbach. (1983). Density of fish and salamanders in relation to riparian canopy and physical habitat in streams of the northwestern United States. *Can. J. Fish. Aquat. Sci.* 40:1173–1185.

Hemphill, N. and S.D. Cooper. (1983). The effect of physical disturbance on the relative abundance of two filter-feeding insects in a small stream. *Oecologia* 58:378–382.

Hilborn, R. and S.C. Stearns. (1982). On inference in ecology and evolutionary biology: the problem of multiple causes. *Acta Biotheor.* 31:145–164.

Hill, B.H. and T.J. Gardner. (1987). Benthic metabolism in a perennial and intermittent Texas prairie stream. *Southwest. Nat.* 32:305–311.

Hoopes, R.L. (1974). Flooding, as the result of Hurricane Agnes, and its effect on a macrobenthic community in an infertile headwater stream in central Pennsylvania. *Limnol. Oceanogr.* 19:853–857.

Kaushik, N.K. and H.B.N. Hynes. (1971). The fate of the dead leaves that fall into streams. *Arch. Hydrobiol.* 68:465–515.

Lay, J.A. and A.K. Ward. (1987). Algal community dynamics in two streams associated with different geological regions in the southeastern United States. *Arch. Hydrobiol.* 108:305–324.

Lienkaemper, G.W. and F.J. Swanson. (1987). Dynamics of large woody debris in streams in old-growth Douglas-fir forests. *Can. J. For. Res.* 17:150–156.

Likens, G.E., ed. (1972). *Nutrients and Eutrophication.* American Society of Limnology and Oceanography Special Symposium Vol. 1. Allen Press, Lawrence, Kansas.

Lodge, D.M. and P. Kelly. (1985). Habitat disturbance and the stability of freshwater gastropod populations. *Oecologia* 68:111–117.

Matthews, W.J. (1986). Fish faunal structure in an Ozark stream: stability, persistence and a catastrophic flood. *Copeia* 1986:388–397.

McArthur, J.V. and G.R. Marzolf. (1986). Interactions of the bacterial assemblages of a prairie stream with dissolved organic carbon from riparian vegetation. *Hydrobiologia* 134:193–199.

McAuliffe, J.R. (1984). Competition for space, disturbance, and the structure of a benthic stream community. *Ecology* 65:894–908.

McElravy, E.P., G.A. Lamberti, and V.H. Resh. (1989). Year-to-year variation in the aquatic macroinvertebrate fauna of a northern California stream. *J. North Am. Benthol. Soc.* 8:51–63.

Meyer, J.L. and G.E. Likens. (1979). Transport and transformation of phosphorus in a forest stream ecosystem. *Ecology* 60:1255–1269.

Minshall, G.W. and J.N. Minshall. (1978). Further evidence on the role of chemical factors in determining the distribution of benthic invertebrates in the River Duddon. *Arch. Hydrobiol.* 83:324–355.

Minshall, G.W. and R.C. Petersen. (1985). Towards a theory of macroinvertebrate community structure in stream ecosystems. *Arch. Hydrobiol.* 104:49–76.

Minshall, G.W., R.C. Petersen, K.W. Cummins, T.L. Bott, J.R. Sedell, C.E. Cushing, and R.L. Vannote. (1983). Interbiome comparison of stream ecosystem dynamics. *Ecol. Monogr.* 53:1–25.

Molles, M.C. (1985). Recovery of a stream invertebrate community from a flash flood in Tesuque Creek, New Mexico. *Southwest. Nat.* 30:279–287.

Mulholland, P.J. (1981). Organic carbon flow in a swamp-stream ecosystem. *Ecol. Monogr.* 51:307–322.

Murphy, M.L., C.P. Hawkins, and N.H. Anderson. (1981). Effect of canopy modification and accumulated sediment on stream communities. *Trans. Am. Fish. Soc.* 110:469–478.

Naiman, R.J., J.M. Melillo, M.A. Lock, T.E. Ford, and S.R. Reice. (1987). Longitudinal patterns of ecosystem processes and community structure in a subarctic river continuum. *Ecology* 68:1139–1156.

Newbold, J.D., J.W. Elwood, R.V. O'Neill, and W. Van Winkle. (1981). Measuring nutrient spiraling in streams. *Can. J. Fish. Aquat. Sci.* 38:860–863.

Petersen, R.C. and K.W. Cummins. (1974). Leaf processing in a woodland stream. *Freshwater Biol.* 4:343–368.

Peterson, B.J., J.E. Hobbie, and T.L. Corliss. (1986). Carbon flow in a tundra stream ecosystem. *Can. J. Fish. Aquat. Sci.* 43:1259–1270.

Peterson, C.G. (1987). Influences of flow regime on development and desiccation response of lotic diatom communities. *Ecology* 68:946–954.

Pickett, S.T.A. and P.S. White. (1985). *The Ecology of Natural Disturbance and Patch Dynamics.* Academic Press, New York.

Power, M.E. and A.J. Stewart. (1987). Disturbance and recovery of an algal assemblage following flooding in an Oklahoma stream. *Am. Midl. Nat.* 117:333–345.

Power, M.E., W.J. Matthews, and A.J. Stewart. (1985). Grazing minnows, piscivorous bass and stream algae: dynamics of a strong interaction. *Ecology* 66:1448–1456.

Power, M.E., R.J. Stout, C.E. Cushing, P.P. Harper, F.R. Hauer, W.J. Matthews, P.B. Moyle, B. Statzner, and I.R. Wais De Badgen. (1988). Biotic and abiotic controls in river and stream communities. *J. North Am. Benthol. Soc.* 7:456–479.

Rae, J.G. (1987). The effects of flooding and sediment on the structure of a stream midge assemblage. *Hydrobiologia* 144:3–10.

Resh, V.H., A.V. Brown, A.P. Covich, M.E. Gurtz, H.W. Li, G.W. Minshall, S.R. Reice, A.L. Sheldon, J.B. Wallace, and R. Wissmar. (1988). The role of disturbance in stream ecology. *J. North Am. Benthol. Soc.* 7:433–455.

Risser, P.G. (1987). Landscape ecology: state of the art. In: *Landscape Heterogeneity and Disturbance*, M.G. Turner, ed. Ecological Studies, Vol. 64, Springer-Verlag, New York, pp. 3–14.

Ross, S.T., W.J. Matthews, and A.A. Echelle. (1985). Persistence of stream fish assemblages: effects of environmental change. *Am. Nat.* 126:24–40.

Rushforth, S.R., L.E. Squires, and C.E. Cushing. (1986). Algal communities of springs and streams in the Mt. St. Helens region, Washington, U.S.A. following the May 1980 eruption. *J. Phycol.* 22:129–137.

Shugart, H.H. and R.V. O'Niell, eds. (1979). *Systems Ecology.* Dowden, Hutchinson, & Ross, Stroudsburg, Pennsylvania.

Siegfried, C.A. and A.W. Knight. (1977). The effects of washout in a Sierra foothill stream. *Am. Midl. Nat.* 98:200–207.

Silsbee, D.G. and G.L. Larson. (1983). A comparison of streams in logged and unlogged areas of Great Smoky Mountain National Park. *Hydrobiologia* 102: 99–111.

Skinner, W.D. and D.E. Arnold. (1988). Absence of temporal succession of invertebrates in Pennsylvania streams. *Bull. North Am. Benthol. Soc.* 5:63.

Soluk, D.A. (1985). Macroinvertebrate abundance and production of psammophilous Chironimidae in shifting sand areas of lowland rivers. *Can. J. Fish. Aquat. Sci.* 42:1296–1302.

Sousa, W.P. (1984). The role of disturbance in natural communities. *Annu. Rev. Ecol. Syst.* 15:353–391.

Steele, J.H. (1985). A comparison of terrestrial and marine ecological systems. *Nature* (London) 313:355–358.

Stout, J. (1980). Leaf decomposition rates in Costa Rican lowland tropical rainforest streams. *Biotropica* 12:264–272.

Tate, C.M. and M.E. Gurtz. (1986). Comparison of mass loss, nutrients, and invertebrates associated with elm leaf litter decomposition in perennial and intermittent reaches of tallgrass prairie streams. *Southwest. Nat.* 31:511–520.

Towns, D.R. (1981). Effects of artificial shading on periphyton and invertebrates in a New Zealand stream. *N. Z. J. Mar. Freshwater Res.* 15:185–192.

Urban, D.L., R.V. O'Neill, and H.H. Shugart, Jr. (1987). Landscape ecology. *BioScience* 37:119–127.

Vannote, R.L., G.W. Minshall, K.W. Cummins, J.R. Sedell, and C.E. Cushing. (1980). The river continuum concept. *Can. J. Fish. Aquat. Sci.* 37:130–137.

Vannote, R.L. and B. Sweeney. (1980). Geographic analysis of thermal equilibria: a conceptual model for evaluating the effect of natural and modified thermal regimes of aquatic insects. *Am. Nat.* 115:667–695.

Vitousek, P.M. and W.A. Reiners. (1975). Ecosystem succession and nutrient retention: a hypothesis. *BioScience* 25:376–381.

Vogl, R.J. 1980. The ecological factors that produce perturbation-dependent ecosystems. In: *The Recovery Process in Damaged Ecosystems*, J. Cairns, ed. Ann Arbor Science, Ann Arbor, Michigan, pp. 63–94.

Ward, A.K., J.A. Baross, C.N. Dahm, M.D. Lilley, and J.R. Sedell. (1983). Qualitative and quantitative observations on aquatic algal communities and recolonization within the blast zone of Mt. St. Helens, 1980 and 1981. *J. Phycol.* 19:238–247.

Ward, J.V. (1985). Thermal characteristics of running waters. *Hydrobiologia* 125: 31–46.

Webster, J.R. and E.F. Benfield. (1986). Leaf breakdown in aquatic ecosystems. *Annu. Rev. Ecol. Syst.* 17:567–594.

Webster, J.R. and J.B. Waide. (1982). Effects of forest clearcutting on leaf breakdown in a southern Appalachian stream. *Freshwater Biol.* 12:331–344.

Webster, J.R., J.B. Waide, and B.C. Patten. (1975). Nutrient recycling and stability of ecosystems. In: *Mineral Cycling in Southeastern Ecosystems*, F.G. Howell, J.B. Gentry, and M.H. Smith, eds. U.S. Energy Research and Development Administration, Washington, D.C., pp. 1–27.

Winterbourn, M.J., J.S. Rounick, and B. Cowie. (1981). Are New Zealand stream ecosystems really different? *N. Z. J. Mar. Freshwater Res.* 15:321–328.

11
Searching for Specific Measures of Physiological Stress in Forest Ecosystems

RICHARD H. WARING

Abstract

Subtle changes in climate, atmospheric chemistry, or management policies may eventually lead to shifts in ecosystem structure. Stresses may occur before shifts in structure are evident. Insights regarding the development of stress can be obtained by monitoring decreases in photosynthetic or growth efficiency. Where decreases in efficiency are noted, additional selective measures are suggested to confirm physiological limitations and to help distinguish stress induced by drought from pollution or management policies. Changes in carbon partitioning, nutrient balance, biochemical indices, and stable isotope composition help identify probable sources of stress. Confirmation requires experimentation and regional assessment across confirmed environmental gradients.

Introduction

Ecologists are challenged to assess how changes in environment and management practices affect a host of ecosystems. Unfortunately, undisturbed communities of plants and animals that might serve as benchmarks for ecological studies no longer exist in many parts of the world. Even where undisturbed ecosystems do exist, subtle changes in climate and atmospheric chemistry may alter the rates of key processes. New combinations of environmental stresses also make it likely that major disturbances will affect the composition as well as the function of ecosystems.

In addressing global-scale change, we ecologists seek more general indices of ecosystem stress. Approaches that scan regions for signs of disturbance may pinpoint where additional analyses are warranted. The kind of scanning that might be appropriate, and additional analyses that could provide insights into the origin of stresses, are the subject of this chapter. I choose forest ecosystems as examples because they are dominated by long-lived organisms that are buffered against minor changes in their environment.

Trees and other woody plants maintain their structural integrity, often for centuries. They therefore provide, through analysis of their growth rings or by carbon dating, a historical record of change (Cook et al. 1987; Graumlich et al. 1989; Worbes and Junk 1989).

In this chapter, some indices are introduced that can be used in cross-system analysis of plant responses to stress. Three examples of stresses will be considered: (1) chemical, as induced through the deposition of sulfur; (2) climatic, as might be induced by the accumulation of greenhouse gases in the atmosphere; and (3) structural, as induced by changes caused by management or natural succession. From these examples, a set of diagnostic characteristics will be developed for recognizing chronic stress and pointing to the most probable cause.

Definition of Stress

There are many definitions of stress. Classically, plant communities have been viewed as under stress when the normal direction of succession is reversed (Whittaker 1970). Odum (1969) considered an entire ecosystem to be inefficient following disturbance, because fewer resources (sunlight, water, and nutrients) were captured by the system than in later phases of succession. Grime (1977) described in more detail cycles of vegetation that changed in their efficiency in using resources at various stages in succession.

An alternative view, applied here, looks at the performance of various components within an ecosystem and measures efficiency in terms of changes in biomass per unit of resource actually used. This concept is related to the idea that an organism's environment is only what it senses directly and responds to (Mason and Lagenheim 1957). In assessing vegetation, an obvious measure of overall efficiency is the net annual accumulation of biomass or photosynthate in relationship to the amount of solar radiation absorbed by photosynthetic organs throughout the year (Monteith 1977; Landsberg and Wright 1989). A surrogate definition of efficiency might relate annual biomass accumulation to the amount of photosynthetic tissue displayed (Waring 1983) and its duration of display (Kira and Shidei 1967). These latter indices can be applied to individual plants or to communities. The seasonal penetration of solar radiation through vegetation is a more appropriate index for comparing stands (Lang 1987) or regional vegetation using satellite-derived measures of light absorbance (Tucker 1977; Goward and Dye 1987).

"Stress" for photosynthetic plants can be defined along a scale of growth efficiency on which, below a certain point, death becomes increasingly probable (Waring 1987). The concept of efficiency in performance is a general one that has been applied in diagnosing stresses on higher plants, animals, and microbial organisms (Mould and Robbins 1981; McLaughlin et al. 1982; Schimel 1988; Nordgren et al. 1988).

There is an additional concept of internal balance in allometry and in chemistry that is useful in assessing the possible cause for an observed reduction in growth efficiency. Organisms exhibit shifts in chemistry and in the way they allocate resources daily and seasonally. To discern chronic stress requires integrative measures that encompass decades or longer periods.

Undisturbed ecosystems have numerous components that are under stress because there is considerable competition among plants for light, water, and nutrients; among consumers for space, food, and cover; and among microbes for substrate. Rarely are all components under stress at the same time. Stressed plants make desirable food for many animals, and dead animals make a favorable substrate for microbes. Shifts in the efficiency of any major entity (group of plants, herbivores, or detrivores) should signal a potential shift in the availability of resources and eventual change in community structure.

A major disturbance leading to a change in ecosystem structure is likely when a large percentage of trees in a forest become inefficient in capturing light (Waring 1985). This is particularly true when stressed forests exists over large areas. Fire and insect and disease outbreaks are typical when regional-scale stress occurs (Waring and Schlesinger 1985). A disturbance releases resources to surviving, regenerating, or emigrating organisms. In a normal cycle, therefore: (1) stress enhances the probability of disturbance; (2) disturbance occurs, improving the availability of resources; (3) the efficiency by which organisms obtain key resources increases; and (4) competition again becomes intense and stress permeates throughout the system (Fig. 11.1). Only when this normal cycle is broken by a continuation of disturbance, induced by adverse climatic change or by other factors, are major amounts of resources lost (or added in excess) to the system. The very survival of some systems require episodic disturbances to redistribute or add to the available pool of resources. Of course, if all major components of a region's ecosystems were pushed beyond their evolutionary limits, a permanent shift in the composition and structure might result.

Sulfur Deposition on Forests

Diagnosing the impact of sulfur deposition on forests has proven difficult. Usually other pollutants, in addition to sulfur dioxide and sulfuric acid, are involved. Variation in soils and in management practices further complicates interpretations.

In spite of these complications, it is possible to select areas where impacts differ. The most obvious simplification is to find a single source of sulfur pollutant and follow its accumulation and impact along a gradient. Winner et al. (1978) did just this in a Canadian spruce forest downwind from a source of sulfur dioxide. They confirmed that the first organisms to accu-

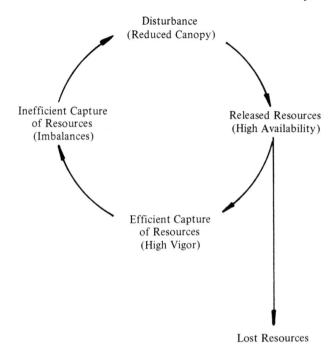

FIGURE 11.1. A normal cycle of stress and disturbances occurs in all ecosystems. A disturbance releases resources. Some resources are lost from the system but, if the disturbance is moderate, most key resources are efficiently recaptured by vigorous plants, animals, and microbes. As competition intensifies, resources are less efficiently captured and stress imbalances permeate throughout the system, setting up conditions again for a major disturbance.

mulate imported sulfur were mosses. Eventually excess sulfur caused these plants to die but until their litter decomposed, little, if any, sulfur appeared in the soil. The unique difference in stable isotope balance of ^{34}S to ^{32}S between source and naturally occurring forms of sulfur in the soil allowed discrimination of these relationships (Fig. 11.2).

The advantage of analyzing stable isotope composition under such circumstances as that found in the Canadian study is that the isotopic signal accumulates slowly and integrates over time the relative contribution of a changing sources of sulfur on various components of an ecosystem. The stable isotopes of nitrogen may also be useful in assessing the impact of N enrichment on ecosystems (Peterson and Fry 1987; Aber et al. 1989).

What happens after the mosses die in a sulfur-polluted forest? A comprehensive series of ecosystem investigations in German spruce forests provided some of the answers (Oren et al. 1988). These studies contrasted two sites where spruce trees were similar in age, size, stocking density and canopy leaf area. Yet one forest showed the first external signs of decline, some yellowing of older needles. The atmospheric inputs, climatic condi-

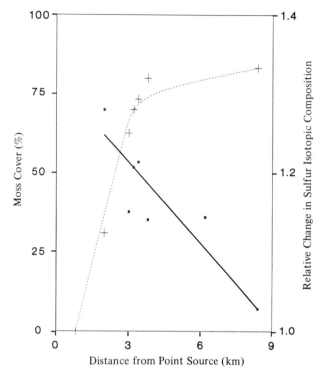

FIGURE 11.2. Moss cover (+) decreases to 0 within 1 km from a refinery producing SO₂. The isotopic composition of sulfur (squares) found in the moss mirrors the contribution of sulfur from the refinery (Winner et al. 1978).

tions, soil chemistry, and tree growth were monitored; photosynthesis, transpiration, nutrient uptake, and translocation were evaluated seasonally; and carbohydrate and elemental analyses were performed on each age-class of needles and other components of the trees.

From this great mass of data, some important insights were distilled (Table 11.1). The impact of sustained acid rain had begun to limit magnesium availability and reduced that element's uptake from the soil. The restrictive uptake of an important nutrient decreased the primary producer's ability to construct wood.

The change in performance was recognized in a number of ways. Certainly a ratio of photosynthesis against growth would show a reduction. But simpler, more integrative measures associated with allocation of resources were also indicative of falling performance. For trees, new leaves are more essential than stemwood. Thus, while the production of leaves was similar in the two spruce forests, the production of stemwood was not (Table 11.1, Fig. 11.3). Because both forests had a similar range in canopy leaf areas, foliage and twig growth, photosynthesis and transpiration, it would be difficult to perceive any early effect of sulfur-enriched acid rain

TABLE 11.1. Physiological assessment of a stressed and unstressed spruce forest.[a]

Variable	Unstressed forest	Stressed forest
Climate	N.D.[b]	N.D.
Photosynthesis	N.D.	N.D.
Transpiration	N.D.	N.D.
Leaf conductance	N.D.	N.D.
Tree water relations	N.D.	N.D.
Soil water status	N.D.	N.D.
Needle growth	N.D.	N.D.
Litterfall	N.D.	N.D.
Twig production	N.D.	N.D.
Branch production	N.D.	N.D.
Leaf area index	N.D.	N.D.
Stemwood growth efficiency		
(g wood m^2 per needle/yr)	100	58
Mycorrhizal on root tips (%)	61	41
Concentration of elements in soil solution		
(μmol/1)		
NO_3-N	N.D.	N.D.
NH_4-N	129	254
Total N	N.D.	N.D.
Phosphorus	N.D.	N.D.
Magnesium	20	9
Calcium	157	47
Hydrogen	48	108
Aluminum	N.D.	N.D.
Concentration of elements in sap		
(mmol/1)		
Nitrogen	N.D.	N.D.
Magnesium	0.44	0.35
Potassium	2.26	1.92
Calcium	1.20	0.89
Concentration of elements in needles		
(μmol/gD)		
Nitrogen	N.D.	N.D.
Magnesium	38	17
Potassium	N.D.	N.D.
Calcium	N.D.	N.D.

[a]After Oren et al. (1988); Osonubi et al. (1988); and Zimmerman et al. (1988).
[b]N.D. signifies no difference statistically between mean values at 0.05 level. When numerical values are presented, differences are significant.

without looking at the efficiency of tree performance, integrated over time.

The integrative effects of nutrient imbalance can be assessed on fresh litter or foliage, and perhaps even on stemwood laid down annually (Contrufo 1983; Bondietti et al. 1989). Additional measurements can confirm where the imbalance originates. In the German study, reduced uptake of magnesium and some other minerals were confirmed by analysis of the mineral composition in the sapstream and in the soil solution and by subsequent fertilization experiments (see Table 11.1).

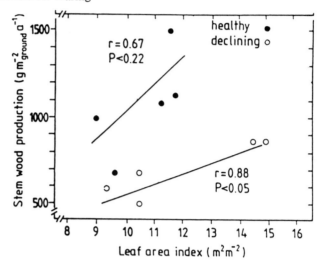

FIGURE 11.3. Spruce trees growing with comparable leaf-area indices produce substantially more stemwood in healthy than in declining forests (Oren et al. 1988).

In addition to a change in the balance of allometric growth relationships, the German spruce study showed a change in the normal balance of major nutrients (Ingestad 1979, 1987). The reduced concentration of a key mineral in the sapstream affected the normal balance of nutrients in the foliage. Through mineral translocation the needs of new leaves and twigs, but not the production of stemwood, were accommodated.

Climatic Change

A change in climate is predicted as a result of the increasing concentration of CO_2 and other greenhouse gases accumulating in the atmosphere. Extremes in climate obviously affect the distribution of vegetation. The increasing levels of carbon dioxide also may influence photosynthesis. Where plants are not exposed to extremes in climate, minor changes in growth rates are difficult to interpret (Graumlich et al. 1989). Climatic effects are also masked when other kinds of stresses are present, as appears may be the case with the decline of red spruce in the eastern United States (Cook et al. 1987).

Drought may be one of the easier symptoms to diagnose. Its effect is usually first felt among the young, less deeply rooted components of a plant community. For this reason, moderate-sized plants are often selected for comparing the predawn plant water potential required to extract water from the soil (Fig. 11.4). Although physiologists can make periodic measurements of plant water stress and integrate these over time (Waring 1985; Myers 1988), the ecosystem ecologists desire a single integrative measure.

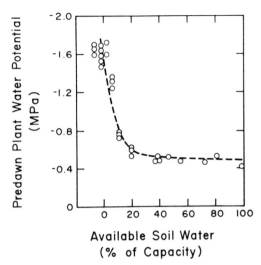

FIGURE 11.4. Drought causes predawn plant water potentials to become increasingly more negative as water is extracted by pine roots from the soil (Sucoff 1972).

One possible approach to evaluating drought is to assess whether the isotopic composition of water changes over time in the sapstream.

When surface water supplies are adequate, the isotopic composition of water closely matches that of recent precipitation (White et al. 1985). On the other hand, when drought occurs, plants with roots able to tap sources of groundwater record a shift in isotopic composition of their sap. The isotopic signature of groundwater closely reflects the mean annual temperature when precipitation falls (Yapp and Epstein 1982. In cases where brackish water is invading freshwater, the isotopic balance also shifts predictably (Sternberg and Swart 1987).

When plants roots draw exclusively from groundwater, the isotopic composition of hydrogen and oxygen in their tissue may reflect changes in the mean annual temperature and relative humidity when photosynthesis occurs. The hydrogen isotopic composition in cellulose tends to mirror that in the sapstream. The oxygen isotopic composition differs because oxygen is produced during photosynthesis and the lighter fraction is differentially lost to the atmosphere as transpiration increases. A long resident time of water in foliage confounds interpretation of the isotopic balance (Cooper and DiNiro 1989). A short residence time is characteristic of plants with small leaves growing in warm environments. Edwards and Fritz (1986) determined the isotopic balance of hydrogen and oxygen in cellulose of modern and fossil wood of conifers growing on a bog in Canada. From these analyses they were able to reconstruct temperature and humidity trends for the past 11,500 years. (Fig. 11.5). Paleobotanical evidence tends to support their interpretations.

The ratio of carbon isotopes in plant tissue has special attraction beyond

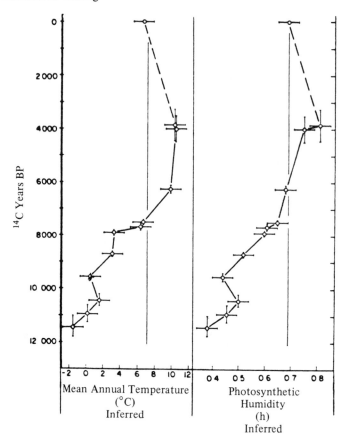

FIGURE 11.5. Current and fossil wood samples from a bog in Canada indicate from the natural abundance of heavier and lighter forms of hydrogen and oxygen isotopes in cellulose that major changes occurred in climate over the last 11,500 years (Edwards and Fritz 1986).

that of carbon dating. The nonradioactive isotopes, ^{12}C and ^{13}C, vary in cellulose and some other constituents of plant tissues depending on the water use efficiency of photosynthesis. When drought occurs, the diffusion of carbon dioxide into leaves is constrained. This forces the carboxylating enzymes to use proportionally more ^{13}C (Francey and Farquhar 1982). If, on the other hand, the enzymatic activity is more constrained than CO_2 diffusion, say by an adverse temperature change, relatively more ^{12}C accumulates in tissue.

Care must be taken in sampling to select open-grown trees or tree that have been long dominant. Otherwise soil respiration injects more of the lighter isotope of carbon in the air and confounds interpretation (Sternberg et al. 1989).

In canopies where the atmosphere is well mixed, an integrative measure

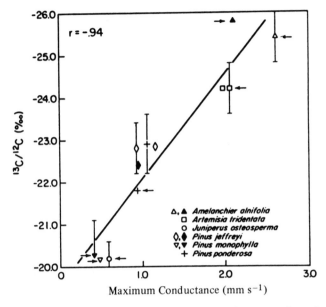

FIGURE 11.6. The carbon isotopic composition of leaves from a variety of arid-zone trees and shrubs become progressively more enriched in ^{13}C compared to ^{12}C as stomatal conductance decreases (DeLucia et al. 1988).

of drought stress is often indicated by the relative enrichment of the heavier isotope ^{13}C. Normally the comparison is best done on a single species, but trends appear in comparisons among a variety of plants with carbon-3 photosynthetic pathways, whether they grow in arid (Fig. 11.6) or in humid environments (Table 11.2).

The historical effects of drought may also be preserved in the isotopic composition of cellulose in the annual growth rings of perennial plants. It is important, however, to separate climatic change from other kinds of

TABLE 11.2. Comparison of $\delta^{13}C$ values of leaf tissue for bog and wet forest habitats in the Hawaiian Islands.[a]

	Island	Bog	Wet forest	$\Delta\delta^{13}C$
Argyroxiphium grayanum	Maui	−25.82	−29.18	3.36
Dubautia laxa	Kauai	−25.36	−28.97	3.51
Dubautia plantaginea	Hawaii	−24.46	−27.89	3.43
Dubautia raillardioides	Kauai	−25.58	−27.05	1.57
Deschampsia nubigena	Kauai	−24.35	−29.14	4.79
Styphelia tameiameiae	Kauai	−26.45	−28.17	1.72
Metrosideros polymorpha	Kauai	−23.74	−27.77	4.03
Metrosideros polymorpha	Maui	−25.57 (2)	−27.71	2.14

[a]After Rundel et al. (1989).

stresses that might induce stomatal closure, such as pollutants (Winner and Mooney 1985).

In summary, changes in temperature, humidity, and the availability of water may be expressed in the isotopic balance of hydrogen and oxygen conducted through the sapstream and deposited in the cellulose of annual growth rings. The efficiency by which plants extract water or carbon dioxide may, with appropriate care to sampling, be assessed to interpret the impact of climatic change on resident species or upon their predecessors [see book edited by Rundel et al. (1989) for many applications of stable isotopes in ecological research].

Management- and Succession-Induced Changes

The composition of organisms in an ecosystem is drastically altered through management policies affecting the frequency of fire, the application of herbicides, the selective harvesting of trees, browsing of shrubs, or grazing of grass. Changes in the physiological performance observed in any surviving plants can mirror stresses or improvements induced through management policies.

In much of the temperate zone, logging and preservation are two counter-management policies practiced on forests. Both policies may induce stresses on forest ecosystems. Preservation policies that include protection from wildfire have led to extensive areas that are more susceptible than normal to insect attack (Christiansen et al. 1987). On the other hand, logging has reduced the degree to which trees cover the landscape. Small gap openings are less likely in young forests, whereas in old forests the gaps become sufficient to favor new groups of species. Even rather subtle changes in canopy openings affect the efficiency of plant processes. These changes can influence the susceptibility and sensitivity of plants to herbivores.

An interesting example comes from southeastern Alaska, where logging of old-growth hemlock has led to a decrease rather than the expected increase in deer population. How is this explained when the favorite food of the deer, Alaskan blueberry, is more abundant following disturbance? A study of the blueberry's performance across a range of environments illustrates major changes in its ability to grow and to support deer browsing (Rose 1990).

In the original forest where large gaps in the canopy allowed blueberry to grow moderately well, plants produced leaves low in tannins (Fig. 11.7). Moreover, restricted growth associated with a shaded environment allowed nitrogen to accumulate, providing an improved source of digestible protein to deer (Fig. 11.8A), associated with a parallel increase in free amino acids (Fig. 11.8B).

Disturbance or lack of disturbance can cause important changes in the

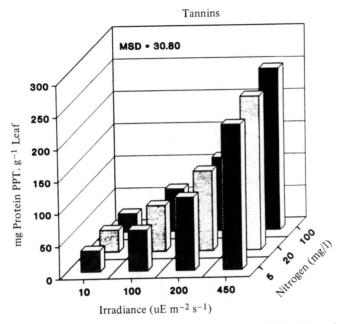

FIGURE 11.7. The concentration of tannins in the leaves of *Vaccinium* increases with irradiance (Rose 1990).

palatability or susceptibility of plants to herbivores or pathogens (Waring and Schlesinger 1985). The Alaskan study on blueberry demonstrates that these changes occur within a species as well as from species replacement. The study also confirmed that specific leaf area (Fig. 11.9) is a good index to changes in the light environment (Araki 1971; Nygren and Kellomaki 1983).

High levels of free amino acids in leaves is a general index of biochemical stress, not a specific one to low light conditions (Vessey and Layzell 1987; Turner and Lambert 1986; Zedler et al. 1986).

Integrative Measures of Stress

In the series of examples, a number of different kinds of indices have been introduced. Some fall into the category of general indices of stress, such as a decrease in the production of stemwood per unit of foliage or a large increase in the levels of free amino acids. Other indices help distinguish one kind of stress from another (Table 11.3). When several different indices indicate inefficiency in capturing resources or physiological imbalances, a disturbance that will reduce the vegetation's ability to absorb solar radiation is likely.

Once stress is recognized, ecosystem ecologists can speculate on the most

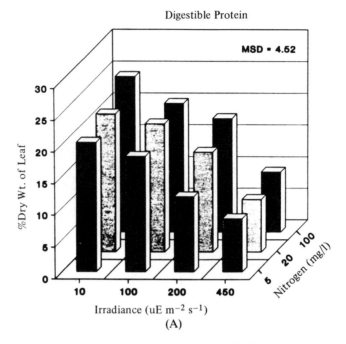

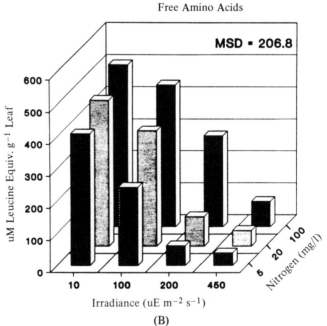

FIGURE 11.8. As irradiance decreases, the concentration of digestible protein (A) and free amino acids (B) increase in the leaves of *Vaccinium* plants grown in controlled environments (Rose 1990).

Specific Leaf Area

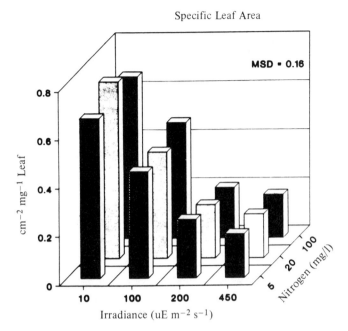

FIGURE 11.9. As irradiance decreases, the specific leaf area of *Vaccinium* leaves increases by more than threefold (Rose 1990).

likely cause by observing patterns across a known gradient in climate or chemical deposition. If climatic change is actually involved, then alterations in the stable isotopic composition of oxygen and hydrogen in leaves and other tissues should reflect this. If chemical changes are important, variation in the balance of essential minerals should develop. If the sources of chemicals originate from anthropogenic activities this may, in some cases, also be diagnosed by observing shifts in the isotopic composition of sulfur or nitrogen along spatial or temporal gradients.

It is more likely that a combination of factors causes stress. In such cases, experimentation may help unravel the relative importance of various factors. Irrigation and addition of balanced mixes of fertilizers should reverse symptoms of stress in situations where drought and nutrient imbalance are important. Such treatments should have less effect where temperature change or other factors dominate.

In assessing the predicted impact of global climatic change we might search first those ecosystems assumed most susceptible. By definitions outlined here, stressed systems should show changes in the efficiency of their performance before changes in community structure occur. The relative importance of various factors is a more difficult issue, but experimentation and careful selection of environmental gradients may lead to renewed understanding of how ecosystems respond to changing environments.

TABLE 11.3. Signals of stress in terrestrial ecosystems referenced to perennial plants.

Category	Integrative signal
General stress:	Reduced biomass production/unit leaf area departure from normal allometric relationships elevated levels of free amino acids
Shade stress:	Increase in area/unit weight of leaves reduction in tannins and phenolics
Water stress:	Extremes in predawn plant water stress enrichment in ^{13}C compared with ^{12}C
Climatic stress:	Shifts in $^{18}O/^{16}O$ and in $^{2}H/^{1}H$ Ratios Depletion in ^{13}C compared with ^{12}C
Mineral imbalance:	Departure from normal N : P : S etc. relationships in foliage, litter, sapwood, and soil
Air pollutants:	Foreign sources of sulfur and nitrogen sensed through change in stable isotope ratios of sulfur and nitrogen

Acknowledgments. I appreciate the opportunity to have participated in all three Cary Conferences and to have been invited to present this paper. I acknowledge helpful review comments from W.H. Schlesinger, B. Yoder, and C. Rose.

REFERENCES

Aber, J.D., K.J. Nadelhoffer, P. Steudler, and J.M. Melillo. (1989). Nitrogen saturation in northern forest ecosystems. *BioScience* 39:378–386.

Araki, M. (1971). The studies on specific leaf areas of forest trees: I. The effects of RLI, season, density and fertilization on the specific leaf area of larch (*Larix leptolepis* Gord.) leaves. *J. Jpn. For. Soc.* 53:359–367.

Bondietti, E.A., C.F. Baes, III, and S.B. McLaughlin. (1989). The potential of trees to record aluminum mobilization and changes in alkaline earth availability. In: *Biological Markers of Air-Pollution Stress and Damage in Forests.* National Academy Press, Washington, D.C., pp. 281–292.

Christiansen, E., R.H. Waring, and A.A. Berryman. (1987). Resistance of conifers to bark beetle attack: searching for general relationships. *For. Ecol. Manage.* 22: 89–106.

Contrufo, C. (1983). Xylem nitrogen as a possible diagnostic nitrogen test for loblolly pine. *Can. J. For. Res.* 13:355–357.

Cook, E.R., A.H. Johnson, and T.J. Blasing. (1987). Forest decline: modeling the effect of climate in tree rings. *Tree Physiol.* 3:27–40.

Cooper, L.W. and M.J. DeNiro. (1989). Covariance of oxygen and hydrogen isotopic composition in plant water: species effects. *Ecology* 70:1619–1628.

DeLucia, E.H., W.H. Schlesinger, and W.D. Billings. (1988). Water relations and the maintenance of Sierran conifers on hydrothermally altered rock. *Ecology* 69: 303–311.

Edwards, T.W.D. and P. Fritz. (1986). Assessing meteoric water composition and relative humidity from ^{18}O and ^{2}H in wood cellulose: paleoclimatic implications for southern Ontario, Canada. *Appl. Geochem.* 1:715–723.

Francey, R.J. and G.D. Farquhar. (1982). An explanation of $^{13}C/^{12}C$ variation in tree rings. *Nature* (London) 297:28–31.

Goward, S.N. and D. Dye. (1987). Evaluating North American net primary productivity with satellite observations. *Adv. Space Res.* 7:165–174.

Graumlich, L.J., L.B. Brubaker, and C.C. Grier. (1989). Long-term trends in forest net primary productivity: Cascade Mountains, Washington. *Ecology* 70:405–410.

Grime, J.P. (1977). Evidence for the existence of three primary strategies in plants and its relevance to ecological and evolutionary theory. *Am. Nat.* 111:1169–1194.

Ingestad, T. (1979). Mineral nutrient requirements for *Pinus silvestris* and *Picea abies* seedlings. *Physiol. Plant.* 45:373–390.

Ingestad, T. (1987). New concepts on soil fertility and plant nutrition as illustrated by research on forest trees and stands. *Geoderma* 40:237–252.

Kira, T. and T. Shidei. (1967). Primary production and turnover of organic matter in different forest ecosystems of the western Pacific. *Jpn. J. Ecol.* 10:70–87.

Landsberg, J.J. and L.L. Wright. (1989). Comparisons among *Populus* clones and intensive culture conditions, using an energy-conversion model. *For. Ecol. Manage.* 27:129–147.

Lang, A.R.G. (1987). Simplified estimate of leaf area index from transmittance of the sun's beam. *Agric. For. Meteorol.* 41:179–186.

Mason, H.L. and J.H. Langenheim. (1957). Language analysis and the concept of environment. *Ecology* 38:325–339.

McLaughlin, S.B., R.K. McConathy, D. Duvick, and K.L. Mann (1982). Effects of chronic air-pollution stress on photosynthesis, carbon allocation, and growth of white pine. *For. Sci.* 28:60–70.

Monteith, J.L. (1977). Climate and efficiency of crop production in Britain. *Philos. Trans. R. Soc. Lond. B* 281:277–294.

Mould, E.D. and C.T. Robbins. (1981). Nitrogen metabolism in elk. *J. Wildlife Manage.* 45:323–334.

Myers, B.J. (1988). Water stress integral—a link between short-term stress and long-term growth. *Tree Physiol.* 4:315–324.

Nordgren, A., E. Baath, and B. Soderstrom. (1988). Evaluation of soil respiration characteristics to assess heavy metal effects on soil microorganisms using glutamic acid as a substrate. *Soil Biol. Biochem.* 20:949–954.

Nygren, M. and S. Kellomaki. (1983). Effect of shading on leaf structure and photosynthesis in young birch, *Betula pendula* Roth. and *B. pubescens* Ehrn. *For. Ecol. Manage.* 7:119–132.

Odum, E.P. (1969). The strategy of ecosystem development. *Science* 164:262–270.

Oren, R., E.-D. Schulze, K.S. Werk, J. Meyer, B.U. Schneider, and H. Heilmeier. (1988). Performance of two *Picea abies* (L.) Karst. stands at different stages of decline. I. Carbon relations and stand growth. *Oecologia* 75:25–37.

Osonubi, O., R. Oren, K.S. Werk, E.-D. Schulze, and H. Heilmeier (1988). Performance of two *Picea abies* (L.) Karst. stands at different stages of decline. IV. Xylem sap concentrations of magnesium, calcium, potassium and nitrogen. *Oecologia* 77:1–6.

Peterson, B.J. and B. Fry. (1987). Stable isotopes in ecosystem studies. *Annu. Rev. Ecol. Syst.* 18:293–320.

Rose, C. (1990). Application of the Carbon/Nutrient Balance Hypothesis to Predicting the Nutritional Quality of Blueberry Foliage to Deer in Southeastern Alaska. Ph.D. Dissertation, Oregon State University, Corvallis, Oregon.

Rundel, P.W., J.R. Ehleringer, and K.A. Nagy, eds. (1989). *Stable Isotopes in Ecological Research*. Ecological Series 68, Springer-Verlag, New York.

Schimel, D. (1988). Calculation of microbial growth efficiency from ^{15}N immobilization. *Biogeochemistry* 6:239–243.

Sternberg, L.L. and P.K. Swart. (1987). Utilization of fresh water and ocean water by coastal plants of southern Florida. *Ecology* 68:1898–1905.

Sternberg, L.S.L., S.S. Mulkey, and S.J. Wright. (1989). Ecological interpretation of leaf carbon ratios: influence of respired carbon dioxide. *Ecology* 70:1317–1324.

Sucoff, E. (1972). Water potential in red pine: soil moisture, evapotranspiration, crown position. *Ecology* 53:681–686.

Tucker, C.J. (1977). Spectral estimation of grass canopy variables. *Remote Sens. Environ.* 6:11–26.

Turner, J. and M.J. Lambert. (1986). Nutrition and nutritional relationships of *Pinus radiata*. *Annu. Rev. Ecol. Syst.* 17:325–350.

Vessey, J.K. and D.B. Layzell. (1987). Regulation of assimilate and partitioning in soybean. *Plant Physiol.* 83:341–348.

Waring, R.H. (1983). Estimating forest growth and efficiency in relation to canopy leaf area. *Adv. Ecol. Res.* 13:327–354.

Waring, R.H. (1985). Imbalanced ecosystems: assessments and consequences. *Forest Ecol. Manage.* 12:93–112.

Waring, R.H. (1987). Characteristics of trees predisposed to die. *BioScience* 37:569–574.

Waring, R.H. and W.H. Schlesinger. (1985). *Forest Ecosystems: Concepts and Management*. Academic Press, Orlando, Florida.

White, J.W.C., E.R. Cook, J.R. Lawrence, and W.S. Broecker. (1985). The D/H ratios of sap in trees: implications for water sources and tree ring D/H ratios. *Geochim. Cosmochim. Acta* 49:237–249.

Whittaker, R.H. (1970). *Communities and Ecosystems*, 1st Ed. Macmillan, New York.

Winner, W.E., J.D. Bewley, H.R. Krouse, and H.M. Brown (1978). Stable sulfur isotope analysis of SO_2 pollution impact on vegetation. *Oecologia* 36:351–361.

Winner, W.E. and H.A. Mooney. (1985). Ecology of SO_2 resistance. V. Effect of volcanic SO_2 on native Hawaiian plants. *Oecologia* 66:387–393.

Worbes, M. and W.J. Junk. (1989). Dating tropical trees by means of ^{14}C from bomb tests. *Ecology* 70:503–511.

Yapp, C.J. and S. Epstein. (1982). Climatic significance of the hydrogen isotope ratios in tree cellulose. *Science* 297:636–639.

Zedler, B., R. Plarre, and G.M. Rothe. (1986). Impact of atmospheric pollution on the protein and amino acid metabolism of spruce *Picea abies* trees. *Environ. Pollut.* 40:193–212.

Zimmermann, R., R. Oren, E.-D. Schulze, and K.S. Werk. (1988). Performance of two *Picea abies* (L.) Karst. stands at different stages of decline. II. Photosynthesis and leaf conductance. *Oecologia* 76:513–518.

Part IV   Biogeochemical Cycles

12
A Cross-System Study of Phosphorus Release from Lake Sediments

NINA CARACO, JONATHAN J. COLE, AND GENE E. LIKENS

Abstract

Phosphorus plays an important role in the control of primary production in aquatic systems. Research during the past several decades has shown that a major loss of P from the surface photic waters of aquatic systems is by sinking of particles to the sediments. Not surprisingly, therefore, the extent to which sediments recycle P to overlying waters is critical in maintaining system productivity and controlling eutrophication of aquatic systems.

Research on P cycling in individual aquatic systems has indicated that dissolved oxygen content of waters is a major factor controlling P release from sediments. Although control by oxygen is widely accepted, this view is not substantiated by the large variation in P release among lakes. Looking among lakes it becomes clear that factors that are relatively invariant in single lakes (e.g., major ion concentrations) are critical in controlling sediment P release.

Introduction

Ecosystems are complex natural units, the structure and function of which depend on countless interactions between biotic and abiotic components. Because of this complexity, intensive studies of ecosystems may be desirable to gain understanding of ecosystem function and structure. In support of this view, there are several outstanding examples in which intensive studies of single systems have led to important advances in the understanding of aquatic and terrestrial ecosystems (e.g., Likens et al. 1977; Taylor 1989). However, a shortcoming of the "single-system" approach, and of ecology in general, is our lack of ability to translate the understanding from single systems to other systems and ultimately to a more general understanding of ecosystem process. Further, by focusing on single systems, ecologists by definition must limit dramatically their observations. Within this limited view, important controls of ecosystem processing can be overlooked.

Another approach to ecosystem research, which incorporates information from diverse ecosystems in an attempt to achieve broader understanding, is the "cross-system approach." By studying how ecosystem processes vary across diverse ecosystems, new insights can be gained about the important control points. In this chapter we present an example in which data from single systems have been used to develop understanding of an ecosystem process. Using these same data, we will show that the cross-system approach reveals a new, and previously unexpected, view of what these important control points are. The example we use is the control of recycling of phosphorus (P) from the sediments of aquatic systems. Before delving into this comparison, however, we will first provide a brief review of P release from sediments.

Sediment P Release in Aquatic Systems: A Brief Review

It is well known that primary production in many aquatic ecosystems depends on the supply and availability of phosphorus (Vallentyne 1972; Schindler 1977; Heckey and Kilham 1988). Phosphorus can enter a lake through any number of diverse sources, including atmospheric loading, runoff, and groundwater seepage (Schindler et al. 1976; Likens et al. 1985; Cole et al. 1990, Caraco et al. 1988; Shaw 1989). The productivity of a lake depends to a large extent on the magnitude of the sum of all these loading terms (Vollenweider 1968). Reactions within a lake are, however, also critical in controlling lake productivity. A within-lake reaction that has been recognized for a long time as important is the exchange of P between lake sediments and the overlying water column (Einsele 1936; Mortimer 1941, 1942; Hasler and Einsele 1948; Barrett 1953).

The importance of sediment exchange of P was first recognized nearly a century ago by fish farmers who discovered that the type of "underlying soil" in ponds could have a dramatic effect on fish productivity (Neess 1949). During the past several decades eutrophication (nutrient enrichment leading to increased biological productivity) has become a widespread problem in lakes, and attention to P exchange between the water column and sediments has become more intense (e.g., Holdren and Armstrong 1980; Bostrom et al. 1982; Baccini 1985; Stauffer 1985; Quigley and Robbins 1986; Schindler et al. 1987; Carlton and Wetzel 1988; Enell et al. 1989). It is now recognized that loading of P from sediments can be so important that management steps taken to prevent P release from sediments (dredging, sediment "sealing," etc.) may, in some cases, be more effective in eutrophication abatement than management schemes to remove P inputs to the lake from the watershed (Cooke et al. 1986; Henderson-Sellers and Markland 1987).

Because P release from sediments can affect productivity of aquatic systems significantly, the factors that control sediment P release have been the focus of much research. Of the numerous factors that have been implicated as having some control over P release from sediments (Table 12.1), dissolved oxygen content of waters is generally thought to be the most important (see Fig. 12.1, for a widely accepted model of the link between dissolved oxygen and P release from sediments). This belief is based on observations in many lakes that show, relatively consistently, an increase in P release from sediments of lakes concurrent with the onset of seasonal anoxia in waters overlying these sediments (Einsele 1936; Mortimer 1941, 1942, 1971; Burns and Ross 1971; Sen Gupta 1973; Larsen et al. 1979). Conversely, the relationship between P release from sediments and many other variables proposed to control P release has not been consistent. In

TABLE 12.1. Factors that have been proposed to control release of P from bottom sediments.

Factors	Reference
Oxygen concentration	Andersen 1974; Baccini 1985; Banoub 1975; Burns and Ross 1971; Callender 1982; Clasen and Bernhardt 1982; Davies et al. 1975; Einsele 1936; Gachter and Imboden 1985; Hayes and Phillips 1958; Holdren and Armstrong 1980; Kamp-Nielson 1974; Khalid et al. 1978; Krom and Berner 1980; Lean et al. 1986; Li et al. 1972; Mortimer 1941, 1942, 1971; Nurnberg 1984; Olsen 1964; Tessenow 1972.
Nitrate concentration	Andersen 1982; Jannson 1982, 1987.
Sulfate concentration	Curtis 1989; Hasler and Einsele 1948; Hawk et al. 1989; Sugawara et al. 1957.
Salt concentration	Fox et al. 1986; Carritt and Goodgal 1954.
pH or alkalinity	Andersen 1975; Broberg 1987; Barrett 1953; Kamp-Nielson 1974; Stauffer 1985; Curtis 1989; Hawke et al. 1989; Ostrofsky et al. 1989.
DOC or colloids	Banoub 1975; Jackson and Schindler 1975; Ohle 1937.
Calcium	Hawke et al. 1989; Hingston 1981.
Temperature	Holdren and Armstrong 1980; Kamp-Nielsen 1975.
Turbulence or bioturbation	Davies et al. 1975; Gallep 1979; Holdren and Armstrong 1980; Reynoldson and Hamilton 1981.
Sediment Fe or Fe : P ratio	Baccini 1985; Callender and Hammond 1982; Lean et al. 1986; Stauffer 1985; Syers et al. 1970.
Sediment Al	Richardson 1985.
Sediment Mn	Banoub 1975.
Sediment POC	Baccini 1985.
Sediment CO_3 or CO_3 POC ratio	Barrett 1953; Ohle 1937.
P content of decomposing material	Fenchel and Blackburn 1979; Gachter and Mares 1985; Schindler and Fee 1974; Tezuka 1989.

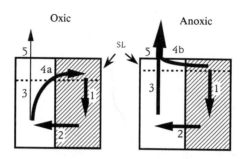

FIGURE 12.1. Schematic of classical view of dependence of P release on dissolved oxygen content of overlying waters. Left-hand panel represents dominant reactions of P when overlying waters and sediment–water interface are oxic; right-hand panel represents these reactions after oxygen has been depleted at sediment-water interface and overlying waters. Sediment (for both left and right panels) is divided into two layers: a surface, iron-oxide-rich layer, denoted SL, and deeper sediments; these two layers are separated by the dashed lines. Stippled area represents P in particulate phases whereas the clear area represents dissolved phases. Numbered arrows show direction of net transformation between these particulate and dissolved phases of P and the movement of P vertically within sediments. Numbered arrow key: (1) particulate P in the form of iron-oxide bound P or organic P is buried into deeper sediments; (2) P solubilized via decomposition or dissolution at depth in the sediments; (3) dissolved P at depth moves up to the iron-oxide-rich surface layer by diffusion or turbulent mixing; (4a) under oxic conditions most of dissolved P from deeper sediments precipitates in surface layer as iron oxide-bound P; (4b) under anoxic conditions surface iron-oxide layer begins to dissolve (from chemical reduction) releasing P bound to iron oxide. Additionally, as surface iron-oxide-rich layer is depleted, little of P from deeper in the sediments is trapped; and (5) as a result of reactions in the surface, iron-oxide-rich layer (4a or 4b) under oxic conditions (left panel), only a small amount of P is released into solution; under anoxic conditions (right panel), a large amount of P is released into solution.

fact, opposing results are reported for the effects of pH, microbial activity, sediment turbation, sediment organic content, and nitrate concentration (Jackson and Schindler 1975; Baccini 1985; Cooke et al. 1986; Jansson 1987; Bostrom et al. 1988; Carlton and Wetzel 1988).

As a result of the perceived importance of dissolved oxygen, most studies on P release from sediments (see Table 12.1) and many aquatic texts that discuss P release from sediments focus on dissolved oxygen (Hutchinson 1947; Fenchel and Blackburn 1979; Stumm and Morgan 1981; Wetzel 1983; Burgis and Morris 1987). We believe, however, that this focus may be misplaced and that variables other than oxygen may be of critical importance to the control of P release. The importance of these other variable(s) are underestimated because of the dominance of single system studies in limnology.

Approach

A variety of approaches ranging from sediment core manipulations to in situ approaches can be used to assess P release from sediments of aquatic systems. In situ approaches, when they can be used, have several obvious advantages over laboratory or even field manipulations and are in many ways preferable. We, therefore, chose to use an in situ approach, which takes advantage of the seasonal vertical stratification which occurs in many lakes.

Lakes often experience seasonal thermal stratification. Solar heating of waters from above causes lakes to become vertically stratified into two layers during summer: warm surface waters (epilimnion) and cooler bottom waters (hypolimnion). In surface waters there is generally a large net biological uptake of dissolved P; thus, release of dissolved P from sediments is not reflected directly in an increase in nutrient concentration in the surface water. Further, even during summer months, there are several other inputs of P to surface waters, including atmospheric sources and groundwater seepage (Winter 1977; Likens et al. 1985; Caraco et al. 1988; Shaw 1989). Because of these complicating factors, changes in surface water P concentration are not interpreted readily. In bottom aphotic waters, on the other hand, decomposition far exceeds assimilation. Further, because bottom waters are shut off from atmospheric and surface seepage inputs, P release from sediments is the main source of dissolved P to bottom waters. Thus, accumulation of dissolved P in hypolimnetic waters often is a relatively good indicator of the release of P from sediments (and settling particles) in aquatic systems.

The direct measurement of the accumulation of P in bottom waters (in situ approach, earlier) can be used in single-system studies to test if dissolved oxygen were a critical variable controlling P release. That is, dissolved P accumulation rate in bottom waters of a lake is compared before and after bottom waters become anoxic (generally early and late summer accumulation, respectively; e.g., Mortimer 1941, 1942). We use a variation of this approach, which can also be used in single-system studies but is more amenable to detecting changes in sediment immobilization in cross-system comparisons. This approach compares P release to that expected when sediments (and particles in bottom waters) have no ability to immobilize P (Richards et al. 1965; Burns and Ross 1971; Fisher et al. 1982; Smith et al. 1987). Because the major source of P to sediments of most aquatic systems is settling of organic matter that was produced in surface waters, the amount of P expected to be released from sediments in the absence of sorption is simply the carbon released from decomposition of this freshly sedimented material times the $P:C$ ratio of this decomposing material (Table 12.2). In other words, in the absence of sediment P immobilization, the $P:C$ release ratio is equal to the $P:C$ ratio of decomposing organic matter (Chapter 13, Smith, this volume). The $P:C$ ratio of freshly sedimented organic matter in most aquatic systems varies between 3 and 10

TABLE 12.2. Major pathways of decomposition in aquatic systems. Note that all these pathways lead to an accumulation of both DIC[a] and TDP. The stoichiometry of the accumulation of these end products depends on the parent material, not the pathway (except for methanogenesis).[b] Under most conditions the accumulation of DIC can be used as a tracer of the amount of decomposition. In extremely alkaline systems, DIC accumulation will be uncoupled from decomposition because of precipitation of carbonates. In these systems decomposition may be traced as the accumulation or depletion of electron acceptors or reduced by-products, respectively (oxidation equivalents). X represents the C : P ratio of decomposing material.

$$[X] (CH_2O) - [y] NH_2 - 1PO_4 + [1.3X]O_2 \rightarrow [X]CO_2 + [2.2X]H_2O + [y]NO_3 + PO_4$$

$$[X] (CH_2O) - [y] NH_2 - 1PO_4 + [4X/5]NO_3 \rightarrow [X]CO_2 + [2/5X]N_2 + [y]NH_4 + PO_4$$

$$[X] (CH_2O) - [y] NH_2 - 1PO_4 + [X]/2]SO_4 \rightarrow [X]CO_2 + [X/2]S^{2-} + [y]NH_4 + PO_4$$

$$[X] (CH_2O) - [y] NH_2 - 1PO_4 + \rightarrow [X/2]CO_2 + [X/2]CH_4 + [y]NH_4 + PO_4$$

$$[X] (CH_2O) - [y] NH_2 - 1PO_4 + [2X]H^+ \rightarrow [X]CO_2 + [2x]H_2 + [y]NH_4 + PO_4$$

[a]Although CO_2 is the oxidized decomposition product of organic C, CO_2 will speciate into other organic forms (HCO_3, CO_3) depending on pH. In practice we measure the total amount of dissolved inorganic carbon (DIC).

[b]When decomposition proceeds by methanogenesis the use of CO_2 as a tracer of decomposition leads to a twofold underestimation. Thus, for every mole of CH_4 that accumulates, we assume that 2 moles of organic C were decomposed. Although we can easily measure the methane that accumulates in bottom waters, this measurement does not include the methane that escapes by ebullition (Strayer and Tiedje 1978). Even if all of the methane produced had escaped by ebullition, we would underestimate decomposition by only twofold. This potential error is small compared to the large natural variation observed in P : C release ratios from sediments (111.C). Further, much of the methane does not bubble out; some accumulates, and the real error is less than twofold. Finally, as methanogenesis would be more prevalent when oxygen was depleted, the occurrence of methanogenesis would lead to the bias of overestimating P : C release under anoxic conditions and thus overestimating the importance of oxygen as a controlling variable of sediment P release.

mmol P/mol C (Moeller 1985; Caraco 1986; Uehlinger and Bloesch 1987); when P : C release ratios are less than 3–10 (depending on the system), it indicates that immobilization of P is occurring in sediments. Variations in P : C release ratios can be analyzed from both a within-system and a cross-system perspective to reveal the importance of control points of P immobilization in sediments.

Data Set

The data for this analysis are derived from 23 aquatic systems for which P : C release ratios are known for both oxic and anoxic conditions. Data for 12 of the 23 systems come from data we collected. Methods used for sampling the lakes and chemical analyses are discussed elsewhere (see Caraco et al. 1989). Briefly, however, the accumulation ratio was calculated by taking vertical profiles of dissolved P and DIC in lakes on two to six occasions during summer stratification. During this time net decomposition occurred and portions of bottom waters changed from oxic to anoxic conditions. Thus, P : C accumulation ratios were available for both sets of conditions (oxic and anoxic). The data for the remaining 11 systems are from the literature; sources of these data are also discussed elsewhere (Caraco et al. 1989).

Results

THE SINGLE-SYSTEM PERSPECTIVE (WITHIN-SYSTEM VARIATION)

Many variables can change seasonally within a lake. The relative importance of these seasonal variables to the control of P release can be tested by within-system studies. For example, to test the oxygen control model, we compare the P : C release ratio into overlying oxic waters to the P : C release ratio into waters devoid of oxygen. Within a given lake this ratio should be greater under anoxic than oxic conditions (Fig. 12.2). Further, if this

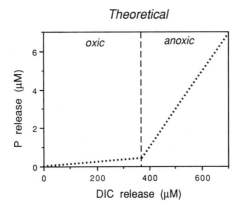

Theoretical

FIGURE 12.2. Theoretical pattern of sediment P release expected if oxygen were the key controlling variable (Fig. 12.1). DIC, dissolved inorganic carbon accumulated in bottom waters of lakes during summer stratification; DIC can be used as tracer for decomposition in all but extremely hardwater systems. In these systems the sum of electron acceptors can be used (Table 12.2). Similarly, TDP is dissolved P accumulated in bottom waters of lakes. Dotted vertical line represents switch from oxic to anoxic waters as decomposition proceeds in bottom waters. If oxygen were a critical controlling variable, sediment immobilization would decrease as waters change from oxic to anoxic. Because TDP accumulation is affected negatively by sediment immobilization (TDP accumulation = decomposition−sediment immobilization), the P : C release ratios (slope on graph) should increase as waters change from oxic to anoxic conditions.

expectation were true for a majority of lakes, it would suggest that oxygen is a major factor controlling P release from sediments.

Lake Lacawac, Pennsylvania, is an example of a system that clearly corresponds to the pattern expected of the oxygen control model (Fig. 12.3A). The P : C release ratio from sediments when overlying water was oxic was 0.5 mmol P/mol C, while the release ratio into anoxic waters was 16 mmol P/mol C. This change represents a 32-fold increase in P release. Further, because P : C ratios of decomposing material (as estimated from P : C ratios of suspended particles in surface waters) in this lake roughly are 4 mmol P/mol C, they suggest that under oxic conditions some 90% of the P released by decomposition was bound in sediments ("trapped"; Fig. 12.1). Under anoxic conditions, on the other hand, the high P : C release ratio (greater than the ratio of decomposing material) suggests release of previously sorbed P.

Squam Lake, New Hampshire, also behaves in correspondence to the expectation of the oxygen-control model (see Figs. 12.1 and 12.2). That is, the P : C release ratio into oxic waters is lower than the P : C release ratio into anoxic waters (Fig. 12.3B). In Squam Lake, however, this increase is smaller (approximately twofold; 0.8 and 1.5 mmol P/mol C under oxic

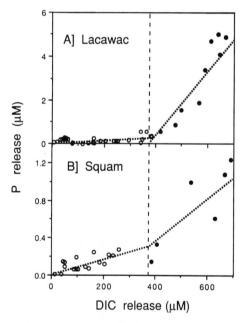

FIGURE 12.3. Observed patterns of dissolved P accumulation in bottom waters of two lakes: (A) Lake Lacawac, Pennsylvania; and (B) Squam Lake, New Hampshire. Both lakes correspond to expectations of the oxygen control model (see Fig. 12.2 and text). Note, however, the different magnitude of P release in the two systems.

and anoxic conditions, respectively). Further, whereas in Lake Lacawac the P : C release ratio under anoxic conditons suggested net release of P sorbed previously, in Squam Lake net sorption continued even after bottom waters become anoxic. That is, in contrast to the results for Lake Lacawac, the P : C release ratio was lower than the P : C ratio of decomposing material (about 3.5 mmol P/mol C) under both oxic and anoxic conditions.

Despite some differences in the response of these two lakes to depletion of dissolved oxygen, both show some decrease in sediment uptake (increase in P : C release ratio) when overlying waters change from oxic to anoxic. This result demonstrates, at least, that the behavior of some lakes does correspond to the expectations of the oxygen control model. Further, the data in Fig. 12.4 demonstrate that a majority of lakes may, in fact, show P release patterns that correspond to these expectations. Of the systems for which we had data, 18 of 23 had higher P : C release ratios under anoxic than oxic conditions. From the within-system perspective, the consistency of this result suggests that oxygen is an extremely important variable in the control of P release from lake sediments. After all, it might be argued, given that there are potentially many other variables changing seasonally within a lake, if oxygen were not of critical importance, such a consistent relationship would not occur. The problem with this argument is that, by restricting comparisons to within-lake seasonal changes, the importance of variables that *do not* change within the temporal viewing period cannot be tested. One way to elucidate the importance of these factors is by comparing the differences in P release among systems.

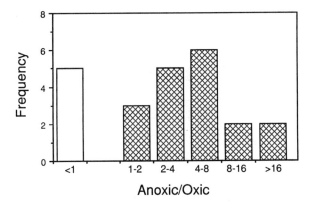

FIGURE 12.4. Change in P accumulation in 23 different aquatic ecosystems as waters change from oxic to anoxic conditions. When oxygen is a critical controlling variable (Fig. 12.1), the slope of TDP versus DIC accumulation in bottom waters should be greater under anoxic than under oxic conditions (Fig. 12.2). ANOXIC/OXIC is the ratio of these slopes; values greater than 1, therefore, suggest compliance with the oxygen-control model. Note that 18 of 23 systems (stippled bars) behave in correspondence to the oxygen control model; 5 systems (open bars) do not.

THE CROSS-SYSTEM PERSPECTIVE (AMONG-SYSTEM VARIATION)

If oxygen were the key variable controlling P binding by sediments, a large part of the variation observed in the P : C release ratio between systems would be eliminated if we were to divide that release into oxic and anoxic components. If the oxic and anoxic phases of P release from the sediments of lakes were not considered separately, the P : C release ratio would vary by 350 fold (0.1 to 35 mmol P/mol C) for the 23 systems for which we have data. Surprisingly, this tremendous variation in P : C release ratio is not reduced dramatically when release is divided into oxic and anoxic phases (Fig. 12.5). By dividing release into these two components we find a decrease in the range in observed P : C release ratios under oxic conditions. However, the range in observed release ratio is still extremely large (100 fold). Further, the range in P : C release ratios under anoxic conditions is 350 fold; a value identical to the range when oxic and anoxic P release are considered together. Put another way, the oxygen status of water (oxic versus anoxic) explains a relatively small fraction of the total variation in the P : C release ratios observed among systems. A one-way analysis of variance (ANOVA) indicated that 18% of the variation in P : C release could be explained by oxygen status, while 82% remained unexplained.

Thus, the oxygen control model did not explain well the among-system variation in P release. This observation, coupled with the fact that the model seems to explain well the within-system changes in P release, suggested that some variable that is relatively invariant within single systems

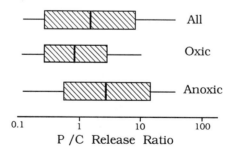

FIGURE 12.5. Variation in P : C release ratio from sediments of aquatic systems as expressed by box-whisker plot. Release into overlying oxic and anoxic waters was calculated for 23 different aquatic systems. Oxic and anoxic P : C release ratios are considered together (All, $n = 46$) and are also analyzed separately (oxic, $n = 23$; anoxic, $n = 23$). "Box" in plot covers lower and upper 25% quartile of data; "whiskers" cover entire data range; dark line in box represents mean value of these data. Note that separation of data into oxic and anoxic components does not decrease greatly the large variation in P : C release ratios observed among different aquatic systems.

(at least on short time scales) is critical in the control of P release from sediments. We investigated the importance of several of these potentially important variables, including lake trophic status, P : C ratio of material within the system, sediment iron content, lake pH, salt concentration of waters in the system, and sulfate concentration of waters (Caraco et al. 1989). The only variable which showed a significant relationship to P : C release ratio into overlying anoxic waters was sulfate concentration of waters. Further, sulfate concentration also showed the best relationship to P release into oxic waters (Caraco et al. 1989).

To compare the importance of this variable to dissolved oxygen content of bottom waters, we performed a one-way ANOVA between sulfate concentration and the P : C release ratio. We found that sulfate concentration explained 30% of the variation in P : C release ratio observed between systems, or almost double the variation explained by oxygen alone (18%). Further, using a two-way ANOVA, including both sulfate and oxygen, we were able to explain 65% of the variation in P : C release ratio, leaving only 35% of the 350-fold variation unexplained (Table 12.3). These data suggest, therefore, that sulfate concentration of waters may be a major determinant of P uptake by sediments. That is, the result suggests that systems with low sulfate concentration have sediments with a high capacity to take up P and prevent its release to overlying waters, compared to systems with higher sulfate concentrations. The relationship between P immobilization in sediments and sulfur content of waters may be due to the interaction of the sulfur and iron cycles in sediments of lakes (Fig. 12.6). There are three implications of this interaction:

1. Increased atmospheric deposition of S from anthropogenic activities could cause alterations in aquatic systems that previously were considered insensitive to the effects of sulfur emissions (Caraco et al. 1989).

TABLE 12.3. Two-way analysis of variance of P release from sediments of 23 aquatic systems.[a]

	Sum of squares	Significance level
Oxygen	466	0.001
Sulfate	779	0.000
Interaction	458	0.003
Residual	918	

[a]In this ANOVA oxygen status was either oxic, waters overlying sediments have dissolved oxygen concentration >0.5 mg/l, or anoxic, waters overlying sediments have dissolved oxygen concentration ≤ 0.5 mg/l. Sulfate concentration was considered in three categories: <60 μM, 60–300 μM, >300 μM. A model that included sulfate and oxygen explained much (65%) of variation in P release among different aquatic systems.

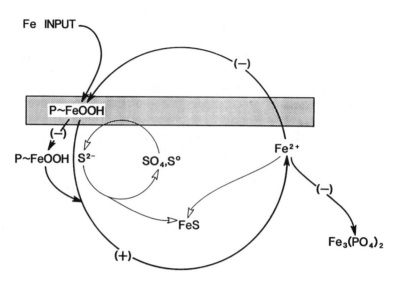

FIGURE 12.6. Hypothesized mechanism for sulfur control of P release from sediments. P immobilization in sediments is proportional to amount of iron oxide in sediments. Sulfide affects this iron oxide pool in two ways: first, sulfide binds Fe (as iron sulfides) and inhibits reformation of iron oxides at the oxic–anoxic border; second, sulfide may enhance dissolution of iron oxides, as chemical reduction of iron oxides by sulfide may occur more rapidly (or more completely) than dissolution by microbial reduction. Both processes, in systems with high amounts of S, may remove the FeOOH layer to outpace regeneration leading to P release. This interaction of S and Fe could be driving the correlation between sulfate concentration of waters and P release.

Presently, it is considered that the effects of atmospheric pollution by emissions of SO_2 are limited to "acid-sensitive regions," which are susceptible to pH drops because of low buffering of the aquatic ecosystem or its watershed. The fact that changing sulfate concentrations in themselves can affect dramatically aquatic system functioning suggests that a larger region may be affected by atmospheric S pollution.

2. It is well established that biological production in freshwaters is, in general, P limited. In coastal marine systems, on the other hand, P is often present in concentrations above that limiting to phytoplankton growth and N often is limiting to phytoplankton production (Ryther and Dunstan 1971; Howarth and Cole 1985; D'Elia et al. 1986; Caraco et al. 1987; Caraco 1988). This discrepancy has not been explained fully. We suggest that the relatively higher P concentrations in coastal marine systems (compared to N and Si; Valiela 1984; Heckey and Kilham 1988) result, in part, from the lower immobilization of P in sediments of marine systems, with high sulfate concentrations, as compared to sediments of freshwater systems, with lower sulfate concentrations (Caraco et al. 1990).

3. As mentioned previously, it is believed generally that dissolved oxygen is of critical importance in controlling P release from sediments of aquatic systems. This belief has, in part, justified the management practice of hypolimnetic aeration (Taggart 1984; Cooke et al. 1986). In this practice, bottom waters of lakes are aerated so that oxygen levels remain relatively high throughout summer stratification. The belief is that, because P retention of sediments is far greater under oxic as compared to anoxic conditions, immobilization of P in sediments should increase, and eutrophication can be controlled, by keeping bottom water oxic. This procedure is extremely expensive, and in fact has had only mixed success in significantly inhibiting P release from sediments (Taggart 1984; Gachter 1988). We suggest that this lack of success is because in many systems the difference between P release under oxic and anoxic conditions actually is quite small (see Figs. 12.4 and 12.5). It would be interesting to test the success of hypolimnetic aeration experiments on the basis of sulfate concentration of waters.

Significance of Different Perspective. When observations are made within single systems, the importance of variables that are invariant over short time scales will be underestimated. Conversely, the importance of variables that change dramatically over a seasonal cycle will be overestimated with this approach. In this chapter we demonstrated this difference with the example of P release from sediments. When analyzing within-systems variation, dissolved oxygen content at the sediment–water interface can be shown to be the critical control of P release from lake systems. When we compared among systems, sulfate content of waters appeared to be a factor that was critical in the control of P release from sediments. Not surprisingly, oxygen content at the sediment–water interface in lakes can change dramatically over a seasonal viewing period, whereas sulfate content of waters is relatively invariant within single systems over short time scales.

The difference in perspective gained from single- and cross-systems studies suggests that when the results from single-system studies dominate, opportunities may be missed to gain understanding about control points of ecosystem processes. We believe that cross-system studies are underused in an attempt to derive ecological insights. New insights into the important control points of ecological function can be gained, therefore, by comparing among systems. Such new insights are valuable in themselves and may also be critical for inspiring ecological experimentation.

REFERENCES

Andersen, J.M. (1974). Nitrogen and phosphorus budgets and the role of sediments in six shallow Danish Lakes. *Arch. Hydrobiol.* 74:528–550.
Andersen, J.M. (1975). Influence of pH on release of phosphorus from lake sediments. *Arch. Hydrobiol.* 76:411–419.

Andersen, J.M. (1982). Effect of nitrate concentration in lake water on phosphate release from the sediment. *Water Res.* 16:1119–1126.

Baccini, P. (1985). Phosphate interaction at sediment-water interfaces. In: *Chemical Processes in Lakes*, W. Stumm, ed. Wiley, New York, pp. 189–205.

Banoub, M.W. (1975). Experimental studies on material transactions between mud and water of the Gradensee (Bodensee). *Verh. Int. Verein. Limnol.* 14:1263–1271.

Barrett, P.H. (1953). Relationship between alkalinity and adsorption of added phosphorus in fertilized trout lakes. *Trans. Am. Fish. Soc.* 82:78–90.

Bostrom, B., M. Jansson and G. Forsberg. (1982). Phosphorus release from lake sediments. *Arch. Hydrobiol. Beih. Ergebn. Limnol.* 18:5–59.

Bostrom, B., J.M. Andersen, S. Fleischer, and M. Jansson. (1988). Exchange of phosphorus across the sediment–water interface. *Hydrobiologia* 170:229–244.

Broberg, O. (1987). Phosphorus, Nitrogen and Carbon in the Acidified and Limed Lake Barsjon, SW Sweden. Ph.D. Dissertation, University of Uppsala, Sweden.

Burgis, M.J. and P. Morris. (1987). *The Natural History of Lakes*. Cambridge University Press, Cambridge.

Burns, N.M. and C. Ross. (1971). Nutrient relationships in a stratified eutrophic lake. *Proc. 14th Conf. Great Lakes Res. Res.* 1971:749–760.

Callender, E. (1982). Benthic phosphorus regeneration in the Potomac River. *Estuary Hydrobiol.* 92:431–446.

Callender, E. and D.E. Hammond. (1982). Nutrient exchange across the sediment-water interface in the Potomic River. *Estuarine Coastal Shelf Sci.* 15:395–413.

Caraco, N. (1986). Phosphorus, Iron, and Carbon Cycling in a Stratified Coastal Pond. Ph.D. Dissertation, Boston University, Massachusetts.

Caraco, N.F. (1988). What is the mechanism behind the seasonal switch between N and P limitation in estuaries? *Can. J. Fish. Aquat. Sci.* 45:381–382.

Caraco, N.F., J.J. Cole, and G.E. Likens. (1989). Evidence for sulfate-controlled P release from sediments of aquatic systems. *Nature* (London) 341:316–318.

Caraco, N.F., J.J. Cole, and G.E. Likens. (1990). A comparison of phosphorus immobilization in sediments of freshwater and marine systems. *Biogeochemistry* 9:277–290.

Caraco, N., A. Tamse, O. Boutros, and I. Valiela. (1987). Nutrient limitation of phytoplankton growth in brackish coastal ponds. *Can. J. Fish. Aquat. Sci.* 44: 473–476.

Caraco, N.F., J.J. Cole, G.E. Likens, M.D. Mattson, and S. Nolan. (1988). A very imbalanced nutrient budget for Mirror Lake, New Hampshire, USA. *Verh. Int. Verein. Limnol.* 23:170–175.

Carlton, R.G. and R.G. Wetzel. (1988). Phosphorus flux from lake sediments: Effects of epipelic algal oxygen production. *Limnol. Oceanogr.* 33:562–570.

Carritt, D.E. and S. Goodgal. (1954). Sorption reactions and some ecological implications. *Deep-Sea Res.* 1:224–243.

Clasen, J. and H. Bernhardt. (1982). A bloom of the chrysophycea *Synura uvella* in the Wharbach Reservoir as indication for the release of phosphates from the sediments. *Arch. Hydrobiol. Beih. Ergebn. Limnol.* 18:61–68.

Cole, J.J., N.F. Caraco, and G.E. Likens. 1990. Atmospheric contribution to the phosphorus budget of Mirror Lake, NH *Limnol. Oceanogr.* 35:1230–1237.

Cooke, G.D., E.B. Welch, S.A. Peterson, and P.R. Newroth. (1986). *Lake and Reservoir Restoration*. Butterworth, Boston.

Curtis, P.J. (1989). Effects of hydrogen ion and sulphate on the phosphorus cycle of a precambrian shield lake. *Nature* (London) 337:156-158.

Davies, R.B., D.L. Thurlow, and F.E. Brewster. (1975). Effects of burowing tubificid worms on the exchange of phosphorus between lake sediments and overlying water. *Verh. Int. Verein. Limnol.* 19:382-394.

D'Elia, C.E., J.G. Sanders, and W.R. Boynton. (1986). Nutrient enrichment studies in a coastal plain estuary: phytoplankton growth in large-scale, continuous cultures. *Can. J. Fish. Aquat. Sci.* 43:397-406.

Einsele, W. (1936). Ueber die Beziehugen des Eisenkreislaufs zum Phophatkreislauf im eutrophen See. *Arch. Hydrobiol.* 29:664-686.

Enell, M., S. Fleischer, and M. Jansson. (1989). Phosphorus in sediments — conference Report. *Ambio* 18:137-138.

Fenchel, T. and T.H. Blackburn. (1979). *Bacteria and Mineral Cycling.* Academic Press, New York.

Fisher, T.R., P.R. Carlson, and R.T. Barber. (1982). Sediment nutrient regeneration in three North Carolina estuaries. *Estuarine Coastal Shelf Sci.* 14:101-116.

Fox, L.E., S.L. Sager, and S.C. Wofsy. (1986). The chemical control of soluble phosphorus in the Amazon Estuary. *Geochim. Cosmochim. Acta* 50:783-794.

Gachter, R. (1988). Effects of oxygenation on phosphorus retention by lake sediments. In: *Abstracts, 2nd International Workshop on Phosphorus in Sediments*, Fiskebackskil, Sweden.

Gachter, R. and D.M. Imboden. (1985). Lake restoration. In: *Chemical Processes in Lakes*, W. Stumm, ed. Wiley, New York, pp. 365-388.

Gachter, R. and A. Mares. (1985). Does settling seston release soluble reactive phosphorus in the hypolimnion of lakes. *Limnol. Oceanogr.* 30:364-371.

Gallep, G.W. (1979). Chironomid influence on phosphorus release in sediment-water microcosms. *Ecology* 60:547-556.

Hasler, A.C. and W.G. Einsele. (1948). Fertilization for increasng productivity of natural inland waters. *Trans. North Am. Wildlife Conf.* 13:527-555.

Hawke, D., P.D. Carpenter, and K.A. Hunter. (1989). Competitive adsorption of phosphate on geothite in marine electrolytes. *Environ. Sci. Technol.* 23:187-191.

Hayes, F.R. and J.E. Phillips. (1958). Lake water and sediments IV. Radiophosphorus equilibrium with mud, plants, and bacteria under oxidized and reduced conditions. *Limnol. Oceanogr.* 3:459-475.

Hecky, R.E. and P. Kilham. (1988). Nutrient limitation of phytoplanton in freshwater and marine environments: a review of recent evidence of the effects of enrichment. *Limnol. Oceanogr.* 33:796-822.

Henderson-Sellers, B. and H.R. Markland. (1987). *Decaying Lakes — The Origins and Control of Cultural Eutrophication.* Wiley, New York.

Hingston, F.G. (1981). A review of anion adsorption. In: *Inorganics at Solid-Liquid Interfaces*, M.A. Anderson and A.J. Robin, eds. Ann Arbor Science, Ann Arbor, Michigan.

Holdren, G.C. and D.E. Armstrong. (1980). Factors effecting phosphorus release from intact lake sediment cores. *Environ. Sci. Technol.* 14:79-87.

Howarth, R.W. and J.J. Cole. (1985). Molybdenum availability in natural waters. *Science* 229:653-655.

Hutchinson, G.E. (1947). *Treatise on Limnology I. Geography, Physics, and Chemistry.* Wiley, New York.

Jackson, T.A. and D.W. Schindler. (1975). The biogeochmistry of phosphorus in an experimental lake environment: evidence for the formation of humic-metaphosphate complexes. *Verh. Int. Verein. Limnol.* 19:211–221.

Jansson, M. (1982). Does high nitrate supply induce internal phosphorus loading in lakes. *Vatten* 38:360–362.

Jansson, M. (1987). Anaerobic dissolution of iron-phosphorus complexes in sediments due to the activity of nitrate-reducing bacteria. *Microbiol. Ecol.* 14: 87–89.

Kamp-Nielsen, L. (1974). Mud-water exchange of phosphate and other ions in undisturbed sediment cores and factors affecting the exchange rates. *Arch. Hydrobiol.* 73:218–237.

Kamp-Nielsen, L. (1975). Seasonal variation in sediment-water exchange of nutrient ions in Lake Esrom. *Verh. Int. Verein. Limnol.* 19:1057–1065.

Khalid, R.A., W.H. Patrick, Jr., and R.P. Gambrell. (1978). Effect of dissolved oxygen on heavy metals, phosphorus, and nitrogen in an estuarine sediment. *Estuarine Coastal Mar. Sci.* 6:21–35.

Krom, M.D. and R.A. Berner. (1980). Adsorption of phosphate in anoxic marine sediments. *Limnol. Oceanogr.* 25:797–806.

Larsen, D.P., J. VanSickle, K.W. Malueg, and P.D. Smith. (1979). The effect of wastewater phosphorus removal on Shagawa Lake, Minnesota: phosphorus supplies, lake phosphorus and chlorophyll a. *Water Res.* 13:1259–1272.

Lean, D.R.S., D.J. McQueen, and V.A. Story. (1986). Phosphate transport during hypolimnetic aeration. *Arch. Hydrobiol.* 108:269–280.

Li, W.C., D.E. Armstrong, J.D.H. Williams, R.F. Harris, and J.K. Syers. (1972). Rate and extent of inorganic phosphate exchange in lake sediments. *Soil Sci. Soc. Am. Proc.* 36:279–285.

Likens, G.E., J.S. Eaton, N.M. Johnson, and R.S. Pierce. (1985). Mirror Lake— Physical and chemical characteristics. E. Flux and balance of water and chemicals. In: *An Ecosystem Approach to Aquatic Ecology: Mirror Lake and its Environment*, G.E. Likens, ed. Springer-Verlag, New York, pp. 135–155.

Likens, G.E., F.H. Bormann, R.S. Pierce, J.S. Eaton, and N.M. Johnson. (1977). *Biogeochemistry of a Forested Ecosystem.* Springer-Verlag, New York.

Moeller, R. (1985). Paleolimnology—A2. Contemporary sedimentation. In: *An Ecosystem Approach to Aquatic Ecology: Mirror Lake and Its Environment*, G.E. Likens, ed. Springer-Verlag, New York, pp. 355–366.

Mortimer, C.H. (1941). The exchange of dissolved substances between mud and water in lakes (Parts I and II). *J. Ecol.* 29:280–329.

Mortimer, C.H. (1942). The exchange of dissolved substances between mud and water in lakes (Parts III and IV). *J. Ecol.* 30:147–201.

Mortimer, C.H. (1971). Chemical exchanges between sediments and water in the Great Lakes—speculations on probable regulating mechanisms. *Limnol. Oceanogr.* 16:387–404.

Neess, J.C. (1949). Development and status of pond fertilization in central Europe. *Trans. Am. Fish. Soc.* 76:335–358.

Nurnberg, G. (1984). The prediction of internal phosphorus load in lakes with anoxic hypolimnia. *Limnol. Oceanogr.* 29:111–124.

Olsen, S. (1964). Phosphate equilibrium between reduced sediments and water. *Verh. Int. Verein. Limnol.* 15:333–341.

Ostrofsky, M.L., D.A. Osborne, and T.J. Zebulske. (1989). Relationship between anaerobic sediment phosphorus release and sedimetary phosphorus species. *Can. J. Fish. Aquat. Sci.* 46:416–419.

Peng, T.H. and W.S. Broecker. (1987). C/P ratios in marine detritus. *Global Biogeochem. Cycles* 1:155–161.

Quigley, M.A. and J.A. Robbins. (1986). Phosphorus release processes in nearshore Lake Michigan. *Can. J. Fish. Aquat. Sci.* 43:1201–1207.

Reynoldson, T.B., Jr. and H.R. Hamilton. (1981). Spatial heterogeneity in whole lake sediments towards a loading estimate. *Hydrobiologia* 91:235–240.

Richards, F.A., J.D. Cline, W.W. Broenkow, and L.P. Atkinson. (1965). Some consequences of the decomposition of organic matter in Lake Nittinat, an anoxic fjord. *Limnol. Oceanogr.* 10 (Suppl.):R185–R201.

Richardson, C.J. (1985). Mechanisms controlling phoshorus retention capacity in freshwater wetlands. *Science* 228:1424–1430.

Ryther, J.H. and W.M. Dunstan. (1971). Nitrogen phophorus and eutrophication in coastal marine environments. *Science* 171:1008–1013.

Schindler, D.W. (1977). Evolution of phosphorus limitation in lakes. *Science* 195: 260–262.

Schindler, D.W. and E.J. Fee. (1974). Experimental lakes area: whole-lake experiments in eutrophication. *Can. J. Fish. Res. B. Can.* 31:937–953.

Schindler, D.W., R.H. Hesslein, and M.A. Turner. (1987). Exchange of nutrients between sediments and water after 15 years of experimental eutrophication. *Can. J. Fish. Aquat. Sci.* 44 (Suppl.):26–33.

Schindler, D.W., R.W. Newburg, K.G. Beaty, and P. Campbell. (1976). Natural water and chemical budgets for a small precambrian lake basin in central Canada. *J. Fish. Res. Bd. Can.* 33:2536–2543.

Sen Gupta, R. (1973). A Study on Nitrogen and Phosphorus and Their Interrelationships in the Baltic. Ph.D. Dissertation, University of Gotëborg, Sweden.

Shaw, J.F.H. (1989). Potential Release of Phosphorus from Shallow Sediments to Lakewater. Ph.D. Dissertation, University of Alberta, Canada.

Smith, S.V., W.J. Wiebe, J.T. Hollibaugh, S.J. Dollar, S.W. Hager, B.E. Cole, G.W. Tribble, and P.A. Wheeler. (1987). Stoichiometry of C, N, P, and Si fluxes in a temperate-climate embayment. *J. Mar. Res.* 45:427–460.

Stauffer, R.E. (1985). Nutrient internal cycling and the trophic regulation of Green Lake Wisconsin. *Limnol. Oceanogr.* 30:347–363.

Strayer, R.F. and J.M. Tiedje. (1978). In situ methane production in a small hyper-eutrophic hard-water lake: loss of methane from sediments by vertical diffusion and ebullition. *Limnol. Oceanogr.* 23:1201–1206.

Stumm, W. and J.J. Morgan. (1981). *Aquatic Chemistry, 2d Ed. An Introduction Emphasizing Chemical Equilibria in Natural Waters.* Wiley, New York.

Sugawara, K., T. Koyama, and E. Kamata. (1957). Recovery of precipitated phosphate from lake muds related to sulfate reduction. *Chem. Inst. Fac. Sci. Nagoya Univ.* 5:60–67.

Syers, J.K., R.F. Harris, and D.E. Armstrong. (1970). Phosphate chemistry in lake sediments. *J. Env. Qual.* 2:1–14.

Taggart, C.T. (1984). Hypolimnetic aeration ad zooplankton distribution: a possible limitation to the restoration of cold-water fish production. *Can. J. Fish. Aquat. Sci.* 41:191–198.

Taylor, L.R. (1989). Objective and experiment in long-term research. In: *Long-Term Studies in Ecology: Approaches and Alternatives*, G.E. Likens, ed. Springer-Verlag, New York, pp. 20–70.

Tessenow, V.U. (1972). Solution diffusion and adsorption in the upper layer of lake sediments. I: A long term experiment under aerobic and anaerobic conditions in a steady-state system. *Arch. Hydrobiol.* 38 (Suppl.):353–398.

Tezuka, Y. (1989). The C:N:P ratio of phytoplankton determines the relative amounts of dissolved inorganic nitrogen and phosphorus released during aerobic decomposition. *Hydrobiologia* 173:55–62.

Uehlinger, U. and J. Bloesch. (1987). Variation in the C:P ratio of suspended and settling seston and its significance for P uptake calculations. *Freshwater Biol.* 17: 99–108.

Valiela, I. (1984.) *Marine Ecological Processes*. Springer-Verlag, New York.

Vallentyne, J.R. (1972). Freshwater supplies and pollution: effects of demographic explosion on water and man. In: *The Environmental Future*, N. Polunin, ed. Macmillan, New York, pp. 181–211.

Vollenweider, R.A. (1968). Scientific fundamentals of eutrophication of lakes and flowing waters with particular reference to nitrogen and phosphorus as factors in eutrophication. OECD Tech. Rep. DAS CSI 68 27.

Wetzel, R.G. (1983). *Limnology*. W.B. Saunders, Philadelphia.

Winter, T.C. (1977). Ground-water component of lake water and nutrient budgets. *Verh. Int. Verein. Limnol.* 20:438–444.

13
Stoichiometry of C:N:P Fluxes in Shallow-Water Marine Ecosystems

Stephen V. Smith

Abstract

The difference between primary production and respiration defines the "net trophic status" of an ecosystem. Net metabolism (whether autotrophy or heterotrophy) is ordinarily a small fraction of gross metabolism, so estimates of net metabolism derived as the difference between measurements of primary production and respiration are subject to large error.

An alternative approach to estimating net metabolism, applicable in many large aquatic systems, is to use hydrographic and stoichiometric budgets. The carbon-to-nitrogen-to-phosphorus ratios of net reactions in the system are assumed to be primarily a function of organic metabolism and to be determined by a few simple chemical reactions involving material with the "local Redfield Ratio" (the average C : N : P ratio of organic matter in the system). A hydrographic budget establishes the nonconservative flux of phosphorus: uptake implies ecosystem net organic production, and release implies net consumption. The difference between the nonconservative carbon flux observed minus that estimated by net organic metabolism to account for phosphorus flux defines CO_2 gas exchange. The difference between observed and expected nitrogen flux defines net nitrogen fixation minus denitrification.

Calculations using this procedure have been performed on two examples each from three kinds of marine ecosystems: coral atoll lagoon reefs, coastal embayments isolated from terrigenous inputs, and estuaries. The systems examined, except estuaries, proved to be net autotrophic, but a general survey of the literature suggests that most coastal marine ecosystems are net heterotrophic. Nitrogen fixation minus denitrification dominates the nonconservative nitrogen flux of coastal systems, and the rate of these processes appears to be largely controlled by the rate of net metabolism.

Introduction

There are two underlying themes in this chapter. First, measurements of net ecosystem metabolism are useful for comparison among ecosystems, and such measurements should be made at the time and space scales of primary interest to the investigation at hand. Second, simple aspects of stoichiometry are useful in ecosystem comparisons and in evaluating ecosystem processes. Each of these themes is briefly considered, and a comparative analysis of P and N cycling through several shallow-water marine ecosystems is then used to suggest revisions in present concepts about N cycling and C metabolism.

NET ECOSYSTEM METABOLISM

Net ecosystem metabolism is defined to be primary production minus respiration of biota in the ecosystem. Primary production and respiration can be used to describe the "net trophic status" of a system: a system is said to be net autotrophic if it produces more organic matter than it respires and net heterotrophic if respiration exceeds primary production.

Net ecosystem metabolism is a measure of some combination of three characteristics of that system: changing biomass, exchange of organic matter with adjacent systems, and accumulation or depletion of organic material in the sedimentary record. To the extent that changing biomass and sediment accumulation or loss can be measured, net metabolism becomes a record of the organic exchange between ecosystems.

Such material coupling is, of course, of interest to a variety of ecological and biogeochemical studies. For example, what is the relationship between coastal marine fisheries, exchange of materials with offshore waters, and inputs from land? What is the role of such couplings in the global C cycle? And, as is discussed later in this chapter, how do the C, N, and P cycles interact with one another?

All ecosystems contain an array of biotic communities and abiotic components interacting with one another. Net system rates are the sum, over space and time, of component rates. For any system, measurements that implicitly encompass the smaller scale variations are more likely to be accurate assessments of the summed component rates than are laborious and incomplete arithmetic summations. It follows that measurements at the scale of interest are likely to contain attributes not discerned by arithmetic summation. In discussion of net ecosystem metabolism, I use approaches scaled to the ecosystem rather than to components within the ecosystem.

Table 13.1 lists estimates of annual average primary production, respiration, and net metabolism for several shallow-water marine systems. In general, such systems appear to be net heterotrophic, especially at higher rates of primary production. The data also show net metabolism to be generally small relative to gross metabolism. This result is not unexpected and is consistent with the concept that primary production in most systems is largely supported by recycled nutrients, rather by than externally supplied nutrients.

It follows that proportionally small errors in production and respiration

TABLE 13.1. Total-system annual mean primary production (p), respiration (r), and net system metabolism (p − r) for selected shallow-water sites.

Site	Reference	p	r	(p − r)	p/r
Bissel Pond salt marsh, RI	Nixon and Oviatt (1973)	960[a]	980	− 20	0.98
Georgia coast	Hopkinson (1985)	539	759	− 220	0.71
El Verde Lagoon, Mexico	Flores-Verdugo et al. (1988)	521	599	− 78	0.87
Estero Pargo, Mexico	Day et al. (1988)	345	405	− 60	0.85
Narragansett Bay, RI	Nixon and Pilson (1984)	310	230	80	1.35
Tomales Bay, CA	Smith et al. (1989, unpublished)	300	345	− 45	0.87
North Atlantic Bight	Rowe et al. (1986)	230	280	− 50	0.82
Terminos Lagoon, Mexico	Day et al. (1988)	219	219	0	1.00
South Kaneohe Bay, HI	Smith et al. (1981)	217	294	− 77	0.74
Newport River estuary, NC	Kenney et al. (1988)	211	242	− 31	0.87
MERL control micro-cosms[b]	Frithsen et al. (1985)	177	180	− 3	0.98
South San Francisco Bay, CA	Hammond et al. (1985)	118	110	8	1.07
Spencer Gulf, Australia	Smith and Veeh (1989)	92	84	9	1.10
Average		326	364	− 37	0.94
± SD		233	263	70	0.17
Coastal ocean	Smith and Mackenzie (1987)	267	279	− 12	0.96

[a]Rates in g C m^{-2} yr^{-1}.
[b]MERL, Marine Ecosystems Research Laboratory.

will translate into large proportional errors in the difference between these metabolic processes. Either inaccuracies or imprecisions in measuring individual components of production and respiration or simply missing entire metabolic components will obscure trends in net metabolism obtained by summation of components. If the desired property of the system is this net performance, then a measurement strategy that evaluates the net directly is preferable to a strategy of summation.

Following this theme one step further, consider the case in which production (p), respiration (r), and production minus respiration (p − r) are all three desired properties of the system—for whatever purpose. Only two of these properties need be explicitly known to describe the third. Because we are unlikely ever to know p and r individually with sufficiently high accuracy at the ecosystem scale to distinguish between these quantities (i.e., to prove that p − r ≠ 0), some strategy other than measuring p and r individually and obtaining (p − r) by difference is required to define all three quantities in an internally consistent fashion. An optimum strategy would be to estimate (p − r) directly as the net system property, then to measure either p or r (or the average of both) as an estimate of gross system performance, and finally to calculate either r or p by difference. The least desirable strategy for evaluating the quantity (p − r) is to estimate p and r independently, and then to derive (p − r) by difference.

For two of the systems listed in Table 13.1, Tomales Bay and Spencer Gulf, the quantity (p − r) was obtained directly. The other estimates are based on calculation of the quantity (p − r) by difference between estimates of p and r. I assume, by comparison with global model calculations (Smith and Mackenzie, 1987), that the tendency toward net heterotrophy in coastal systems is correct. For the reasons given previously, however, I do not have great confidence in most of the individual (p − r) estimates in this table.

Ecosystem Stoichiometry

"Ecosystem stoichiometry" is a concept based on the observation that major chemical composition either within or between ecosystems is less variable than taxonomic composition of biota in ecosystems. The mean chemical composition of individual plants and animals does not vary greatly, at least with respect to certain key constituents. Chemical reactions at the system scale appear similarly constrained. For example, photoautotrophic production by one major biochemical reaction pathway (in water) and oxidation along only a few biochemical pathways dominate the turnover of organic matter in aquatic ecosystems. These constraints on the variation of organic composition and reaction pathways have some important consequences.

A specific consequence recognized in aquatic environments was the observation that the composition of plankton in the oceans has a reproducible C : N : P composition (Redfield 1934); this ratio (106 : 16 : 1, by moles) has come to be known as the Redfield ratio. It was further noted that dissolved inorganic C, N, and P in the ocean are highly correlated with regression slopes approximating the Redfield ratio. Not only do N and P covary, but the regression intercept of dissolved inorganic N versus P concentrations in the open ocean is near 0.

Redfield (1958) used these observations to derive another important concept. He argued that the covariation of dissolved inorganic C, N, and P in proportion to the plankton C : N : P ratio reflected control of oceanic composition (for these elements, at least) by the organisms. Moreover, he pointed out that the near-zero N : P intercept suggests that these elements are exhausted almost simultaneously by biotic uptake. Dissolved inorganic C does not approach 0 as N and P are exhausted, so dissolved inorganic C is apparently not limiting to biomass production. The only significant source of P input to the ocean is stream flow, while there is a potentially large delivery of N across the air-sea interface. Redfield argued, before nitrogen fixation had been demonstrated to occur in the ocean, that biota of the ocean could fix as much atmospheric N as required to use available P. He concluded that P ultimately limits the production of biomass in the ocean.

This argument ran counter to observations about nutrient limitation of organisms in the sea. Ryther and Dunstan (1971) cited Redfield (1958) in concluding that nitrogen fixation might "be important in regulating the level or balance of nutrients in the ocean as a whole and over geological time . . . " But "It [nitrogen fixation] is certainly not effective locally or in the short run." Repeated experiments and observations have tended to demonstrate two major factors interpreted as support of the Ryther and Dunstan model. In the first place, inoculations of plankton cultures with N and P often demonstrate that the P inoculations have little effect; N inoculations usually cause biomass increase. Second, linear regression plots of oceanic dissolved inorganic N versus dissolved inorganic P usually show a small, but significant, negative N intercept; N apparently runs out, with some P left.

With this background information in mind, I now introduce a strategy for examining aspects of net metabolism in shallow-water marine environments. This strategy involves hydrographic assessment of net uptake or release of materials and then the calculation of uptake or release fluxes not directly amenable to hydrographic analysis by stoichiometric considerations. I then use the hydrographic-stoichiometric analyses to derive conclusions about interactions among ecosystem C, N, and P fluxes not readily derived from smaller scale experiments or observations. The hydrographic-stoichiometric strategy is then applied in a comparison among several shallow-water marine ecosystems. I believe that my conclusions have general application in the ocean, probably also carry over to other aquatic systems, and may have some bearing in terrestrial systems as well.

System-Level Flux Estimates

Fluxes of dissolved, biogeochemically reactive materials in aqueous systems can be related to net metabolic processes. Let us first consider how to estimate net fluxes of dissolved materials in aquatic ecosystems. I used a modification of one-dimensional mixing calculations, long used by coastal oceanographers to estimate the roles of river flow, mixing, and net uptake or release in controlling the distribution of reactive materials in estuaries. This model has provided robust flux estimates in a variety of systems. I recommend the paper by Officer (1979) as a particularly simple discussion of the use of such one-dimensional mixing diagrams in estuaries. Note, however, that I have worked on systems experiencing low net water flow, in which case the basic equations are modified to include rainfall and evaporation along with river flow (e.g., Smith et al. 1987, 1989).

The relevant equations need not be repeated here. However, the essence of the model for mixing water masses of two end-member compositions is contained in Fig. 13.1. Consider that we have mixtures of these two water types containing differing amounts of a nonreactive dissolved constituent (X) and a constituent of unknown reactivity (Y). If those mixtures of two end-member water types lie along a straight line, then Y, like X, is nonreactive. In this instance, Y is said to be "conservative" with respect to X. A convex-upward distribution of data indicates that there is a net source of Y

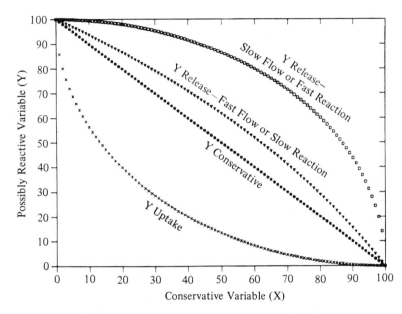

FIGURE 13.1. Schematic diagram shows effects of nonconservative uptake or release of Y on composition of two water masses during mixing. In all cases, variable X is known to be conservative. Note two cases of release of Y: slow flow (or fast reaction) and fast flow (or slow reaction).

between the end-member extremes of X, while a concave upward distribution of data indicates a net sink for Y. If Y is either taken up or released, then it is said to be "nonconservative" with respect to X.

If input and output rates for X can be budgeted, then that budget and the XY distribution data allow estimation of the internal net source or sink for Y. In the ocean, salinity is a good, precisely measurable term for X. Budgeting water input and output, and the salinity of that water, establishes salinity input and output. For the examples discussed here, dissolved inorganic P and dissolved inorganic N (DIP, DIN) have been measured as the Y constituents.

Careful site selection is a legitimate and indeed essential criterion in ecosystem analysis. Selection of sites for total-system analysis should be based on design criteria of the analytical procedures being applied and on "natural history common sense" that the site generally represents the habitats of interest. One can then move from concepts developed in simple systems to specific hypothesis testing in more complicated systems, rather than either extrapolating upward in spatial and temporal scales from concepts that describe incubation chambers to description of ecosystems or attempting initially to develop concepts in physically complex systems. As long as the essential components of the biotic communities of interest are present in the simple systems, this deliberate "bias" toward simple systems should be a powerful method of ecosystem analysis.

In the case of the one-dimensional mixing analysis being used here, the major site selection criteria can be briefly summarized:

1. A relatively simple (linear) geometry allows ready application of a simple hydrographic model; more complex systems not only require more complex (and probably expensive) hydrographic analysis to decipher water flow and nonconservative uptake but are more likely to embroil the results in model-dependent interpretations. If the system is sufficiently simple, even a very large system can be analyzed relatively rapidly and cheaply.

2. Systems with low water flow are generally preferable to systems with high water flow for analysis of non-conservative fluxes, unless flow rate is considered to be directly a controlling variable. As summarized in Table 13.2, coastal systems experience a wide dynamic range of water flow. By contrast, the dynamic range of net ecosystem fluxes is likely to be much smaller. All else being equal, high water flow will tend to mask chemical signals generated by nonconservative fluxes. In shallow-water

TABLE 13.2. Freshwater flow to and from selected shallow-water marine and estuarine systems[a].

System	Area (km^2)	Daily flow/area (mm d^{-1})	Comments
Sites discussed in this chapter			
Canton Atoll, Phoenix Island	40	−7	F ~ 0, E > R
Shark Bay (E. gulf), Australia	4,300	−4	F ~ 0, E > R
Spencer Gulf, Australia	21,700	−3	F ~ 0, E > R
Christmas Atoll, Line Is.	140	5	F ~ 0, E < R
Tomales Bay (inner), CA	18	20	F > > R in winter, E > F > R in summer
Ochlockonee Bay, FL	24	225	F only
Selected other sites			
South River estuary, NC	25	4	F only
Kaneohe Bay, HI	31	9	F > > R, E
Newport River estuary, NC	25	16	F only
Chesapeake Bay, MD, VA	11,300	17	F only
Delaware Bay, DE, NJ, PA	1,940	20	F only
Narragansett Bay, RI	328	28	F only
San Francisco Bay, CA	1,240	40	F > > R, E
Terminos Lagoon, Mexico	2,500	75	F only
Neuse River estuary, NC	400	375	F only
Columbia River estuary, WA, OR	380	1,250	F only
Fourleague Bay, LA	93	5,300	F only

[a]For ease of comparison, water flow (m^3 d^{-1}) has been divided by system area. F denotes stream + groundwater inflow; R is rainfall; E is evaporation. Data from various sources.

ecosystems with evaporation and rainfall being significant terms of the water budget relative to runoff, net nonconservative biogeochemical fluxes are likely to impart strong signals on water chemistry. For example, if a coastal ecosystem shows apparently conservative behavior for a clearly reactive constituent like DIP (no net uptake or release), it is likely that the conservative behavior is an artifact of rapid flow, not the lack of net P reaction in the system (see Fig. 13.1).

3. Multiple inputs of materials (e.g., sewage discharge or stream discharge into an embayment at multiple sites) are likely to complicate interpretation of the results unless these inputs are well described. Besides resulting from biogeochemical reactions within the system, curvature in a plot like Fig. 13.1 can result from mixtures of more than two water masses. Thus the complexity of inputs and the available knowledge about these inputs should be considered in the selection of a site for total-system analysis.

It is, of course, true that such complex inputs — especially if they are to be perturbed according to some predictable schedule — can provide opportunity for major large-scale, total-system experiments, which might otherwise be unattainable. Such opportunity has been used in Lake Washington (Edmondson 1970), Kaneohe Bay (Smith et al. 1981), the Thames River estuary (Mann 1972; Potter 1973), and a few other systems to assess various aspects of ecosystem responses to sewage diversion.

Thus, aquatic ecosystems that are simple, one-dimensional, low-flow systems with readily described water budgets and nutrient inputs are optimal sites for the analysis of nonconservative net fluxes of materials at the scale of ecosystems. Proximity of a study site to a marine laboratory is a poor guarantee that any of these criteria will be met.

Stoichiometric Analysis of C : N : P Reactions

We now move from estimates of net nonconservative fluxes for dissolved, reactive materials to estimates of net system metabolism. Consider the simple "Redfield stoichiometry" for production and consumption of organic matter with a C : N : P molar composition ratio of 106 : 16 : 1 (typical of plankton):

$$106CO_2 + 16HNO_3 + H_3PO_4 + 122H_2O = (CH_2O)_{106}(NH_3)_{16}(H_3PO_4)$$
$$+ 138(Ol_2) \qquad [1]$$

If the material being produced or consumed has a different C : N : P composition ratio, the equation can be rewritten. The C : N : P ratio on either side of this equation is called the "Redfield ratio"; I use the term "local Redfield ratio" to refer to local variations on this oceanic average. At least in the case of organic matter in the ocean, the N : P composition ratio appears to be less variable than the C : P and C : N composition ratios. The reason for this seems to be that the amount of structural C in organisms can vary a great deal, while other aspects of composition are less variable. For example, the median C : N : P ratio for benthic plants has been estimated to be 550 : 30 : 1 (Atkinson and Smith 1983). Thus, the characteristic C : P ratio varies by about fivefold between plankton and benthic plants, while the N : P ratio varies by only about twofold.

If the dominating composition of the system is "Redfield organic matter," and the dominating reactions are production and respiration according to reactions similar to Equation 1, then deviations of water composition from expectations drawn from that equation can yield valuable information about additional reaction pathways. In the deep ocean, C, N, P, and O may all be good tracers of net production minus consumption. Simply put, reactions proceeding solely according to Equation 1 should use or release C, N, and P (and O) according to the Redfield ratio. If the change in any one of these elements is known, expected changes in the other elements can be predicted. Thus the good C : N : P relationships observed by Redfield (1934, 1958) are derived. Carbon suffers the slight complication of reacting along a separate, important pathway ($CaCO_3$ precipitation and dissolution), but changes in total alkalinity can be used to parse out this effect (Smith 1985).

In shallow-water systems that are open to the atmosphere, there are immediately other complications. Both O and C can exchange across the air–water interface via physically mediated gas exchange, and fixed N in the water exchanges with the enormously larger dissolved gaseous N_2 reservoir via nitrogen fixation and denitrification. In general, P appears to be potentially the best pure tracer of net metabolism. Phosphorus can, itself, suffer as a tracer because of competing reactions—for example, coprecipitation with $CaCO_3$ or precipitation in one of a number of $CaHPO_4$ compounds, or redox-mediated adsorption-desorption reactions involving Fe. However, in many shallow-water systems the organic reactions appear dominating. The assumption is made in this chapter that abiotic P reactions do not occur, but the reader is encouraged to consult various references given (especially Atkinson, 1987; Smith et al., 1987) for discussions of this assumption.

The relationship between N and P flux is particularly important and will provide much of the focus for my development here, because of the still-uncertain roles of N versus P in controlling organic metabolism in marine systems. Deviations of N flux from the Redfield ratio are readily attributed to the net effect of nitrogen fixation and denitrification. Changes in the dissolved N_2 reservoir from these reactions are too small to be detected, yet these changes are of major quantitative importance to the fixed N content of water. In qualitative terms, this has been recognized by various authors (Richards 1965; Cline and Richards 1972; Nixon 1981). In general, this qualitative interpretation of deviations in the fixed N content of water from expected $N:P$ stoichiometry has not been converted to a quantitative measure of nitrogen fixation and denitrification (but see Smith 1984; Hammond et al. 1985; Smith et al. 1987, 1989).

The specific algebra for converting net fluxes of C, N, and P to net metabolism according to Equation 1 versus fluxes from the sum of other processes is presented here:

1. The flux of P, in particular DIP, is assumed to be controlled by advection and mixing (estimated according to the hydrographic model that has been presented) plus chemical reactions within the system. This assumption ignores, as insignificant, transfer of DIP across the air–water interface.

2. Some fraction of DIP flux is, in general, not conservative relative to salinity; the nonconservative flux of DIP is assumed to be in response to reactions of the form of Equation 1. The only sink for DIP is assumed to be production of organic matter, and the only source is oxidation of organic matter. The $C:N:P$ composition ratio of the organic matter being produced and consumed is assumed to be the same as the composition ratio of mean organic matter in the system. This ratio is taken to be the local Redfield ratio, if that value has been estimated; otherwise, the general Redfield ratio is used. This assumption considers that, over time, differential uptake or release of P relative to C and N in organic matter averages out to zero. The assumption also considers other (mineral precipitation or sorption) P sources and sinks to be small in comparison to biotic reactions. The organic matter being produced or consumed includes both particulate material and dissolved material.

3. Net metabolism $(p - r)$, expressed in terms of C, $(p - r)_C$, can be directly scaled from the Redfield $C:P$ ratio $(C/P)_R$ and the DIP flux (DIP):

$$(p - r)_C = (C/P)_R \times DIP \qquad [2]$$

In the treatment presented here, other pathways of C flux ($CaCO_3$ precipitation and dissolution, CO_2 gas flux across the air–water interface) will be ignored. Note that these processes do occur, locally, at important rates. Leaving them out of the analysis here does not detract from the the model presented.

4. Net flux of fixed N to or from organic matter $(p - r)_N$ can be similarly scaled:

$$(p - r)_N = (N/P)_R \times DIP \qquad [3]$$

5. Net nonconservative flux of DIN is known from the hydrographic model. Discrepancies between DIN and $(p - r)_N$ occur and can be attributed to the sum of flux pathways other than organic production. If DIN transfer across the air–water interface is ignored, as done for DIP, then the difference between nitrogen fixation and denitrification $(n - d)_N$ is the most plausible set of additional N flux pathways:

$$DIN = (p - r)_N + (n - d)_N \qquad [4]$$

Substituting Equation 3 into Equation 4 and rearranging:

$$(n - d)_N = DIN - (N/P)_R \times DIP \qquad [5]$$

A positive value for $(n - d)_N$ indicates net nitrogen fixation, while a negative value indicates net denitrification. A more complete analysis would also take into account the nonconservative fluxes of dissolved organic N and P. These are discussed only qualitatively in the present treatment. Equation 5 is the central focus for much of the remaining presentation in this paper.

There are, of course, biochemical assay methods for estimating nitrogen fixation and denitrification. These cannot be applied at the ecosystem level (other than by summation), but there is the potential for intercomparison at the scale of incubations. Surprisingly few data of this kind are available. However, one useful comparison is available for Narragansett Bay. That comparison is summarized in Table 13.3. Incubation measurements of benthic denitrification by Seitzinger et al. (1984) averaged 1.4 mmol m^{-2} d^{-1}. Hale (1975) measured benthic nutrient fluxes that can be used according to Equation 5 to estimate a net denitrification rate of 1.0 mmol m^{-2} d^{-1}, within 30% of the bioassays.

For lack of more site-specific data, the following general comparison can also be drawn. Figure 13.2 includes a plot of benthic flux measurements of DIN and DIP fluxes from sediments compiled from a variety of sources. Seitzinger (1988a) summarized 11 studies for which both DIN flux and N$_2$ flux (the major product of denitrification) — but not DIP flux — have been measured. I use here summed DIN + N$_2$ flux as a measure of total oxidation of organic matter in the sediments, and I assume that DIP flux from those sediments can be calculated by Redfield stoichiometry. Without elaborating, this calculation follows the logic presented in Equation 5, but proceeds from measured N fluxes to estimated DIP fluxes.

The resultant calculated points on Fig. 13.2 lie in the same cluster with

TABLE 13.3. Comparison of stoichiometrically calculated denitrification with assays of denitrification in cores from Narragansett Bay, RI

Denitrification assays (from Seitzinger et al. 1984)	1.4 mmol m^{-2} d^{-1}
Benthic fluxes (from Hale 1975)	Δ DIN 0.23 mmol m^{-2} d^{-1}
	Δ DIN 2.0 mmol m^{-2} d^{-1}
Local Redfield ratio (from Nixon and Pilson 1984)	$(N/P)_R$ 13
Stoichiometrically calculated nitrogen fixation minus denitrification rate (from test Equation 5)	$(n - d)_N = \Delta$ DIN $- (N/P)_R$ $\times \Delta$ DIP $= 2.0 - 13 \times 0.23$ $= -1.0$ mmol m^{-2} d^{-1} (negative indicates net denitrification)

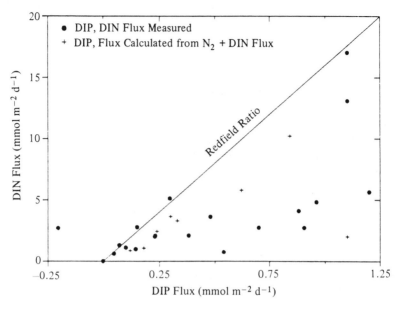

FIGURE 13.2. Comparison of dissolved inorganic phosphorus (DIP) versus dissolved inorganic nitrogen (DIN) flux in benthic incubation chambers (data from literature summary) (solid circles). DIN + N_2 flux data used to calculate expected DIP flux in benthic incubation chambers for which DIP data were unavailable, as discussed in text (data from summary by Seitzinger 1988a) (crosses).

the direct measurements of DIN and DIP flux. In general, sediment denitrification seems to consume somewhat more than half of the N released during sediment oxidation. This conclusion is derived either by measurement of DIN and N_2 fluxes simultaneously or by examination of DIN : DIP flux ratios. It can be seen that assays and stoichiometrically analyzed benthic flux measurements lead to very similar results.

A comparison can also offered between stoichiometrically derived estimates of nitrogen fixation and assay results. Pilson and Betzer (1973) reported that there was no net DIP flux across a reef flat at Enewetak Atoll, while Webb et al. (1975) reported a net DIN export equivalent to 4 mmol N m^{-2} d^{-1}. Based on the lack of measurable net DIP flux, I assume by Redfield stoichiometry that all the DIN flux results from nitrogen fixation. In a different study, also conducted on the Enewetak reef flat, Wiebe et al. (1975) measured nitrogen fixation by acetylene reduction; they found that a random selection of coral-algal rubble from the reef flat fixed nitrogen at a rate of 3 mmol m $^{-2}$ d $^{-1}$, while various *Calothrix* communities (the blue-green algae apparently largely responsible for nitrogen fixation on that reef) fixed N at a rate of 5 mmol m $^{-2}$ d $^{-1}$. Again, the results of the assays and stoichiometric analysis are similar.

I tentatively conclude from these analyses that the stoichiometric derivations of denitrification in sediments and nitrogen fixation on reef flats yield results that are comparable to biochemical assays, at least at the scale of individual incubations. I acknowledge that this data set is too limited for great confidence, but it provides some basis for assessing ecosystem-scale net nitrogen fixation minus denitrification from deviations of N : P flux ratios from Redfield stoichiometry.

Estimates of net nitrogen fixation minus denitrification according to this reasoning are as much "direct" estimates of net ecosystem function as are summations of local assays of nitrogen fixation and denitrification. If the individual assays show no evidence of nitrogen fixation and denitrification at rates exceeding the stoichiometrically calculated net system rates, then there is reason to suspect that the calculated rates might be in error (possibly because of other important terms, such as input of N via atmospheric fallout or alternative (e.g., mineral) P sources or sinks). There are, of course, potential problems (methodological discrepancies, container effects, assumptions that should be, but usually are not, tested on a case-by-case basis) with the assay techniques as well. As long as the stoichiometrically estimated rates lie within the variance of the assay measurements, I believe that the stoichiometrically derived rates for the entire system are more nearly quantitative estimates of net system performance than are the summed results from assays.

I now use a combination of the hydrographic flux model and the stoichiometric flux model to calculate the net nitrogen fixation minus denitrification for a series of shallow-water marine ecosystems. I demonstrate that the calculations derived are within the range to be anticipated for net nitrogen fixation minus denitrification, based on available incubation data. I will compare the results among the several ecosystems analyzed and then offer a general theory for metabolic control on N availability in aquatic ecosystems.

Stoichiometry of C : N : P Fluxes in Shallow-Water Marine Ecosystems

Data satisfactory for the analyses undertaken here are available for only a small selection of ecosystems. It should be emphasized that the shortage of data primarily reflects the small number of appropriate studies, rather than constraints on the number of sites to which the model can be applied. Numerous hydrographic studies using the general salt and water balance model discussed here are available. However, most of these studies are not useful to the present discussion. Some are one-time (or "few-time") studies in clearly time-varying systems. These data provide no sense of seasonal or shorter term variation. Some such studies do not provide sufficiently complete N and P budgets to be used in the stoichiometric analysis; only DIP or DIN data are presented, not both. Some studies do not report stream flow, so patterns of uptake or release cannot be converted to rates. There are examples of flux data from high-flow systems in which no net noncon-servative nutrient flux can be discerned. There are good nutrient data sets for systems that are too complex for analysis by the one-dimensional model employed here.

After going through the apparently relevant papers that I could find, I identified useful data from six sites, five of them studied by me and my immediate associates, for which I can identify or infer the variables neces-sary to estimate the stoichiometry of N and P fluxes. Two of these sites are enclosed coral atoll lagoon reef complexes in the equatorial Pacific Ocean; two are large hypersaline embayments in Australia, neither receiving stream input; and two are North American estuaries. Table 13.4 summarizes the general characteristics of these six sites.

To give some sense of the magnitude of errors that might be associated with each flux estimate, I have assumed that the C : P and C : N ratios used to scale from P flux to organic carbon metabolism and nitrogen fixation minus denitrification might be in error by as much as $\sim 50\%$. This error might include not only uncertainties in the C : N : P ratio of organic matter being produced or consumed but also alternative (e.g., mineral) sources or

TABLE 13.4. Comparative characteristics of sites used for total-system net metabo-lism analyses.[a]

Characteristic	Atoll lagoons		Hypersaline bays		Estuaries	
	Canton	Christmas	Shark	Spencer	Tomales	Ochlock
Area (km²)	40	140	4300	21700	18	24
Mean depth (m)	6	3	7	21	4	1
Stream flow/ (mm d⁻¹)	0	0	0	0	17	225
Groundwater (mm d⁻¹)	0	0	0	0	1	?
Rain (mm d⁻¹)	1	10	1	1	3	?
Evaporation (mm d⁻¹)	8	5	5	4	1	?
Ocean salinity (‰)	36	35	36	36	33.4	30
Average system salinity (‰)	38	32	45	37	33.1	14
Ocean DIP (μM)[b]	0.6	0.3	0.2	0.2	1.2	0.1
Average system DIP (μM)	0.2	0.1	0.1	0.1	2.0	0.4
Ocean DIN (μM)[b]	3.6	2.7	0.5	0.4	8.8	4.8[c]
Average system DIN (μM)	0.7	0.9	0.5	0.4	5.8	0.4[c]

[a]Data from sources cited in text.
[b]DIP, dissolved inorganic phosphorus; DIN, dissolved inorganic nitrogen.
[c]These data do not include NH_4^+. However, NH_4^+ flux does not contribute significantly to net DIN flux in any of the other systems analyzed.

sinks for P. Actual error analysis would, of course, be more complex than I have presented, including contributions of organic nutrients to the fluxes and errors in the water budgets. Stoichiometric models used in the original papers to calculate fluxes differed slightly from the calculations presented here. I constructed an internally consistent model framework for this chapter.

CORAL ATOLL LAGOON REEF COMPLEXES

Canton and Christmas Islands are atolls in the equatorial Pacific Ocean. Smith and Jokiel (1978) described the Canton study in detail, while Smith et al. (1984) presented similar detail about Christmas. Smith (1984) summarized much of what I report here for both sites, although the present analysis has been refined somewhat.

Unlike most atolls, which are relatively open about their perimeter, both of these are characterized by an enclosing island with a single pass, through which ocean water exchanges tidally with lagoon water. Canton Island is in an area of high evaporation. Normally, Christmas Island is a net evaporative system as well; at the time of the study described here, Christmas Island was experiencing high rainfall from the 1982 El Niño, so salinity was depressed. Both studies represent data collected over only a few days; however, I am assuming that the net metabolism of such equatorial systems is not strongly seasonal. The geometry of Canton Island is very amenable to the one-dimensional analysis advocated here. Although Christmas Island does not present such a simple one-dimensional delineation, nevertheless relatively clean nutrient versus salinity gradients were derived.

Table 13.5 summarizes the results of nonconservative DIP and DIN hydrographic flux analysis, and the calculated net nitrogen fixation minus denitrification. Metabolism is assumed to be dominated by benthic organisms, so a benthic plant C : N : P ratio of 550 : 30 : 1 (Atkinson and Smith 1983) is used in the calculations. Plants from reef sites listed in the Atkinson and Smith summary do not differ significantly in composition from this overall median ratio. In both of these reef systems, the ratio of DIN to DIP in water entering these systems is well less than the Redfield ratio, yet the systems deplete dissolved DIP to well below oceanic concentrations. The average nonconservative DIP flux was -0.035 mmol m^{-2} d^{-1}, and the average DIN flux was -0.25 mmol m^{-2} d^{-1}. These figures lead to a net nitrogen fixation minus denitrification rate averaging 0.8 mmol m^{-2} d^{-1} (Table 13.5).

By comparison, nitrogen fixation rates in a variety of reef environments average about 5 mmol m^{-2} d^{-1} (Capone 1983). Denitrification rates for reef systems are not well known, but are likely to be similar to (or perhaps lower than) denitrification rates in other shallow-water sediment systems, 1 to 2 mmol m^{-2} d^{-1} (Seitzinger 1988a; Seitzinger and D'Elia 1985). If this assumption is correct, then the nitrogen fixation rates for these reef systems

TABLE 13.5. Nonconservative fluxes[a] for systems examined.

Variable	Atoll lagoons		Hypersaline bays		Estuaries	
	Canton	Christmas	Shark	Spencer	Tomales	Ochlock
DIP[b]	−0.05	−0.02	−0.004	−0.007	+0.11	+0.13
DIN[b]	−0.40	−0.10	−0.002	−0.001	−0.19	+0.70[c]
C:N:P	550:30:1 (assumed)	550:30:1 (assumed)	550:30:1 (assumed)	300:23:1 (assumed)	99:18:1 (measured)	106:16:1 (assumed)
$(p - r)_C$	28	11	2	2	−14	−11
range	14–41	6–17	1–3	1–3	−5−−16	−7−−21
Average	20		2		−13	
		←———— net production ————→			←— net respiration —→	
$(n - d)_N$	1.1	0.5	0.1	0.2	−2.2	−1.4
Range	0.4–1.9	0.2–0.8	0.1–0.2	0.1–0.2	−1.2−−3.2	−0.3−−2.4
Average	0.8		0.2		−1.8	
		←———— net N$_2$ fixation ————→			←— net denitrification —→	

[a] Fluxes, mmol m^{-2} d^{-1}.

[b] DIP, dissolved inorganic phosphorus; DIN, dissolved inorganic nitrogen.

[c] These data do not include NH$_4^+$. However, NH$_4^+$ flux does not contribute significantly to net DIN flux in any of the other systems analyzed.

[d] The range is based on the assumption that there might be a 50% error in the C:P and C:N ratios of the organic matter that is being produced or consumed.

corrected for denitrification would be 2 to 3 mmol m^{-2} d^{-1}. The important point to note about this number is that it is somewhat less than the average nitrogen fixation rate derived from Capone (1983) for coral reefs. This is to be expected if that average represents some bias toward sampling of areas exhibiting rapid fixation rates.

In the case of Christmas Island, for which the N and P budgets are better defined than they are for Canton, there is an additional piece of information: net export of dissolved organic N is approximately 0.2 mmol m^{-2} d^{-1}, actually exceeding the net import of dissolved inorganic N. There are two important aspects to this observation. First, there must be some nonhydrographic source of N to balance the N budget, regardless of N : P stoichiometric considerations; nitrogen fixation is the most likely source. Second, this dissolved organic N export is considerably less than the estimated nitrogen fixation. This direct assessment of the N budget, independent of the P budget, provides evidence for net nitrogen fixation.

The P flux data can also be used by Redfield stoichiometry to estimate net organic carbon production: 20 mmol m^{-2} d^{-1}. This value is about half the net system production rate estimated from dissolved inorganic C budgets. Those budgets have uncertainty in the gas-exchange rate coefficient but do not depend on an assumed knowledge of C : P ratios of the major organisms in the system. In any event, this rate of net system production is about 4% of the gross production rate of typical coral reef flat communities (Kinsey 1985; Smith 1988). Of course such lagoonal reef systems consist of many components that do not produce organic matter as rapidly as the reef flats, including likely consumer communities such as lagoonal sediments.

The conclusions that emerge for these two reef systems is that they appear to be slight net producers of organic C, deriving most of their N requirements by nitrogen fixation rather than by hydrographic supply. Primary production on coral reefs tends to be extremely high, even in the face of low DIN and DIP concentrations. These systems apparently slowly accumulate nutrients into biomass over time. The instantaneous primary production of these systems is controlled by the accumulated biomass. Addition of DIN or DIP might well stimulate the biomass-specific growth rate of N- or P-limited components of the reef community. Both of these systems draw down the delivered DIP and DIN concentrations significantly. Continuing stimulation of N-limited primary production by DIN additions could, in principle, eventually deplete DIP until net production would cease. Primary production supported by recycling could, of course, continue. Stimulation of P-limited primary production by addition of DIP would be expected to shift the composition of the community toward nitrogen-fixing organisms.

ISOLATED HYPERSALINE COASTAL EMBAYMENTS

Shark Bay and Spencer Gulf are large embayments on the west and south coasts of Australia, respectively. Smith and Atkinson (1983, 1984) and Atkinson (1987) discussed relevant aspects of Shark Bay; a version of the stoichiometric model presented here was given by Smith (1984). Smith and Veeh (1989) presented salinity and nutrient budgets for Spencer Gulf.

Both of these bays are adjacent to desert regions and receive insignificant runoff or groundwater input. This lack of terrigenous input makes these systems unusual in comparison to most bays and estuaries that have been investigated; both systems are net evaporative. Salinity at the head of Shark Bay reaches 60 $\frac{0}{00}$, while salinity in Spencer Gulf reaches 45 $\frac{0}{00}$. Despite the elevated salinities, both embayments support normal shallow-water marine communities dominated by seagrass beds throughout much of their extent. In both systems, water residence time estimated from salt budgets is about one year, and therefore the data should adequately represent an annual cycle. Both systems are sufficiently elongate to make the one-dimensional hydrographic model entirely applicable.

Table 13.5 includes the DIP and DIN hydrographic flux analysis for each of these systems and the calculated net nitrogen fixation minus denitrification rates. In Shark Bay, the local Redfield C : N : P ratio is assumed to be 550 : 30 : 1, because the metabolism of that system appears to be dominated by seagrass. In Spencer Gulf, plankton appear to play a significant part in the net metabolism, because there is some P drawdown in water too deep to support significant seagrass biomass. I therefore assume a local Redfield C : N : P ratio of 300 : 23 : 1, as intermediate between plankton and seagrass metabolism. As with the two atoll lagoon reef systems, both of these hyper-saline embayments show evidence of net nitrogen fixation. The average rate of net nitrogen fixation minus denitrification is 0.2 mmol m^{-2} d^{-1}, substantially lower than the reef rates. The rates would be somewhat affected by the choice of the local N : P ratio (see Table 13.5), but the general patterns and conclusions would not be significantly affected. If denitrification rates average about 2 mmol m^{-2} d^{-1} in these systems (inferred from the summary of Seitzinger 1988a), then nitrogen fixation would slightly exceed this rate.

The seawater entering both of these gulfs is very low in nutrients. In both cases, oceanic water has approximately 0.2 μM DIP and 0.5 μM DIN. In both cases, the systems deplete DIP to analytical detection limits with insignificant change in DIN. The only recourse these systems would have to continued net production would be the production of P-free organic matter!

Atkinson (1987) has argued that the P taken up in Shark Bay is bound with $CaCO_3$, rather than with organic matter. This may be true, and it may affect the ultimate scaling of DIP uptake to net system organic production. Nevertheless, the distribution of DIP flux in this system seems to indicate that the initial uptake pathway is via organic production. The region where P uptake occurs coincides with the area of maximum seagrass abundance. Phosphorus binding with $CaCO_3$ may represent P sorption onto carbonate particles during oxidation of organic matter in the sediments. Such differential cycling during oxidation of organic matter is a complication in the system-scale analyses and bears further consideration.

In the case of Shark Bay, no data on dissolved organic nitrogen (DON) flux are available. For Spencer Gulf, DON export is estimated to be 0.05 mmol m^{-2} d^{-1}, which is large relative to the DIN import; regardless of system stoichiometry, there must therefore be a DON source other than conversion from DIN to DON within the system. The estimated DON flux is about 30% of the estimated nitrogen fixation.

Acknowledging the possibility that at least some of the net DIP flux is associated with differential C versus P release during oxidation and possible P binding with CaCO$_3$, I nevertheless used Redfield scaling to estimate net system production. Organic carbon production for both systems is estimated to be about 2 mmol m^{-2} d^{-1}. In the case of Spencer Gulf, Smith and Veeh (1989) have estimated primary production to be about 20 mmol m^{-2} d^{-1}, so net system production is estimated to be 10% of the primary production.

Seagrass in both systems is apparently highly productive (West and Larkum 1979; Walker and McComb 1988). I therefore derive a conclusion for these embayments similar to that advanced for the reef systems. Primary production is controlled by the ability to accumulate biomass. N or P additions would stimulate growth of N- or P-limited organisms. The N addition might result in a more rapid drawdown of P to oxygen, but that depletion represents an absolute limit on organic production for the system as a whole.

ESTUARIES

Two North American estuaries, Ochlockonee Bay, Florida, and Tomales Bay, California, provide the final two systems for which P and N budgets are presented. Ochlockonee Bay was described by Kaul and Froelich (1984) and Seitzinger (1988b and personal communication), and Tomales Bay by me and my immediate colleagues (especially Smith et al. 1987, 1989; and unpublished data describing a full annual cycle of metabolism). The two bays show quite different flow regimes (see Table 13.4). Ochlockonee Bay receives high stream discharge, while discharge in Tomales is low. In fact, during the summer, stream discharge into Tomales Bay is less than evaporation, and the system becomes slightly hypersaline. Primary production in both systems seems to be largely plankton based, and both systems also have active benthic metabolism (primarily heterotrophic metabolism).

Table 13.5 also summarizes DIP and DIN hydrographic flux analyses for the two systems. In the case of the Tomales Bay N flux, the budget includes both nitrate and ammonium. Ammonium data are not available for Ochlockonee Bay, but I assume this flux to be small. Unlike the previous four systems examined, both estuaries show evidence of net denitrification. The average estimated denitrification rate for the two estuaries is 1.8 mmol $m^{-2} d^{-1}$, close to the average rate reported by Seitzinger (1988a). Seitzinger (1988b) reported a benthic denitrification rate of 1.8 mmol m^{-2} d^{-1} for Ochlockonee Bay, based on assays; the stoichiometrically derived rate for the bay is 1.4 mmol $m^{-2} d^{-1}$. Assays have not been used to estimate the baywide denitrification rate in Tomales Bay, but have been used to ascertain that denitrification does occur in the sediments of that system (Smith et al. 1987).

In the case of Tomales Bay, for which DON flux data are also available, DIN import by this system approximately equals DON export. Nevertheless, denitrification is inferred to occur, so that the export of DIP is balanced. The local $C:N:P$ ratio for Tomales Bay is estimated from suspended load data to be $99:18:1$, not significantly different from the Redfield ratio. I assume that the Redfield ratio ($106:16:1$) adequately describes Ochlockonee Bay.

The observed nonconservative export of DIP indicates net respiration in excess of primary production; the average rate is about 13 mmol C m^{-2} d^{-1}. In the case of Tomales Bay, the primary production rate has been estimated to be approximately 70 mmol $m^{-2} d^{-1}$, so respiration apparently exceeds primary production in that system by approximately 15%.

In the absence of denitrification, a net heterotrophic system should release more DIN than it takes up. Both of these estuaries are interpreted to be net heterotrophic, yet both deplete DIN to near zero. Plankton primary production is demonstrably stimulated by DIN additions in Tomales Bay (Smith et al. 1987); I assume that this would also be true in Ochlockonee Bay. In fact, plankton primary production in bays and estuaries is usually demonstrated in incubations to be N limited (e.g., Howarth 1989), yet estuarine systems tend to be net heterotrophic (see Table 13.1). Various authors (e.g., Seitzinger 1988a) have argued that denitrification may play a major role in accounting for this N-limited primary production.

Smith and Hollibaugh (1989) have presented a somewhat different interpretation of the same observations. We have argued that denitrification is an efficient sink for excess N, responding to DIN buildups by accelerating the rate of denitrification. A small amount of carbon oxidation by denitrification is proportionally very important to the C budget. This model is based on the assumption that nitrate availability at the site of denitrification is a biochemical limit on denitrification. Data from incubation assays support this assumption (Oremland et al. 1984), as does the observation that most nitrate-supporting sediment denitrification appears to be "locally produced" by nitrification of ammonium released during respiration, rather than being nitrate available in the ambient environment (Seitzinger 1988a).

Because of closely juxtaposed aerobic and anaerobic microenvironments, estuarine sediments are probably particularly efficient sites for respiration, nitrification, and denitrification to be coupled. Increased sediment respiration will lead to increased release of ammonium, which can become nitrified and available for denitrification. In a net heterotrophic ecosystem with slow water exchange, the response of denitrification to accelerated rates of DIN release can be manifested by a significant drawdown, both of ambient DIN concentrations and of DIN released during respiration in the sediments.

Figure 13.2 suggests that denitrification in sediments is about 50% efficient in consuming DIN released during organic decomposition. Only about 25% of total respiration in shallow-water systems occurs in the sediments (Nixon 1981), so the process of denitrification in estuaries probably consumes an average of only about 12.5% of the DIN released during respiration. In systems with rapid water exchange, N may cycle only a few times between organic and inorganic forms. In such systems, biotic processes may have insufficient time to alter water composition. The apparent general tendency of estuaries towards net heterotrophy (see Table 13.1), generally high N:P loading ratios from land (Meybeck 1982), and the aforementioned N limitation of primary production suggest that water residence time in estuaries is frequently long enough to allow control of the N cycle by net C metabolism.

Comparison Among Systems

Calculations presented here have included two examples each from three distinctive marine ecosystem types: coral atoll lagoon reef systems, coastal embayments isolated from terrigenous inputs, and estuaries. In a search for communalities among these systems, I have plotted net nonconservative DIP flux (a measure of net organic matter metabolism) versus apparent net nitrogen fixation minus denitrification (Fig. 13.3).

The high correlation evident from these data as they are presented is largely a computational artifact: nonconservative DIP flux (Δ DIP) is entirely a phenomenon reflected in water chemistry, whereas there is a relatively small change in water-column DIN associated with Δ DIP in these shallow-water systems (see Table 13.5). The importance of this graph is that, a priori, this pattern need not have occurred. In the water column of the open ocean, where neither nitrogen fixation nor denitrification is ordinarily a rapid process, DIN and DIP are strongly correlated, with a positive slope that approximates the Redfield ratio. A near-zero slope must occur for Δ DIP versus $(n - d)_N$ in the open ocean.

Perhaps because of bottom-mediated effects associated with sharp redox gradients at the sediment-water interface, the pattern seen in Fig. 13.3 develops for shallow-water systems. In the slowly exchanging systems examined here, the biogeochemical processes of nitrogen fixation and denitrification are apparently sufficiently rapid to remove all but minor variations

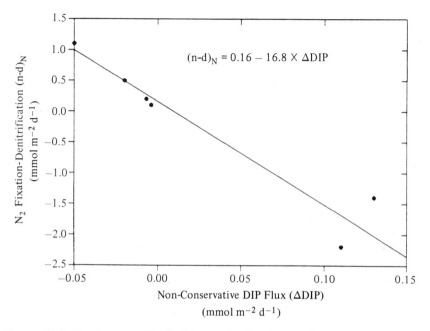

$$(n-d)_N = 0.16 - 16.8 \times \Delta DIP$$

FIGURE 13.3. Total-system dissolved inorganic phosphorus (DIP) versus calculated $(n - d)_N$ from Table 13.5.

in water-column DIN concentrations. The details of Fig. 13.3 would vary slightly to accommodate varying $N : P$ ratios used to estimate $(n - d)_N$ for each system (see Table 13.5), but the general pattern would remain.

The contrasting biogeochemical cycles of P and N within a particular ecosystem are shown schematically in Fig. 13.4. Phosphorus and N within the system are represented, simplistically, by dissolved inorganic and dissolved plus particulate organic reservoirs. Dashed lines are used to illustrate hydrographically controlled external sources and sinks; for both N and P, these fluxes are largely controlled by physical considerations rather than by the characteristics of biota in the system. This is, of course, not entirely true; for example, benthic organisms do not wash out from a system as readily as do plankton. Atmospheric fluxes do not exist (or at least are trivial) for P, so external controls on P supply and export are entirely controlled hydrographically. By contrast, biogeochemical processes can exert substantial control on atmospheric fluxes for N.

Without having modeled the linkages between the C and N fluxes numerically, my concept of C control on the N cycle is as follows. If DIN builds up (through excess input or net respiration), denitrification will increase and tend to decrease the DIN. If the level of DIN is insufficient to support organic production minus washout of organic matter and respiration, then nitrogen fixation will take on increased importance. Nitrogen fixation and

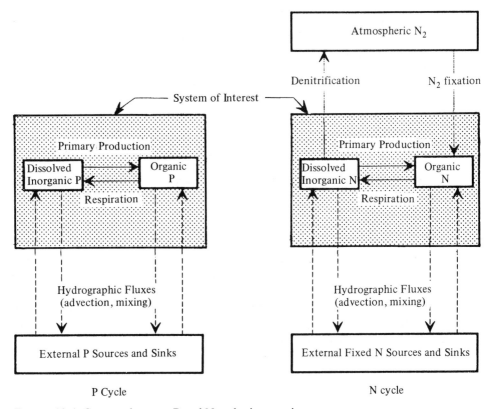

FIGURE 13.4. Contrast between P and N cycles in aquatic ecosystems.

denitrification can be locally high, biochemically controlled, and patchy within ecosystems. For example, the site of denitrification will tend to be the sediments, as organic matter accumulates there, is consumed, and sets up a redox gradient between oxic and anoxic conditions. Nitrogen fixation, initially and locally favoring nitrogen-fixing and P-limited organisms, will contribute fixed N to the cycle of the entire system as organic matter produced in conjunction with the nitrogen fixation decomposes. I believe that incubation assays and even more sophisticated microcosm experiments tend to decouple these locally biochemically controlled processes from one another and hence do not describe the summed function of the systems.

The summed rates of nitrogen fixation minus denitrification for entire systems are likely to represent a dynamic control between input, output, and net system metabolism. Systems that are net heterotrophic (because of excess input of labile organic matter over inorganic nutrients that can be taken up) or systems heavily loaded with DIN relative to DIP should show denitrification in excess of nitrogen fixation. Net autotrophic systems receiving low DIN : DIP loading ratios (compared to the local Redfield ratio) are likely to be net nitrogen-fixing systems. We have termed this concept of feedbacks in the N cycle responding to net system carbon metabolism "carbon-controlled nitrogen cycling" (Smith et al. 1989; Smith and Hollibaugh 1989).

Seitzinger (1988a) has suggested that denitrification dominates over nitrogen fixation in most coastal marine ecosystems. Both global geochemical arguments (Smith and Mackenzie 1987) and the summary data in Table 13.1 suggest that the coastal marine environment is net heterotrophic. This region of the ocean should lose more N to the atmosphere than it fixes from the atmosphere. The two estuarine systems examined here, as well as those systems summarized by Seitzinger (1988a), are consistent with this pattern. The two lagoonal reef systems are central ocean sites, removed from terrestrial inputs and exposed to surface seawater with a low ratio of DIN to DIP (relative to the Redfield ratio). These systems should fix more N than they lose to the atmosphere; they appear to do so, as expected. Finally, the two hypersaline embayments appear to be exceptions to the generality of net heterotrophy and net denitrification in nearshore systems. This apparently exceptional behavior is actually to be expected, because these systems receive no inputs from land, and the seawater has a low DIN to DIP ratio. These systems should, and do, function metabolically more like coral reefs than like estuaries.

Conclusions

1. The stoichiometric approach to total ecosystem analysis is a demonstrably useful way to assess net community metabolism in aquatic ecosystems. The discussion presented here has been largely limited to N and P fluxes, with some attention to C flux. Similar reasoning can be applied to other elements. In general, for Redfield-like reactions that produce and consume organic matter of known C : N : P ratios, I believe that P flux most nearly represents net production minus consumption of that organic matter. Deviations of N and C fluxes from simple Redfield ratios relative to P fluxes are attributed to other important flux pathways in the ecosystem.
2. Limited biochemical assay data for denitrification and nitrogen fixation compared with stoichiometric flux estimates at similar scales yield comparable results. It therefore seems feasible and fruitful to apply the stoichiometric flux model at scales and degrees of system heterogeneity not directly amenable to biochemical assays. There is a cautionary note, however. As confidence is gained in using such first-order stoichiometric calculations for representations of net system metabolism, emphasis should also be placed on understanding second-order complications to this model. In particular, P flux to and from mineral phases is an impor-

tant consideration that merits further attention. Perhaps alternative stoichiometries (e.g., utilizing micronutrients) will eventually prove more robust than the P-based stoichiometry discussed here.

3. Nitrogen flux in large aquatic ecosystems with slow water exchange cannot be balanced by hydrographic considerations alone. In systems for which DON data are available, this conclusion is demonstrable even without stoichiometric assessment. Small instantaneous imbalances are accumulated through time until they become readily measured chemical signals in these large ecosystems with slow water exchange. Mechanisms accounting for the imbalances are likely to occur, but be less readily detected, in more rapidly exchanging systems, including subsets (e.g., individual communities) within the large systems examined. Flux of N across the air–water interface, probably by nitrogen fixation and denitrification at entirely plausible rates, can be invoked to balance the N budgets of these systems.

4. DIN concentration in slowly exchanging systems tends to remain low, despite both the biotic potential for nitrogen fixation and denitrification and the importance of these processes as demonstrated by budgetary analysis. A ready interpretation of this observation is that, at the ecosystem scale, N is supplied or lost across the air–water interface in response to carbon metabolism. Experimentally demonstrable limitation of primary production by N availability may be less relevant to controls on net ecosystem metabolism than is the inferred global limitation of biomass production in the ocean by P availability. At the scale of ecosystems, nitrogen fixation and denitrification are interpreted to be negative feedbacks in the nitrogen cycle responding to carbon metabolism.

Acknowledgments. I particularly thank the Institute for Ecosystem Studies for inviting me to participate in the third Cary Conference. My thanks to Marlin Atkinson, Steven Dollar, Gordon Tribble, and two anonymous reviewers for critical comments on drafts of this manuscript. My studies of ecosystem metabolism, particularly the relationship between C, N, and P fluxes in ecosystems, have been primarily funded by the National Science Foundation. This paper is Contribution Number 809 of the Hawaii Institute of Marine Biology and Contribution Number 2248 of the Hawaii Institute of Geophysics.

References

Atkinson, M.J. (1987). Low phosphorus sediments in a hypersaline marine bay. *Estuarine Coastal Shelf Sci.* 24:335–347.

Atkinson, M.J. and S.V. Smith. (1983). C : N : P ratios of benthic marine plants. *Limnol. Oceanogr.* 26:1074–1083.

Capone, D.G. (1983). Benthic nitrogen fixation. In: *Nitrogen in the Marine Envi-*

ronment, E.J. Carpenter and D.G. Capone, eds. Academic Press, New York, New York, pp. 105–137.

Cline, J.D. and F.A. Richards. (1972). Oxygen-deficient conditions and nitrate reduction in the Eastern Tropical North Pacific Ocean. *Limnology* 17:885–900.

Day, J.W., Jr., C.J. Madden, F. Ley-Lou, R.L. Wetzel, and A.M. Navarro. (1988). Aquatic primary productivity in Terminos Lagoon. In: *Ecology of Coastal Ecosystems in the Southern Gulf of Mexico: The Terminos Lagoon Region*, A. Yáñez-Arancibia and J.W. Day, Jr., eds. Universidad Nacional Autónoma de México, México City, pp. 221–236.

Edmondson, W.T. (1970). Phosphorus, nitrogen and algae in Lake Washington after diversion of sewage. *Science* 196:690–691.

Flores-Verdugo, F.J., J.W. Day, L. Mee, and R. Briseño-Dueñas. (1988). Phytoplankton production and seasonal biomass variation of seagrass, *Ruppia maritima* L., in a tropical Mexican lagoon with an ephemeral inlet. *Estuaries* 11:51–56.

Frithsen, J.B., A.A. Keller, and M.E.Q. Pilson. (1985). Effects of inorganic nutrient additions in coastal areas: a mesocosm experiment. Data Report Vol. 1, Marine Ecosystems Research Laboratory Series Report No. 3, University of Rhode Island, Kingston.

Hale, S.S. (1975). The role of benthic communities in the nitrogen and phosphorus cycles of an estuary. In: *Mineral Cycling in Southeastern Ecosystems*, F.G. Howell, J.B. Gentry, and M.H. Smith, eds., Energy Research and Development Agency Series, CONF-740513, NTIS, Springfield, Virginia, pp. 291–308.

Hammond D.E., C. Fuller, D. Harmon, B. Hartman, M. Korosec, L.G. Miller, R. Rea, S. Warren, W. Berelson, and S.W. Hager. (1985). Benthic fluxes in San Francisco Bay. *Hydrobiologia* 129:69–90.

Hopkinson, C.S. (1985). Shallow-water benthic and pelagic metabolism: evidence for heterotrophy in the nearshore Georgia Bight. *Mar. Biol.* 87:19–32.

Howarth, R.W. (1989). Nutrient limitation of net primary production in marine ecosystems. *Annu. Rev. Ecol. Syst.* 19:89–110.

Kaul, L.W. and P.N. Froelich, Jr. (1984). Modeling estuarine nutrient geochemistry in a simple system. *Geochim. Cosmochim. Acta* 48:1417–1433.

Kenney, B.E., W. Litaker, C.S. Duke, and J. Ramus. (1988). Community oxygen metabolism in a shallow tidal estuary. *Estuarine Coastal Shelf Sci.* 27:33–43.

Kinsey, D.W. (1985). Metabolism, calcification and carbon production. I. System-level studies. In: *Proceedings of the Fifth International Coral Reef Symposium, Tahiti*, Vol. 4, C. Gabrie and B. Salvat, eds., Antenne Museum-Ephe, Moorea, French Polynesia, pp. 505–526.

Mann, K.H. (1972). Case history: River Thames. In: *River Ecology and Man*, R.T. Oglesby, C.A. Carson, and J.A. McCann, eds. Academic Press, New York, New York, pp. 215–232.

Meybeck, M. (1982). Carbon, nitrogen, and phosphorus transport by world rivers. *Am. J. Sci.* 282:401–450.

Nixon, S.W. (1981). Remineralization and nutrient cycling in coastal marine ecosystems. In: *Estuaries and Nutrients*, B.J. Neilson and L.E. Cronin, eds. Humana Clifton, New Jersey, pp. 111–138.

Nixon, S.W. and C.A. Oviatt. (1973). Ecology of a New England salt marsh. *Ecol. Monogr.* 43:463–498.

Nixon, S.W. and M.E.Q. Pilson. (1984). Estuarine total system metabolism and organic exchange calculated from nutrient ratios: an example from Narragansett Bay. In: *The Estuary as a Filter*, V.S. Kennedy, ed. Academic Press, New York, New York, pp. 261–290.

Officer, C.B. (1979). Discussion of the behavior of nonconservative dissolved constituents in estuaries. *Estuarine Coastal Shelf Sci.* 9:91–94.

Oremland, R.S., C. Umberger, C.H. Culbertson, and R.L. Smith. (1984). Denitrification in San Francisco Bay intertidal sediments. *Appl. Environ. Microbiol.* 47: 1106–1112.

Pilson, M.E.Q. and S.B. Betzer. (1973). Phosphorus flux across a coral reef. *Ecology* 54:581–588.

Potter, J.H. (1973). Management of the tidal Thames. In: *Mathematical and Hydraulic Modelling of Estuarine Pollution*, A.L.H. Gameson, ed. Her Majesty's Stationery Office, London.

Redfield, A.C. (1934). On the proportions of organic derivatives in sea water and their relation to the composition of plankton. James Johnstone Memorial Volume, Liverpool, pp. 177–192.

Redfield, A.C. (1958). The biological control of chemical factors in the environment. *Am. Sci.* 46:205–222.

Richards, F.A. (1965). Anoxic basins and fjords. In: *Chemical Oceanography*, Vol. 1, 1st Ed., J.P. Riley and G. Skirrow, eds. Academic Press, New York, New York, pp. 611–645.

Rowe G.T., S. Smith, P. Falkowski, T. Whitledge, R. Theroux, W. Phoel, and H. Ducklow. (1986). Do continental shelves export organic matter? *Nature* (London) 324:559–561.

Ryther, J.H. and W.M. Dunstan. (1971). Nitrogen, phosphorus, and eutrophication in the coastal marine environment. *Science* 171:1008–1013.

Seitzinger, S.P. (1988a). Denitrification in freshwater and coastal marine ecosystems: ecological and geochemical significance. *Limnol. Oceanogr.* 33:702–724.

Seitzinger, S.P. (1988b). Nitrogen biogeochemistry in an unpolluted estuary: the importance of benthic denitrification. *Mar. Ecol. Prog. Ser.* 41:177–186.

Seitzinger, S.P. and C.F. D'Elia. (1985). Preliminary Studies of Denitrification on a Coral Reef. Symposia Series for Undersea Research, NOAA Undersea Research Program 3H: 199–208.

Seitzinger, S.P., S.W. Nixon, and M.E.Q. Pilson. (1984). Denitrification and nitrous oxide production in a coastal marine environment. *Limnol. Oceanogr.* 29: 73–83.

Smith, S.V. (1984). Phosphorus versus nitrogen limitation in the marine environment. *Limnol. Oceanogr.* 29:1149–1160.

Smith, S.V. (1985). Physical, chemical and biological characteristics of CO_2 gas flux across the air-water interface. *Plant Cell Environ.* 8:387–398.

Smith, S.V. (1988). Mass balance in coral reef-dominated areas. In: *Coastal-Offshore Ecosystem Interactions*, B.-O. Jansson, ed. Springer-Verlag, Berlin, pp. 209–226.

Smith, S.V. and M.J. Atkinson. (1983). Mass balance of carbon and phosphorus in Shark Bay, Western Australia. *Limnol. Oceanogr.* 28:625–639.

Smith, S.V. and M.J. Atkinson. (1984). Phosphorus limitation of net production in a confined aquatic ecosystem. *Nature* (London) 307:626–627.

Smith, S.V. and J.T. Hollibaugh. (1989). Carbon controlled nitrogen cycling in a marine 'macrocosm': an ecosystem-scale model for managing cultural eutrophication. *Mar. Ecol. Prog. Ser.* 52:103–109.

Smith, S.V. and P. Jokiel. (1978). Water composition and biogeochemical gradients in the Canton Atoll lagoon. *Atoll Res. Bull.* 221:15–53.

Smith, S.V. and F.T. Mackenzie. (1987). The ocean as a net heterotrophic system: implications from the carbon biogeochemical cycle. *Global Biogeochem. Cycles* 1:187–198.

Smith, S.V. and H.H. Veeh. (1989). Mass balance of biogeochemically active materials (C, N, P) in a hypersaline gulf. *Estuarine Coastal Shelf Sci.* 29:195–215.

Smith, S.V., J.T. Hollibaugh, S.J. Dollar, and S. Vink. (1989). Tomales Bay California: a case for carbon-controlled nitrogen cycling. *Limnol. Oceanogr.* 34:37–52.

Smith, S.V., W.J. Kimmerer, E.A. Laws, R.E. Brock, and T.W. Walsh. (1981). Kaneohe Bay sewage diversion experiment: perspectives on ecosystem responses to nutritional perturbation. *Pacific Sci.* 35:279–395.

Smith, S.V., S. Chandra, L. Kwitko, R.C. Schneider, J. Schoonmaker, J. Seeto, T. Tebano, and G.W. Tribble. (1984). Chemical stoichiometry of lagoonal metabolism: preliminary report on an environmental chemistry survey of Christmas Island, Kiribati. University of Hawaii SeaGrant Report No. UNIHI-SEA-GRANT-CR-84-02, University of Hawaii, Honolulu.

Smith, S.V., W.J. Wiebe, J.T. Hollibaugh, S.J. Dollar, S.W. Hager, B.E. Cole, G.W. Tribble, and P.A. Wheeler. (1987). Stoichiometry of C, N, P, and Si fluxes in a temperate-climate embayment. *J. Mar. Res.* 45:427–460.

Walker, D.I. and A.J. McComb. (1988). Seasonal variation in the production, biomass and nutrient status of *Amphibolis antarctica* (Labill.) Sonder ex Aschers. and *Posidonia australis* Hook. f. in Shark Bay, western Australia. *Aquat. Bot.* 31:259–275.

Webb, K.L., W.D. DuPaul, W. Wiebe, W. Sottille, and R.E. Johannes. (1975). Enewetak (Eniwetok) Atoll: aspects of the nitrogen cycle on a coral reef. *Limnol. Oceanogr.* 20:198–210.

West, R.J. and A.W.D. Larkum. (1979). Leaf productivity of the seagrass *Posidonia australis* in eastern Australian waters. *Aquat. Bot.* 7:57–65.

Wiebe, W.J., R.E. Johannes, and K.L. Webb. (1975). Nitrogen fixation in coral reef community. *Science* 188:257–259.

14
Gradient Analysis of Ecosystems

PETER M. VITOUSEK AND PAMELA A. MATSON

Abstract

Comparative studies have been useful for understanding the biogeochemistry of terrestrial ecosystems, particularly where comparisons can be carried out along gradients of an underlying controlling factor. Examples of the application of gradient approaches include the analysis of state factors developed by Hans Jenny, analyses of the regulation of streamwater chemistry along elevational and successional gradients, examination of variation in natural abundance of stable isotopes as they integrate ecosystem processes during soil development, and the development of regional and global budgets for biogenic trace gases. Gradient studies of terrestrial ecosystems are strengthened by a connection to intensive, long-term studies of a single ecosystem at some point along the gradient.

Introduction

Comparative analyses are widely used in studies of nutrient cycling; examining how a process varies from place to place offers insights that are difficult to obtain on a single site, no matter how intensively studied. Such comparisons often lead to testable hypotheses for controls of processes that regulate nutrient cycling and nutrient losses from ecosystems, even when the sites are chosen arbitrarily or fortuitously (where data happen to be available). However, we believe that such comparisons are still more fruitful when they are carried out along gradients of factors that can control nutrient cycling or loss. In this way, comparative studies can evaluate how processes in different sites are related as well as how and why they differ.

This approach is widely used in plant community ecology. In some cases, sites for vegetation sampling are selected along gradients of putative controlling factors (e.g., Whittaker 1956), and the results are plotted against those factors (direct gradient analysis); in others, similarities and differences in the composition of the vegetation itself are used to construct axes of variation/relatedness, and these are then correlated with environmental factors (indirect gradient analysis) (Curtis and McIntosh 1951, and many other authors). Both approaches yield a regional understanding of variation in vegetation; both also provide a rich base of information on vegetation patterns that can be used to guide experimental studies of processes.

Gradient analysis has been employed in ecosystem-level studies as well, although usually not as explicitly. Perhaps the first example is also the best. Hans Jenny defined five major 'state factors' (climate, parent material, relief, organisms, and time) that control variation among soils. He then carried out extensive studies of the influence of those factors by identifying soil sequences in which four of the five are more or less constant, and the effect of the fifth is therefore paramount. His results provided much of the early conceptual basis for soil science, and many remain clear illustrations of what a gradient approach can do for ecosystem theory, analysis, and understanding.

A number of more recent studies have also used and extended gradient approaches. Notable among these are the Van Cleve et al. (1983) examination of vegetation, soil properties, nutrient distribution, and nutrient cycling along topographic and successional gradients in central Alaska; the Pastor et al. (1984) evaluation of nitrogen availability and nitrogen cycling along a soil texture gradient in Wisconsin; Jordan's (1985) ordination analysis of patterns of nutrient distribution and cycling in tropical forest ecosystems; and the Lajtha and Schlesinger (1988) analysis of phosphorus dynamics along soil sequences in a New Mexico desert.

In this chapter, we illustrate the utility of gradient analyses of nutrient cycling and loss by summarizing several specific examples, most drawn from our own work. The cases discussed include applications of gradient analyses to (i) interpreting nutrient losses in streamwater, (ii) analyzing the regulation of abundances of stable isotopes in terms of ecological processes, and (iii) extrapolating fluxes of trace gases on coarse spatial scales. Following these illustrations, we discuss some of the weaknesses of a gradient approach, emphasizing particularly the importance of basing gradient studies on intensive, long-term study sites. Finally, we discuss the potential application of gradient studies to analyses of global-scale biogeochemistry.

Examples

Streamwater Chemistry in New Hampshire

Vitousek (1975, 1977) measured concentrations of major ions in 47 first-order streams draining small watersheds in the White Mountains of New Hampshire. The streams were selected along gradients of elevation (ranging from northern hardwoods below 750 m to alpine tundra above 1500 m) and disturbance history (clearcutting more than 30 years previously versus no history of anthropogenic disturbance). A "direct gradient analysis" of results (done by plotting mean annual concentrations versus elevation) demonstrated that sulfate and chloride concentrations were correlated with each other and inversely correlated with elevation. To the extent that chloride ion is biologically inert and absent from rocks in the area, this pattern suggested control by atmospheric inputs and by greater evapotranspiration at low elevations. Concentrations of Ca, Mg, Na, Si, and, to a lesser extent, K, were strongly correlated with each other and inversely correlated with elevation, but the slope of the relationship was much steeper than that for sulfate and chloride. This pattern could be interpreted as reflecting inputs of rock weathering per unit water flux. Finally, N and, to a lesser extent, K concentrations were uncorrelated with elevation; they could be interpreted as reflecting the accumulation of N and K within watersheds dominated by aggrading vegetation.

These results also were evaluated using Bray–Curtis ordination (Bray and Curtis 1957), an indirect technique. The first axis of variation was strongly correlated with elevation, and the second was correlated with disturbance history (Fig. 14.1; Vitousek 1975). As occurs in many ordinations of vegeta-

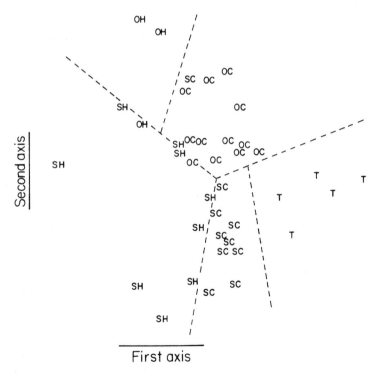

FIGURE 14.1. Bray-Curtis ordination of streamwater chemistry for first-order streams in White Mountains of New Hampshire. Each point represents a stream; the symbols are: T = tundra; OC = old coniferous; SC = successional coniferous; OH = old hardwood; SH = successional hardwood. Redrawn from Vitousek (1975).

tion, the indirect ordination yielded axes that were diagonal with respect to the direct ordination (Loucks 1962).

Most of the streams included in this study were located within 20 km of the Hubbard Brook Experimental Forest (Likens et al. 1977) and on the same bedrock. Consequently, it was possible to relate patterns observed along environmental gradients to the results of the intensive, long-term study there. Without such a comparison, it would have been difficult or impossible to relate concentrations of elements in streamwater to nutrient outputs or to distinguish year-to-year from site-to-site variation. Conversely, the comparative study yielded a regional perspective on the intensive site; it also yielded testable hypotheses for the control of nutrient losses in streamwater.

Soil Development and [15]N Natural Abundance in Hawai'i

The Hawaiian Islands are a valuable natural laboratory for gradient studies of ecosystem structure and function. In these systems it is often possible to develop a sequence in which one of the state factors identified by Jenny (1941, 1980) varies markedly, while the others are more or less invariant. The influence of time is perhaps most obvious in Hawai'i; the islands are volcanoes that result from the movement of the Pacific plate over a stationary convective plume in Earth's mantle. Volcanic activity is now concentrated in the southeast of the archipelago, and the islands are disappearing beneath the sea in the northwest. Surface ages on the islands range from <1 to $>5{,}000{,}000$ years, and their distribution is well mapped and dated (MacDonald et al. 1983; Lockwood et al. 1988).

This age sequence is useful because the other major factors indentified by Jenny can often be held constant. Climate is extraordinarily variable — it includes a temperature gradient from lowland tropical to alpine, and moisture gradients from $>11{,}000$ mm/yr to <250 mm/yr of precipitation. However, this variation is continuous and predictable as a function of elevation and exposure to the prevailing northeast trade winds. Likewise, on young volcanoes the chemistry of parent material is virtually constant, local relief is absent, and, until invasions by exotic species took hold, the biota were also remarkably constant. One tree species, *Metrosideros polymorpha* in the Myrtaceae, is found (and often dominates) from sea level to treeline, from $>11{,}000$ to <500 mm annual precipitation, and (in moist to wet sites) as the earliest woody colonist on young lava flows to the oldest substrates.

The existence of these gradients has been useful in studies of the regulation of nutrient cycles. For example, patterns of nitrogen pools and transformations have been described along a gradient of soil age (to several thousand years) in the montane rainforest climatic zone (Vitousek et al. 1983, 1987; Vitousek and Matson 1988). We also used the soil age and climate gradients to examine patterns in ^{15}N in vegetation and soils. Studies of ^{15}N natural abundance on these gradients were begun to explain a surprising result in a comparison between nitrogen-fixing and nonnitrogen-fixing species in early primary succession. The natural abundance of ^{15}N can be used to estimate nitrogen fixation (Shearer and Kohl 1986) where non-nitrogen-fixing plants derive their nitrogen from a source that differs in ^{15}N from atmospheric N_2, the source for nitrogen fixation. ^{15}N natural abundance is expressed in units of δ^{15}N, the parts per thousand difference in ^{15}N/^{14}N between a sample and a standard. Atmospheric N_2 is the primary standard; pools that are enriched in ^{15}N relative to the atmosphere have positive δ^{15}N values, while ^{15}N depleted pools have negative δ^{15}N. Most often, soil nitrogen is enriched in ^{15}N relative to the atmosphere.

In Hawai'i, we found significant differences in δ^{15}N between nonnitrogen fixers and *Myrica faya*, the nitrogen fixer, in a number of sites, as was expected. However, they were in the opposite direction to the usual pattern — *Myrica* had δ^{15}N values near 0‰ (as would be expected for a fixer), but the nonfixers were ^{15}N *depleted* (Vitousek et al. 1989). Soil nitrogen was also ^{15}N depleted in young sites.

This unusual pattern was interesting, but it was not obvious how to explain it. However, the gradients offered the opportunity to examine δ^{15}N across a broader range of sites. We found no difference in natural abundance of ^{15}N with elevation on young soils, but there was a significant increase (^{15}N enrichment) in both foliage and soils in older substrates (Fig. 14.2).

This pattern suggested that:

1. Nitrogen inputs early in primary succession (mostly in precipitation) are ^{15}N depleted, as observed elsewhere (cf. Heaton 1987).
2. Nitrogen is in short supply early in soil development (Walker and Syers 1976), so almost all nitrogen entering such systems is retained. In support of this suggestion, nitrification is absent, nitrogen immobilization rapid, nitrogen limitation to primary production strong, and nitrous oxide flux absent early in soil development in this age sequence (Matson and Vitousek 1987; Vitousek et al. 1987 Vitousek and Matson 1988).
3. Later in soil development, the quantity and availability of nitrogen increases. Nitrification increases, immobilization decreases, and nitrous oxide flux (and presumably other N losses) increase in older sites. These losses preferentially remove the lighter ^{14}N isotope, and consequently the δ^{15}N of the residual nitrogen increases.

Clearly, a simple examination of the pattern of δ^{15}N along the soil age gradient cannot test this hypothesis, but it is very useful for raising and constraining it.

TRACE GAS FLUXES FROM TERRESTRIAL ECOSYSTEMS

Gradient approaches to ecosystems can also be useful on a global scale for understanding losses of nutrients to the atmosphere. For example, tropical forest ecosystems are believed to be the most important natural source of

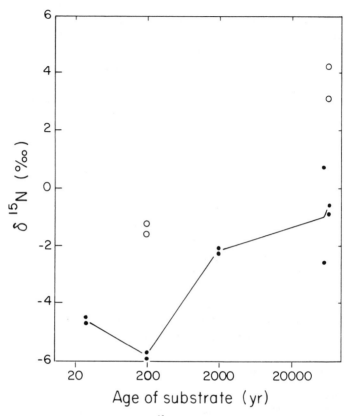

FIGURE 14.2. Natural abundance of ¹⁵N in foliage of *Metrosideros polymorpha* (solid symbols) and soils (open symbols) from soil age sequence in Hawai'i. Data from Vitousek et al. (1989).

the greenhouse gas nitrous oxide (N_2O) (McElroy and Wofsy 1986). Why should that be so? Comparative studies of nutrient cycling in tropical versus temperate forests demonstrate that nitrogen appears to function as an excess nutrient in most lowland tropical forests. The C : N : P ratios of foliage, biomass, and litter differ substantially with latitude; for example, the C : N : P ratio of litterfall in a large sample of lowland tropical forests averages 810 : 23 : 1, versus 630 : 11 : 1 in temperate forests (Vitousek and Sanford 1986). These differences are consistent with a relative deficiency of phosphorus as well as a relative excess of nitrogen in "average" tropical forests; that alone should lead to greater losses of nitrogen from tropical forests.

However, tropical forests are very different from each other; it is no more reasonable to talk of a "typical tropical forest" than it would be to talk of a "typical temperate forest." The average values just given conceal a great deal of variation. Variation among tropical forests has been described in terms of gradients of temperature (lowland to alpine) and precipitation (wet forest to desert) (Beard 1955); a third major gradient is soil fertility (Vitousek and Sanford 1986).

The pattern of C, N, P, and Ca circulation along gradients of soil fertility in tropical forests is fairly well described. Forests on fertile soils such as alfisols and andosols circulate large amounts of N, P, and other elements at low carbon-to-element ratios; forests on infertile oxisols and ultisols circulate large amounts of N but little P or Ca; and forests on leached sands circulate very little N and relatively little P or Ca at high C-to-element ratios (Vitousek and Sanford 1986). Forest primary production appears to be P- or Ca limited on oxisols and N limited on sandy soils (Cuevas and Medina 1988).

Nitrous oxide fluxes might be expected to correlate with nitrogen cycling along this gradient of soil fertility; nitrous oxide is produced and a small fraction emitted during nitrification and denitrification, and the amount of N moving through inorganic nitrogen pools in the soil should at least set an upper limit to nitrous oxide flux (Firestone and Davidson 1989). In fact, we observed that nitrogen transformations in tropical forest soils are correlated with soil fertility (Vitousek and Matson 1988), and that nitrous oxide flux is highly correlated with indices of nitrogen transformations across a fairly wide range of sites (Fig. 14.3; Matson and Vitousek 1987). This correlation suggests that patterns of nitrogen cycling could control variations in nitrous oxide fluxes among tropical forests and between tropical and temperate forests.

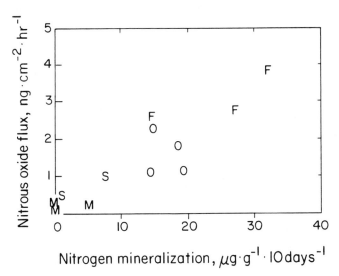

FIGURE 14.3. Nitrous oxide emission as function of net nitrogen mineralization for tropical forest ecosystems on fertile coils (F), oxisols and ultisols (O), spodosols and psamments (S), and montane sites (M). From Matson and Vitousek (1987).

These correlated patterns of nutrient cycling and nitrous oxide flux along a soil fertility gradient may also be useful in building regional and global estimates of flux. The known extent of different soil types within the tropics (Sanchez 1981) can be multiplied by the average of flux measurements in each type to yield a global estimate. Such an approach results in a lower overall flux from tropical forests than that obtained by previous estimates based on the mean of a few sites (Matson and Vitousek, 1990); it does not change the fact that tropical forests are the most important source of nitrous oxide globally.

A more sophisticated approach to estimating fluxes of trace gases along gradients has been used in grassland ecosystems. Jenny (1941) identified the North American Great Plains as a useful system in which to study the influence of climate on ecosystem characteristics, because temperature varies from north to south and precipitation from east to west there. The result is a complete temperature–precipitation interaction, all involving sites that once were dominated by perennial grassland. Jenny (1941) also identified the importance of toposequences of soils (catenas) within the Great Plains, the repeating sequence of gradients from flat interstream areas to erosional areas on slopes to depositional areas near streams. Soil properties vary predictably and more or less continuously along these sequences; primary production and nutrient cycling can vary as much across a catena as they do (on average) across the Great Plains as a whole.

The ecosystems group at the Natural Resource Ecology Laboratory of Colorado State University has used nested gradients (toposequences within broad-scale climatic gradients) to examine and to model nutrient cycling and losses in the region. Ecosystem characteristics vary as a function of slope position along catenas (Schimel et al. 1985), and the mechanisms controlling these differences can be worked out in small-scale experimental studies. The understanding of biogeochemical processes developed in such studies has been summarized in a simulation model (Parton et al. 1988b) and then extended to regional biogeochemistry as it is influenced by both fine- and coarse-scale variations in climate and soils. Finally, the simulation model has been extended to estimate fluxes of nitrogen-containing trace gases from the Great Plains as a whole (Parton et al. 1988a).

Weaknesses of the Gradient Approach

The gradient approach to ecosystem-level comparisons has two major weaknesses. Most notably, it is difficult enough to carry out a good, complete, whole-ecosystem study on a single site. Conducting two for comparative purposes is still more difficult, and trying to do any more reaches toward the impossible. Accordingly, for practical reasons, complete ecosystem studies are very unlikely to be carried out along gradients. Rather, process-level measurements and integrating techniques (such as stable isotopes) are much more appropriately applied to gradient studies. Of course, when these are done alone, it can be difficult to interpret process and integrating measurements and determine their ecosystem-level significance.

A solution to this problem is that wherever possible, gradients should be anchored on a long-term, whole-ecosystem study. The association benefits both: the gradient study then can make use of the context provided by a well-studied site, while the whole-system study benefits by an understanding of regional variations in ecosystem properties and by the testable hypotheses about controlling mechanisms that arise in the comparative studies.

The second weakness is really more of a caveat: it is important that the "right" ecosystem characteristic be studied along the "right" gradient. Most often, the purpose of gradient or other comparative analyses of ecosystems is to develop and test an understanding of controls on ecosystem properties and processes. Ecosystems vary in space and time, and it is quite possible to show correlation of a particular property with a particular gradient without there being a very clear mechanism connecting them and, therefore, to show an association without gaining any real insight in the process. Scientific judgment is therefore as important in gradient studies as in all other kinds of ecosystem studies.

Implications for Global Studies

Gradient studies are inherently useful for gaining an understanding of the spatial distribution of ecosystem properties or processes; they can therefore be used to develop regional or global estimates of nutrient pools and exchanges. A number of regional and global estimates of nutrient pools or fluxes based on gradients have been completed, including the Brown and Lugo (1982, 1984) and Post et al. (1982, 1985) analyses of carbon and nitrogen pools as a function of climatic gradients, and the analyses by Matson and Vitousek (1990) of nitrous oxide flux as a function of soil fertility. It is possible that fluxes of other trace gases also could be based on ecosystem-level processes measured across environmental gradients. For example, biogenic isoprene and terpenes are produced in large quantities in many tropical and temperate forests. Terpenes in particular are carbon-based defensive compounds; their production may be greatest where plant growth is limited by nutrients (and therefore fixed carbon is relatively abundant), and where plant parts are long lived and require substantial investment for defense (McKey et al. 1978; Coley et al. 1985). Fluxes of volatile hydrocarbons could therefore vary in predictable ways along gradients of soil fertility. A significant amount of research would be required to test this possibility; if there is a reasonable causal connection, however, it would be useful in understanding coarse-scale patterns of hydrocarbon flux and hence tropospheric CO and O_3 formation.

Gradient studies are particularly useful where the controlling gradient(s) can be characterized remotely (Matson et al. 1989). The absorption of photosynthetically active radiation by plants is one property that can be characterized from space (Sellers 1985), and seasonal variation in that absorption yields a very clear picture of the moisture gradient in tropical ecosystems (Tucker et al. 1985). It may also become possible to determine aspects of canopy nutrient chemistry remotely (Wessman et al. 1988), allowing remote characterization of gradients in soil fertility.

If remotely derived information on controlling factors as they vary along gradients can be obtained, it will represent a powerful new tool for analyzing regional and global patterns in biosphere–atmosphere and biosphere–hydrosphere exchange. New technologies may also prove useful in addressing another concern of gradient studies, or indeed any regional or global study—once regional fluxes have been calculated, can they be checked? Integrated land–water interactions can be validated with watershed studies (Likens et al. 1977). It is now becoming possible to validate studies of land–atmosphere interaction on similar scales using tower- and aircraft-based gas flux measurements (Desjardins et al. 1985; Matson and Harriss 1988; Sellers et al. 1988); these will allow validation (or rejection) of regional estimates based on extrapolations along gradients. We hope that they will also provide more grist for the mill of comparative ecosystem studies.

REFERENCES

Beard, J.S. (1955). The classification of tropical American vegetation-types. *Ecology* 36:89–100.
Bray, J.R. and J.T. Curtis. (1957). An ordination of the upland forest communities of southern Wisconsin. *Ecol. Monogr.* 27:325–349.
Brown, S. and A.E. Lugo. (1982). The storage and production of organic matter in tropical forests and their role in the global carbon cycle. *Biotropica* 14:161–187.
Brown, S. and A.E. Lugo. (1984). Biomass of tropical forests: A new estimate based on forest volumes. *Science* 233:1290–1293.
Coley, P.D., J.P. Bryant, and F.S. Chapin III. (1985). Resource availability and plant antiherbivore defense. *Science* 230:895–899.
Cuevas, E. and E. Medina. 1988. Nutrient dynamics within Amazonian forests. II. Fine root growth, nutrient availability, and leaf litter decomposition. *Oecologia* 76:222–235.
Curtis, J.T. and R.P. McIntosh. (1951). An upland forest continuum in the prairie-forest border region of Wisconsin. *Ecology* 32:476–496.
Desjardins, R.L., J.L. MacPherson, P. Alvo, and P.H. Schuepp. (1985). Measurements of CO_2 and turbulent heat exchange over forests from aircraft. In: *The Forest-Atmosphere Interaction*, B.H. Hutchinson and B.B. Hicks, eds. Reidel Publishing, Hingham, Massachusetts, pp. 645–658.
Firestone, M.B. and E. Davidson. (1989). Microbiological basis of NO_x and N_2O production and consumption. In: *Exchange of Trace Gases Between Terrestrial Ecosystems and the Atmosphere*, M.O. Andreae, and D.S. Schimel, eds. Wiley, New York, pp. 7–21.

Heaton, T.H.E. (1987). $^{15}N/^{14}N$ ratios of nitrate and ammonium in rain at Pretoria, South Africa. *Atmos. Environ.* 21:843–852.

Jenny, H. (1941). *Factors of Soil Formation.* McGraw-Hill, New York.

Jenny, H. (1980). *Soil Genesis with Ecological Perspectives.* Springer-Verlag, New York. 560 pp.

Jordan, C.F. (1985). *Nutrient Cycling in Tropical Forest Ecosystems: Principles and Their Application in Management and Conservation.* Wiley, New York.

Lajtha, K., and W.H. Schlesinger. (1988). The biogeochemistry of phosphorus cycling and phosphorus availability along a desert soil chronosequence. *Ecology* 69:24–39.

Likens, G.E., F.H. Bormann, R.S. Pierce, J.S. Eaton, and N.M. Johnson. (1977). *Biogeochemistry of a Forested Ecosystem.* Springer-Verlag, New York.

Lockwood, L.P., P. Lipman, L.D. Peterson, and F.R. Warshauer. (1988). Generalized Ages of Surface Lava Flows of Mauna Loa Volcano, Hawaii. U.S. Geological Survey Miscellaneous Publications Map I-1908, U.S. Government Printing Office, Washington, D.C.

Loucks, O.L. (1962). Ordinating forest communities by means of environmental scalars and phytosociological indices. *Ecol. Monogr.* 32:137–166.

MacDonald, G.A., A.T. Abbott, and F.L. Peterson. (1983). *Volcanoes in the Sea: The Geology of Hawaii,* 2nd Ed. University of Hawaii Press, Honolulu.

Matson, P.A. and R.C. Harriss. (1988). Prospects for aircraft-based gas exchange measurements in ecosystem studies. *Ecology* 69:1318–1325.

Matson, P.A. and P.M. Vitousek. (1987). Cross-system comparison of soil nitrogen transformations and nitrous oxide fluxes in tropical forests. *Global Biogeochem. Cycles* 1:163–170.

Matson, P.A., and P.M. Vitousek. 1990. Ecosystem approach to a global nitrous oxide budget. *BioScience* 40:667–672.

Matson, P.A., P.M. Vitousek, and D.S. Schimel. (1989). Regional extrapolation of trace gas flux based on soils and ecosystems. In: *Exchange of Trace Gases Between Terrestrial Ecosystems and the Atmosphere,* M.O. Andreae and D.S. Schimel, eds. Wiley, New York, pp. 97–108.

McElroy, M.B. and S.C. Wofsy. (1986). Tropical forests: Interactions with the atmosphere. In: *Tropical Rain Forests and the World Atmosphere,* G.T. Prance, ed. Westview Press, Boulder, Colorado, pp. 33–60.

McKey, D., P.G. Waterman, J.S. Gartlan, and T.T. Struhsaker. (1978). Phenolic content of vegetation in two African rain forests: ecological implications. *Science* 202:61–64.

Parton, W.J., A.R. Mosier, and D.S. Schimel. (1988a). Rates and pathways of nitrous oxide production in a shortgrass steppe. *Biogeochemistry* 6:45–58.

Parton, W.J., J.W.B. Stewart, and C.V. Cole. (1988b). Dynamics of C, N, P, and S in grassland soils: a model. *Biogeochemistry* 5:109–131.

Pastor, J., J.D. Aber, C.H. McClaugherty, and J.M. Melillo. (1984). Aboveground production and N and P cycling along a nitrogen mineralization gradient in Blackhawk Island, Wisconsin. *Ecology* 65:256–268.

Post, W.M., W.R. Emanuel, P.J. Zinke, and A.G. Stangenberger. (1982). Soil carbon pools and world life zones. *Nature* (London) 298:156–159.

Post, W.M., J. Pastor, P.J. Zinke, and A.G. Stangenberger. (1985). Global patterns of soil nitrogen storage. *Nature* (London) 317:613–616.

Sanchez, P.A. (1981). Soils of the humid tropics. In: *Blowing in the Wind: Deforest-ation and Long-Range Implications* (Department of Anthropology, ed.). College of William and Mary, Williamsburg, Virginia, pp. 347–410.

Schimel, D., M.A. Stillwell, and R.G. Woodmansee. (1985). Biogeochemistry of C, N, and P in a soil catena of the shortgrass steppe. *Ecology* 66:276–282.

Sellers, P.J. (1985). Canopy reflectance, photosynthesis, and transpiration. *Int. J. Remote Sens.* 6:1335–1372.

Sellers, P.J., F.B. Hall, G. Asrar, D.E. Strebel, and R.E. Murphy. (1988). The first ISLSCP field experiment (FIFE). *Bull. Am. Meterol. Soc.* 69(1):22–27.

Shearer, G. and D. Kohl. (1986). N_2-fixation in field settings: estimations based on natural ^{15}N abundance. *Aust. J. Plant Physiol.* 13:699–756.

Tucker, C.J., J.R.G. Townshend and T.E. Goff. (1985). African land-cover classi-fication using satellite data. *Science* 227:369–375.

Van Cleve, K., L. Oliver, R. Schlentner, L.A. Viereck, and C.T. Dyrness. (1983). Productivity and nutrient cycling in taiga forest ecosystems. *Can. J. For. Res.* 13:747–766.

Vitousek, P.M. (1975). The Regulation of Element Concentrations in Mountain Streams in the Northeastern United States. Ph.D. Dissertation, Dartmouth Col-lege, Hanover, New Hampshire.

Vitousek, P.M. (1977). The regulation of element concentrations in mountain streams in the northeastern United States. *Ecol. Monogr.* 47:65–87.

Vitousek, P.M. and P.A. Matson. (1988). Nitrogen transformations in tropical forest soils. *Soil Biol. Biochem.* 20:316–367.

Vitousek, P.M. and R.L. Sanford, Jr. (1986). Nutrient cycling in moist tropical forest. *Annu. Rev. Ecol. Syst.* 17:137–167.

Vitousek, P.M., G. Shearer, and D.H. Kohl. (1989). Foliar ^{15}N natural abundance in Hawaiian rainforest: patterns and possible mechanisms. *Oecologia* 78:383–388.

Vitousek, P.M., K. Van Cleve, N. Balakrishnan, and D. Mueller-Dombois. (1983). Soil development and nitrogen turnover on recent volcanic substrates in Hawaii. *Biotropica* 15:268–274.

Vitousek, P.M., L.R. Walker, L.D. Witeaker, D. Mueller-Dombois, and P.A. Matson. (1987). Biological invasion by *Myrica faya* alters ecosystem development in Hawaii. *Science* 238:802–804.

Walker, T.W. and J.K. Syers. (1976). The fate of phosphorus during pedogenesis. *Geoderma* 14:1–19.

Wessman, C.A., J.D. Aber, D.L. Peterson, and J.M. Melillo. (1988). Remote sensing of canopy chemistry and nitrogen cycling in temperate forest ecosystems. *Nature* (London) 335:154–156.

Whittaker, R.H. (1956). Vegetation of the Great Smoky Mountains. *Ecol. Monogr.* 26:1–80.

Part V   Additional Views

15
Variance and the Description of Nature

CARLOS M. DUARTE

Abstract

I examine here the epistemological and operational consequences of the use of ecological variability in comparative analyses. Ecological variability is used in comparative analyses to describe the possible states of ecological properties and to search for patterns in nature. Success in these goals requires (1) an explicit recognition of the structure and development of ecological variability and (2) appreciation of the important epistemological constraints inherent to the different methods used in the study of ecological variance. Ecological variance is often hierarchical in nature, containing a nested array of scales of variation, each scale contributing peculiar components to the global variability and presenting patterns that may differ across scales. The approach used to examine ecological variability needs, therefore, to be able to cope with the different types of patterns possible (e.g., continuous relationships, boundary conditions, threshold relationships, etc.) and to allow for changes in the existing patterns when sampling across scales. Thus, I contend that increased flexibility in the search for patterns in ecological variability is essential for successful comparative analysis.

Introduction

One of the goals of comparative ecology is to describe the range of states possible for a property by quantifying its variability. The description of the variability (e.g., as statistical variance) in a particular property may be a goal in itself, but far more often ecologists seek to extract patterns or trends (i.e., useful structure) from this variability. Comparative ecology often categorizes nature by describing and explaining the variance in relevant ecological properties. Realization of this common general goal is rare because comparative ecology is usually used as a tool to address specific questions. However, the use of ecological variability to describe nature has important epistemological implications, largely derived from the hierarchical nature of ecological variability and the diversity of patterns possible, that should be articulated to avoid failure and to expedite progress in comparative ecology.

My goals here are to examine the assumptions and limitations implicit in the use of comparative ecology and to describe generalities in the structure of ecological variance that may help the design and interpretation of comparative studies. To achieve my goals, I first discuss the role of comparative ecology and its practical advantages and disadvantages in relation to other approaches. I then examine general characteristics of the hierarchical structure and sources of ecological variability and provide some basic guidelines for proceeding with the study of ecological variability and the search for patterns in nature.

This exercise is not aimed to contribute to the philosophy of science, but rather is focused on the discussion of the practical implications of the use of comparative ecology, relative to other approaches, for problem-solving in ecology from the standpoint of a practicing ecologist. Thus, my primary concern is that the ideas expressed here be operationally, but not necessarily philosophically, justifiable. The terms variance, variation, and variability are used here indistinctly to refer to the range of states possible for a certain property, of which statistical variance is only one of many possible descriptors (e.g., ranges, frequency distributions, coefficients of variation, etc.)

The Role of Comparative Analysis in Ecology

Comparative ecology is an imprecise term, for all approaches in ecology are comparative in essence (e.g., Peters et al., Chapter 4, this volume). For instance, experimental ecology is also comparative, because the outcomes of treatments are compared to assess their significance. I shall avoid any further ambiguities on what I refer to as comparative ecology by restricting it to comparative analyses of naturally existing, as opposed to artificially generated, variance in ecological properties. These types of comparative analyses are common to all natural sciences but clearly have not been developed to their full potential in ecology. This may be attributable to some confusion as to the role of comparative ecology and its relationship with other approaches.

Comparative studies are descriptive and exploratory in nature and provide patterns or generalities that may lead to general ecological laws, contrasting with other, more explanatory, approaches (e.g., most of theoretical and experimental ecology). Both comparative and explanatory studies can be structured in an hypothetico-deductive manner. However, hypotheses are logical constructs and cannot, therefore, reach beyond the logical consequences of existing knowledge. Thus, hypothetico-deductivism is not likely to foster progress in areas where the state of knowledge is often insufficient for postulating nontrivial hypotheses (see Mentis 1988). Further, because hypothetico-deductivism attempts to falsify, rather than support, hypotheses (Popper 1959), this approach has little use in examining some important concepts in ecology, notably core theories (e.g., evolutionary theory, continuum theory), that cannot be readily falsified by any simple test (Lakatos 1968; Mentis 1988).

Comparative ecology has the additional advantage of being amenable to inductive approaches, which attempt to develop knowledge, generate hypotheses, or derive general statements from specific data (Mentis 1988), whereas explanatory ecology has been more confined to hypothetico-deductivism (Mentis 1988). Comparative and explanatory ecology are, however, mutually dependent, for comparative analyses often provide the empirical bases for explanatory theories and their experimental tests and these, in turn, suggest new relevant comparisons.

Comparative and experimental approaches to ecology are similar in that both describe nature through the study of variation in ecological properties. There is, however, a fundamental difference between them, because comparative ecology focuses on naturally *existing* variability, whereas experimental ecology examines variability that is generated artificially by manipulating particular factors. Because an infinite number of factors may underlie *existing* natural variation, comparative ecology can only suggest explanations. Moreover, the patterns found in a particular system may not hold elsewhere, for there is no guarantee of uniformity in nature, although repeated observation of patterns may reinforce our confidence in their generality.

Experimental approaches perturb particular factors in otherwise (nominally) identical conditions to examine whether those perturbations influence significantly the property of interest. Because the variability in the property examined is produced *a posteriori*, experiments offer more confident identification of the sources of biological variability than comparisons based on existing variability, and are, therefore, better suited to examine mechanisms. Thus, identification of cause and effect is often impossible in comparative analyses, whereas experimental manipulation may sometimes resolve this problem.

There are, however, inherent limitations to what experiments can do. Experimental studies have only two possible outcomes: either the treatment(s) influence significantly (i.e., within the limits of resolution) the values of the property studied, or they do not. In neither case do experiments indicate the relative importance of the factors in nature, for factors untested for, or not amenable to experimental manipulation, may have a disproportionate importance in nature. The underlying reason for this limitation is that the structure of experimentally generated variance and naturally existing variability is fundamentally different, largely because experimentally generated variance often lacks the hierarchical structure characteristic of natural variance (see following).

The range of questions susceptible to experimental testing is also limited, particularly when the mechanisms of interest involve multiple influences. For instance, experimental testing of the ideas that turbulence controls primary production and the dominant phytoplankton life forms in the ocean (Margalef 1978) can only be partial, for the covariation of turbulence, light, and nutritional environment found in the sea cannot be easily reproduced in the laboratory (e.g., Estrada et al. 1987). Littoral slope is closely correlated to the maximum biomass of submerged macrophyte biomass (Duarte and Kalff 1986), but littoral slope is linked to so many processes (e.g., wave energy dissipation within the littoral zone, sediment dynamics and characteristics, weed bed extension) that experimental manipulations of littoral slope can only provide a partial test of its importance (Duarte 1987). These constraints often impel ecologists to experimental designs closely approaching *in situ* conditions (e.g., Duarte and Kalff 1988). However, as the experimental conditions approach those in nature, uncontrolled variability reduces the demonstrative power of the interpretation of the results (Duarte and Kalff 1988), and problems of pseudoreplication affect the validity of the inferences (Hurlbert 1984).

Thus, neither comparative nor experimental ecology would satisfy scientists searching for unambiguous explanations of ecological phenomena, for these may still lie beyond the reach of ecology. Ecologists must content themselves with rules or laws (derived from comparisons) that describe patterns in ecological variability and a set of probable explanations or mechanisms (experimentally tested) to account for them.

Comparative Ecology and the Structure of Ecological Variability

Comparison is one of the oldest forms of ecological inquiry, for the questions that stimulated the curiosity of naturalists were derived from observed differences among organismal characteristics across broad geographical areas. However, comparative ecology involves more than an intuitive or verbal comparison of observations. A further step in comparative ecology involves the formulation of classifications (Barnes 1967). Classifications attempt to describe patterns in variability by proposing nominal categories that capture the fundamental difference(s) underlying the observed variability. Classifications may come to dominate the research programs, as was the case of lake typology (cf. Brinkhurst 1974).

Classification systems are useful in organizing our thoughts and observations, but they can be constraining because the criterion chosen to represent "the fundamental difference underlying the variability" is subjective, and because the rationale from observed differences to classification is often reversed, leading to circularities such as "lake x is unproductive because it is oligotrophic." These constraints are partially solved by using quantitative comparisons that provide a more objective basis for the comparisons, allow quantitative tests of specific hypotheses, and reveal regularities that may represent basic ecological patterns. For instance, comparative studies of lake trophic status flourished when typologies were replaced by comparisons based on chlorophyll *a* concentrations (e.g., Dillon and Rigler 1974; Canfield 1981).

Quantitative comparisons provide an estimation of variability in the property studied. In an analogy with analysis of variance, the variance observed can be partitioned into variance due to measurement error, random variation operationally defined as the compartment where no pattern can be discerned, and pattern (Fig. 15.1). Most comparative studies try to describe the last compartment, for they attempt to extract patterns from the differences or similarities revealed by the comparison. The identification of pattern requires that the structure of the variance in ecological properties be recognized to avoid inappropriate methods and unwarranted expectations. Variability in a particular property (e.g., plant biomass) cannot be quantified without a reference frame (e.g., individual plant, stand, forest). The need for reference frames in comparative studies derives from the hierarchical nature of ecosystems (cf. Allen and Starr 1982; Margalef 1982; O'Neill 1988). The choice of the scale of interest to attack the problem is a critical step in comparative analyses because it constrains our success in identifying patterns. Variability in the process examined may differ from one level of observation to another, because variance at one level of observation may be transferred to other levels in the hierarchy, or be dissipated within a particular level. In addition, each level has inherent sources of variability (e.g., demographic interactions at a plant stand level) that compound to, and may overwhelm, variability at lower levels.

Patterns in variability may differ substantially among levels of observa-

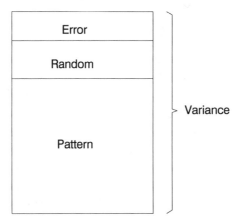

FIGURE 15.1. Conceptual representation of the composition of ecological variability.

tion, for, just as experiments cannot identify variability resulting from factors other than the treatments, comparative analyses cannot reveal variability from factors that remain uniform across the elements compared. For instance, the comparison of physiological properties of isolated plants, which cannot account for variability in plant performance caused by interactions with its neighbors, may fail to reveal patterns at the population level. Ecological variability can, therefore, be conceived as structured in a nested array of scales of variation, each contributing peculiar components to the global variability and presenting patterns that may differ from level to level.

The nested structure of ecological variance suggests that it may be partitioned into a series of components, or sources, much as in a nested experimental design (cf. Sokal and Rohlf 1969). Ecological variability involves temporal and spatial components that are often asymmetrical, one dominating over the other, and that may reflect different phenomena. The temporal and spatial components can be partitioned into a series of scales of increasing variance, for these scales can incorporate the variability in the lower scales and often add some variability from processes inherent to higher scales. This structure can be described for particular problems by combining nested analysis of variance (ANOVA) to examine the contribution of discrete components of variation (e.g., regional, within-lake) to total variability, with spectral analyses, to examine the scale-dependence of the variance along continuous components (e.g., time and space). This exercise identifies the dominant sources of variance, where success in the search for patterns would be most promising, and permits the comparison of the importance of different components and scales of variability. This approach has been used to partition the variability of phytoplankton biomass in the NW Mediterranean (Masó and Duarte 1989) into temporal and spatial components (Fig. 15.2). In addition to partitioning phytoplankton variability into discrete components, spectral analyses of oceanic phytoplankton biomass have revealed the scale-dependence of its variability within these components (e.g., Denman et al. 1977; see Fig. 15.2).

The choices of the unit of analysis, the scale of study, and the components of variance are important determinants of success in comparative ecology. This is often neglected in practice, for the reference frame is often lost by extrapolating patterns across different components or scales. For instance, the demonstration of strong relationships between lake chlorophyll *a* and phosphorus concentrations among lakes has been extensively documented (e.g., Dillon and Rigler 1974; Canfield 1981). However, there is no guarantee that patterns from these cross-lake studies are transferable to explain variability in chlorophyll *a* concentrations within lakes. The temporal reactions of particular lakes to reduced phosphorus loading have shown that the transferability of patterns from one scale (cross-lake) to another (single-lake) involves great uncertainty (Vollenweider 1987). The problems of transferability of patterns in cross-lake variability in phyto-

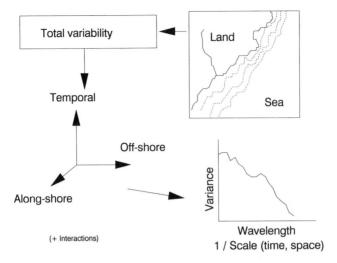

FIGURE 15.2. Schematic representation of the structure of phytoplankton variability in the sea.

plankton biomass to temporal variability in single lakes may arise from a shift of the "meaning" of measured phosphorus concentrations involved in that process. In cross-lake studies, phosphorus concentration is correlated to a number of other factors (e.g., electric conductance, mean depth, catchment geology; Duarte and Kalff 1989), which are important to algal growth and add power to the correlation between total phosphorus and chlorophyll *a* concentrations (Duarte and Kalff 1989). The power of phosphorus concentrations to explain variation in chlorophyll concentration may be partially lost in single-lake studies where the covariation of phosphorus concentrations with other relevant properties (e.g., lake depth or geology) is partially lost.

Failure to consider the restrictions imposed by the level of analysis may lead to unwarranted expectations and to disagreement among authors about regulating factors. For instance, a hierarchical conception of the littoral zone removes apparent disagreement about the factors controlling the biomass of submerged macrophytes in lakes (Duarte and Kalff 1990). A substantial fraction of the variability in submerged macrophyte biomass detected in worldwide studies (Fig. 15.3) can be attributed to differences in the latitudinal position of the lakes (Duarte et al. 1986). Not surprisingly, models based on latitudinal differences have little power to explain regional differences in macrophyte biomass (Canfield and Duarte 1988), which appear, instead, to be attributable to differences in the availability of dissolved inorganic carbon (e.g., Adams et al. 1978; Duarte and Kalff 1990; see Fig. 15.3). Chemical variability within single lakes is more limited,

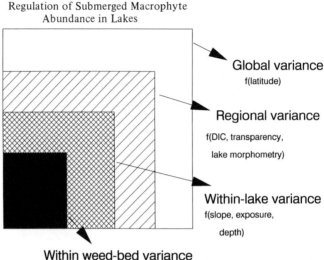

Regulation of Submerged Macrophyte
Abundance in Lakes

Global variance
f(latitude)

Regional variance
f(DIC, transparency,
lake morphometry)

Within-lake variance
f(slope, exposure,
depth)

Within weed-bed variance
f(sediment heterogeneity, neighbor interactions, sampling error)

FIGURE 15.3. Schematic representation of the structure of variability of submerged macrophyte biomass in lakes and factors related to biomass variability at different scales.

and the fraction of variability in submerged macrophyte biomass from within-lake differences apparently derives from local differences in littoral morphometry (Duarte and Kalff 1986, 1990). Differences at yet smaller scales (i.e., within macrophyte beds) can be attributed to sediment heterogeneity (e.g., Anderson and Kalff 1986). Therefore, the statements that submerged macrophyte biomass is primarily related to latitudinal, chemical, sediment, or morphometric differences do not conflict because they reflect the scale of comparison (Duarte and Kalff 1990).

Variation at the largest possible scale (often worldwide comparisons) may comprise many orders of magnitude in the property studied. As a result, even powerful patterns in broad comparisons may leave a residual error greater than an order of magnitude (e.g., maximum density–organismal size relationships; Duarte et al. 1987). However, the large error associated with many large-scale patterns may render them useless at narrower levels. At any rate, the variability observed at the largest scale sets limits to that possible at smaller scales. Therefore, any patterns evident at the largest scale are likely to represent a substantial fraction of the total variance. Moreover, because they involve greater variance, large-scale studies are less sensitive to measurement error than studies at smaller scales. Thus, large-scale comparative analyses will more likely succeed in finding patterns in ecological variance.

Because the structure of ecological variability is shaped by the processes that originate it (e.g., Allen 1985), the assumptions on how variability is generated are at the root of a successful search for patterns in nature. Yet, these assumptions are seldom articulated, for most studies in comparative ecology attempt to explain ecological variability by a series of direct correlations or regression equations with independent factors measured simultaneously (e.g., Rigler 1982; Peters 1986). The expectation of finding such relationships is based on the hope that a substantial portion of the variance in the response variable arises from variation in independent, often environmental, factors (Fig. 15.4). Thus, environmental variability is conceived as a template upon which ecological variability is expressed (Fig. 15.4). This widely held conception of the genesis of ecological variability is rarely, if ever, explicitly addressed in comparative studies. Although adequate to most comparative studies, the unsuitability of this assumption to specific problems will result in failure to identify patterns. The search for direct correlations between ecological variables and environmental factors assumes that signal and response coexist in time and space, that is, changes in environmental factors are translated into changes in organismal growth and abundance without any temporal or spatial lag. This assumption, however, is often inappropriate, because the transfer of environmental perturbations

$$Property = f(factors)$$
$$thus,$$
$$Var(property) = f[var(factors)]$$

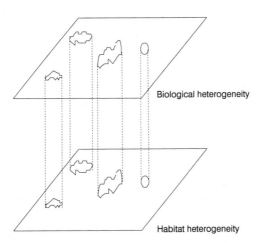

FIGURE 15.4. Conceptual implications of use of direct correlation and regression in search for patterns in ecological variability. Environmental variability is, in these approaches, conceived as a template on which ecological variability is expressed.

at one level to higher levels in the hierarchy may involve some lag (Allen and Hoekstra 1986). For instance, changes in algal growth in response to environmental changes are subject to lags in the transfer of environmental perturbations (e.g., changes in nutrient concentrations) to the cellular level (i.e., increased cell division rate), and from this to the population level (i.e., increased algal biomass; Duarte 1990). Time lags also explain delays between changes in organismal growth and growth in organisms that depend on them (e.g., Vaqué et al. 1989; Rose and Leggett 1990).

Lagged responses account for apparent paradoxes derived from comparative analyses, such as the common observation of negative spatial relationships between phytoplankton biomass and free nutrient concentrations in the sea (i.e., high phytoplankton biomass in waters with low free nutrient levels and vice versa; Cruzado and Kelley 1973; Thomas 1979; Margalef 1982). A way to avoid the bias resulting from temporal uncoupling between signal and response is to aggregate data either temporally or spatially (e.g., Vollenweider 1987), using average values for the comparative studies. For instance, nutrient–chlorophyll relationships based on seasonal or annual averages are stronger than those for single dates (e.g., McCauley and Kalff 1981).

The concept that variance in the response variable arises from variability in environmental conditions also neglects the possibility that ecological variability may develop in the absence of environmental heterogeneity. Patchiness in some plant communities, such as seagrass communities (Duarte and Sand-Jensen 1990a,b) and freshwater submerged macrophytes at the deep vegetation limit (e.g., Rørslett 1987), cannot be explained by any observable differences in environmental conditions before patch establishment (Duarte and Sand-Jensen 1990a,b). These plant communities inhabit transient environments where fluctuations in the factors constraining colonization (sediment movement, and light, respectively; Duarte and Sand-Jensen, 1990a,b; Rørslett 1987) offer time windows when plant colonization is possible (Fig. 15.5). Low frequency or amplitude of fluctuations in the disturbance agent will result in sufficient windows to yield uniform plant covers (i.e., low heterogeneity), whereas frequent fluctuations above the threshold level for colonization will result in the absence of suitable windows for colonization, yielding bare substrata (i.e., uniform sediments; Fig. 15.5). Thus, spatial heterogeneity requires particular fluctuation frequencies in the disturbance agent, similar to the symmetry-breaking instabilities described in chemistry (Turing 1952), where complex heterogeneities are produced in homogeneous fluids on the introduction of fluctuations at characteristic wavelengths (cf. Allen 1985). The ecological variance resulting from this type of process is, therefore, dependent on the variance structure of the disturbance agent rather than on any simultaneous differences in growth conditions between vegetated and unvegetated sediments.

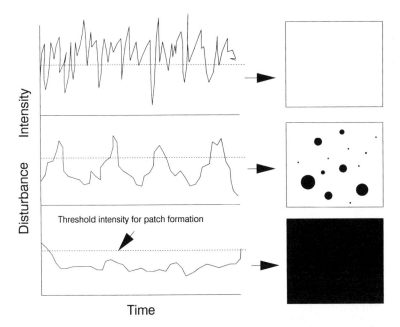

FIGURE 15.5. Schematic representation of the influence of fluctuations in disturbance agent (e.g., current velocity) for macrophyte colonization and development of plant cover.

Description of Ecological Variability and the Search for Pattern

The most effective comparative analyses are quantitative in nature, for the epistemological power of comparative analyses relies largely on the assessment of variability in the response variable and the subsequent extraction of patterns from the observed variability. The quantification of variability and pattern identification are, therefore, critical steps in comparative ecology, for the choice of methods influence the type of patterns that can be disclosed. For instance, most statistical methods (e.g., correlation analyses, principal components analyses) can only demonstrate linear associations between the variables examined.

QUANTIFICATION OF ECOLOGICAL VARIABILITY: AN ARITHMETIC OR MULTIPLICATIVE WORLD?

Variability is often expressed as the standard deviation or the coefficient of variation of the variable studied. This choice is far from trivial, for it depends on whether the variability studied is arithmetic or multiplicative.

There are a priori reasons to favor either an arithmetic or multiplicative treatment of specific problems. Comparisons based on ratios are candidate multiplicative variables, and logarithmic transformation simply transforms these ratios into differences.

The variance of many relative properties (e.g., organismal abundance and size, nutrient concentrations in water) is scaled to the mean values, whereas the variances of some absolute properties (e.g., temperature) are independent of the mean. In comparing the variability of relative properties, it is necessary to account for differences in their means (scaling). Consequently, quantitative comparisons of the variability of relative properties should be based on the coefficient of variation, whereas standard deviations are well suited for comparisons of the variability of absolute properties (see Sokal and Rohlf 1969).

The need to scale relative properties and their variabilities is evident in allometric comparisons (e.g., Peters 1983), because a distance of 1 km or a mass change of 1 kg has different implications for mice and elephants. Consequently, studies of the implications of plant (Hara 1988) and animal (Haldane 1955; Lewontin 1966) size use coefficients of variation to represent size variability; and logarithmically transformed variables to search for patterns (e.g., Peters 1983). Therefore, whether multiplicative or arithmetic data analyses are used should depend on the nature of the problem at hand.

SEARCHING FOR PATTERNS: ENVELOPES, THRESHOLDS, AND CONTINUOUS MODELS

Success in revealing patterns (i.e., useful structure) in the properties compared depends greatly on the assumptions made about the structure of the variability studied. These assumptions influence our choice of methods and statistical devices to describe the patterns which, in turn, restrict the types of patterns that can be described. The power of comparative ecology is recognizing the possible (Rigler 1982), describing the probable (Peters 1987), and revealing the unexpected (i.e., patterns that could not be deduced from our current understanding). Therefore, pattern-seeking in comparative ecology must remain an open, exploratory activity able to accommodate to any particular variance structure.

Scrutiny of the literature shows correlation and regression analyses to be the favored tools for pattern recognition in comparative ecology (Rigler 1982; Peters 1986, 1987). Although regression analysis is indeed a very useful tool, overemphasis on regression analysis serves comparative ecology poorly, for it constrains our flexibility to describe ecological variability. In addition to the methodological limitations of regression and correlation analyses (Montgomery and Peck 1982), these analyses impose limits on the way we conceive of ecological variance. Regression models are best suited to describe relationships that are (1) deterministic, where a most probable value of the dependent variable can be established for each value of the independent variable(s) (Fig. 15.6A); and (2) homogeneous, where the same relationship holds across all the measurement space (Breiman et al. 1984).

Ecological patterns often deviate from the deterministic, homogeneous relationships best described by regression analysis. Independent factors often do not determine, but rather impose, limits on the state properties (Fig. 15.6B). Boundary conditions, that constrain the range of the possible rather than determine the probable, apply to many ecological phenomena. For instance, clear, stratified waters often develop deep phytoplankton maxima (DPM; Anderson 1969; Fee 1976). DPM only develop when the mixed layer (Z_m; i.e., the layer that is amenable of wind-driven mixing), is shallower than the compensation depth (Z_c, the depth at which photosynthetic use of the available irradiance equals respiratory demands). DPM formation is otherwise precluded by algal entrainment into the mixed

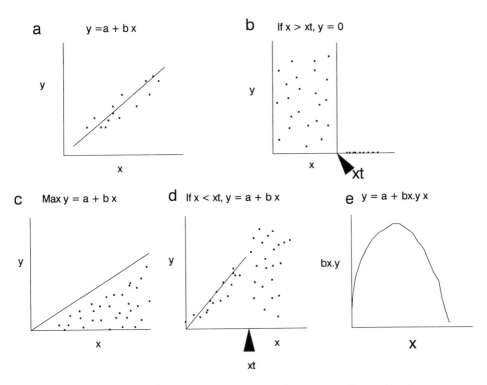

FIGURE 15.6. Schematic representation of different types of relationships between two variables. (A) Homogeneous relationship; (B) threshold relationship; (C) boundary relationship; (D) homogeneous-threshold relationship; (E) continuously changing relationship. Symbols: y, dependent variable; x, independent variable; max y, maximum y for given value of x; xt, critical value of x; a, regression intercept; bx · y, regression slope.

layer and light limitation. Therefore, the presence of DPM can be modeled by a threshold condition of the type (Fig. 15.6B):

If $Zm < Zc$, then DPM may occur, or else DPM are not likely to occur.

Threshold conditions can be defined for a variety of ecological problems, particularly if the problem can be reduced to a dichotomy. For instance, submerged macrophytes are not likely to occur in lake shores steeper than 0.14 m m^{-1} (Duarte 1987). Boundary conditions may also define envelopes enclosing a space encompassing all permissible combinations of the independent and response factors (Fig. 15.6C). For instance, organismal size sets an upper limit to organismal abundance (Duarte et al. 1987) that defines a space of permissible combinations of organismal size and density. The maximum depth at which submerged macrophytes can grow in lakes is a function of water transparency (e.g., Chambers and Kalff 1985) that defines the combinations of depths and water transparencies where macrophytic growth may be expected. Regression analyses applied to envelope-type relationships may produce the feeling of increased precision, by predicting a value of the dependent variable as more probable than others, but the resulting representations of the structure of the relationship examined will be biased and unrealistic (Fig. 15.6C).

Threshold conditions may be used to explore changes in the relationship between two variables over the measurement space (Fig. 15.6D; Breiman et al. 1984). For instance, Rasmussen (1988) found a strong relationship between sediment redox potential and mayfly biomass in lakes up to a redox potential of -120 mV; no relationship existed between the two variables at higher redox potentials. The relationship between maximum submerged macrophyte biomass and littoral slope differs for slopes steeper and shallower than approximately 0.04 m m^{-1} (Duarte and Kalff 1986), and their relationship to habitat conditions differs above and below the depth of maximum biomass (Duarte 1987). In addition, changes in the relationship between two variables may be continuous rather than occur at specific threshold points (Fig. 15.6E). Prairie et al. (1989) have shown that the relationship between algal biomass and phosphorus and nitrogen concentrations changes with changing nitrogen-to-phosphorus ratios, thereby explaining disagreement in published chlorophyll–nutrient relationships. Ecological variance is multilevel in nature, so it is unlikely to fall into a single pattern or relationship. Threshold, boundary, and continuous relationships are probably inherent to most ecological phenomena. Because our ability to describe the structure of ecological variability depends on the ability of our models to combine all these possibilities, a successful search for patterns requires flexibility beyond that provided by regression and correlation alone.

The needed flexibility is provided by regression trees (Breiman et al. 1984). These combine threshold, boundary, and continuous relationships, in a hierarchical, treelike form, to assign an object, through a series of decisions, to the most probable values of the state variable. For instance, the relationship between the biomass of submerged macrophytes in lakes (at a regional scale) and lake (alkalinity, water transparency) and site (slope, depth, exposure) conditions have been described by a regression tree (Fig. 15.7; Duarte 1987). The regression tree proposed (Fig. 15.7) explains a greater proportion (60%) of the variance in submerged macrophyte biomass than direct regression analyses (40%), which yielded a biased equation with more than 13 significant terms (Duarte 1987). In addition, regression trees may incorporate many different tools within its structure, because the rules to split the data can be based on average values, threshold conditions, regression analyses, analyses of covariance, and discriminant analyses.

Regression trees may open new avenues in the search for pattern in comparative ecology, because they are simpler and provide more insight into the predictive structure of variability than regression analyses alone (Breiman et al. 1984), and yet they provide an appropriate frame to incorporate the hierarchical structure of ecological variability. They may also lend themselves to expert systems (e.g., Seip and Ibrekk 1988), thereby providing a bridge between the needs of basic and applied ecology.

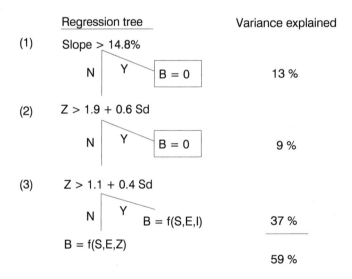

FIGURE 15.7. Regression tree describing patterns in biomass (B) of submerged macrophytes in lakes. Patterns in biomass are modeled through threshold, boundary, and continuous relationships. Condition (1) discriminates shores too steep (S > 14%) to support macrophytes; (2) discriminates depths (Z) beyond those where macrophytes are able to grow (based on water transparency, Sd); (3) discriminates between depths above and below that of maximum macrophyte biomass (based on water transparency, Sd), where the relationship [as multiple regression equations, f(x)] of biomass with habitat conditions differ (alkalinity, A; littoral slope, S; exposure to waves, E; irradiance, I; depth = Z). From Duarte (1987).

Conclusions

Comparative ecology describes nature by studying ecological variability to describe the possible states of ecological properties (e.g., value ranges) and, by searching for patterns within ecological variance, to extract the probable from among the possible. Opening the black box of ecological variability involves first an explicit examination of its structure, which is often hierarchical, corresponding to a nested array of scales of variation each contributing peculiar components to the global variability and presenting patterns that may differ from level to level. Attempts to describe patterns in ecological variability must, therefore, increase in flexibility to accommodate them to the different types of patterns (e.g., continuous relationships, boundary conditions, threshold relationships, etc.) and their changes when sampling across scales.

Acknowledgments. This chapter has benefited from most useful discussions and comments by Daniel E. Canfield, Jr., Mark V. Hoyer, Jacob Kalff, Rob H. Peters, and an anonymous reviewer.

REFERENCES

Adams, M.S., S. Guilizzoni, and S. Adams. (1978). Relationship of dissolved inorganic carbon to macrophyte photosynthesis in some Italian lakes. *Limnol. Oceanogr.* 23:912–913.

Allen, P.M. (1985). Ecology, thermodynamics, and self-organization: towards a new understanding of complexity. In: *Ecosystem Theory for Biological Oceanography*, R.E. Ulanowicz and T. Platt, eds. *Can. Bull. Fish. Aquat. Sci.* 213:3–26.

Allen, T.F.H. and T.W. Hoekstra. (1986). Instability in overconnected and underconnected systems: a matter of relative lag at different hierarchical levels. In: *Proceedings of the International Conference on Mental Images, Values, and Reality*, Vol. I, J.A. Dillon Jr., ed. Society for General Systems Research, Salinas, California, pp. D77–D85.

Allen, T.F.H. and T.B. Starr. (1982). *Hierarchy: Perspectives for Ecological Complexity.* University of Chicago Press, Chicago.

Anderson, G.C. (1969). Subsurface chlorophyll maximum in the northeast Pacific Ocean. *Limnol. Oceanogr.* 14:386–391.

Anderson, R.M. and J. Kalff. (1986). Regulation of submerged aquatic plant distribution in an uniform area of a weed bed. *J. Ecol.* 74:953–961.

Barnes, H. (1967). Ecology and experimental biology. *Helgol. Wiss. Meeresunters.* 15:6–26.

Breiman, L., J.H. Friedman, R.A. Olsen, and C.J. Stone. (1984). *Classification and Regression Trees.* Wadsworth, Belmont, California.

Brinkhurst, R.O. (1974). *The Benthos of Lakes.* MacMillan Press, London.

Canfield, D.E., Jr. (1981). Prediction of chlorophyll a concentrations in Florida lakes: the importance of phosphorus and nitrogen. *Water Resour. Bull.* 19:255–262.

Canfield, D.E., Jr. and C.M. Duarte. (1988). Patterns in the biomass and cover of aquatic macrophytes in lakes: a test in Florida lakes. *Can. J. Fish. Aquat. Sci.* 45:1976–1982.

Chambers, P.A. and J. Kalff. (1985). Depth distribution and biomass of submerged macrophyte communities in relation to Secchi depth. *Can. J. Fish. Aquat. Sci.* 42:701–709.

Cruzado, A. and K.C. Kelley. (1973). Continuous measurements of nutrient concentrations and phytoplankton density in the surface waters of the Western Mediterranean, winter 1970. *Thalassia Jugosl.* 9:19–24.

Denman, K.L., A. Okubo, and T. Platt. (1977). The chlorophyll fluctuation spectrum in the sea. *Limnol. Oceanogr.* 22:1033–1038.

Dillon, P.J. and F.H. Rigler. (1974). The phosphorus-chlorophyll relationship in lakes. *Limnol. Oceanogr.* 19:767–773.

Duarte, C.M. (1987). Biomass and distribution of submerged macrophytes in lakes. Ph.D. Dissertation, McGill University, Montreal, Canada.

Duarte, C.M. (1990). Time lags in algal growth: generality, causes, and consequences. *J. Plankton Res.* 12:873–883.

Duarte, C.M. and J. Kalff. (1986). Littoral slope as a predictor of the maximum biomass of submerged macrophyte communities. *Limnol. Oceanogr.* 31:1072–1080.

Duarte, C.M. and J. Kalff. (1988). The influence of lake morphometry on the response of submerged macrophytes to nutrient additions. *Can. J. Fish. Aquat. Sci.* 45:216–221.

Duarte, C.M. and J. Kalff. (1989). The influence of catchment geology and lake depth on phytoplankton biomass. *Arch. Hydrobiol.* 115:27–40.

Duarte, C.M. and J. Kalff. (1990). Patterns in the submerged macrophyte biomass of lakes and the importance of the scale of analysis in the interpretation. *Can. J. Fish. Aquat. Sci.* 47:357–363.

Duarte, C.M. and K. Sand-Jensen. (1990a). Seagrass colonization: patch formation and patch growth in *Cymodocea nodosa*. *Mar. Ecol. Prog. Ser.* 65:193–200.

Duarte, C.M. and K. Sand-Jensen. (1990b). Seagrass colonization: biomass development and shoot demography in *Cymodocea nodosa* patches. *Mar. Ecol. Prog. Ser.* 67:97–103.

Duarte, C.M., S. Agusti, and R.H. Peters. (1987). An upper limit to the abundance of aquatic organisms. *Oecologia* (Berlin) 74:272–276.

Duarte, C.M., J. Kalff, and R.H. Peters. (1986). Patterns in the biomass and cover of aquatic macrophytes in lakes. *Can. J. Fish. Aquat. Sci.* 43:1900–1908.

Estrada, M., M. Alcaraz, and C. Marrasé. (1987). Effects of turbulence on the composition of phytoplankton assemblages in marine microcosms. *Mar. Ecol. Prog. Ser.* 38:267–281.

Fee, E.J. (1976). The vertical and seasonal distribution of chlorophyll in lakes of the Experimental Lakes Area, northwestern Ontario: implications for primary production. *Limnol. Oceanogr.* 21:767–783.

Haldane, J.B.S. (1955). The measurement of variation. *Evolution* 9:484.

Hara, T. (1988). Dynamics of size structure in plant populations. *Trends Ecol. Evol.* 3:129–133.

Hurlbert, S.H. (1984). Pseudoreplication and the design of ecological field experiments. *Ecol. Monogr.* 54:187–211.

Lakatos, I. (1968). Criticism and the methodology of scientific research programmes. *Proceedings Aristotelian Soc.* 69:159–186.

Lewontin, R.C. (1966). On the measurement of relative variability. *Syst. Zool.* 15:141–142.

Margalef, R. (1978). Life forms of phytoplankton as survival alternatives in an unstable environment. *Oceanol. Acta* 1:493–509.

Margalef, R. (1982). *La Biosfera, entre la Termodinámica y el Juego.* Omega, Madrid.

Masó, M. and C.M. Duarte. (1989). The spatial and temporal structure of hydrographic and phytoplankton biomass heterogeneity along the Catalan Coast (NW Mediterranean). *J. Mar. Res.* 47:813–827.

McCauley, E. and J. Kalff. (1981). Empirical relationships between phytoplankton and zooplankton biomass in lakes. *Can. J. Fish. Aquat. Sci.* 38:458–463.

Mentis, M.T. (1988). Hypothetico-deductive and inductive approaches in ecology. *Funct. Ecol.* 2:5–14.

Montgomery, D.C. and E.A. Peck. (1982). *Introduction to Linear Regression Analysis.* Wiley, New York.

O'Neill, R.V. (1988). Hierarchical theory and global change. In: *Scales and Global Change*, T. Rosswall, R.G. Woodmansee, and P.G. Risser, eds. Wiley, New York.

Peters, R.H. (1983). *The Ecological Implications of Body Size.* Cambridge University Press, New York.

Peters, R.H. (1986). The role of prediction in limnology. *Limnol. Oceanogr.* 31: 1143–1159.

Peters, R.H. (1987). Los objetivos de la investigación y la naturaleza de la ciencia. *Alquibla* 10–11:23–29.

Popper, K.R. (1959). *The Logic of Scientific Discovery.* Hutchinson, London.

Prairie, Y.-T., C.M. Duarte, and J. Kalff. (1989). Unifying nutrient-chlorophyll relationships in lakes. *Can. J. Fish. Aquat. Sci.* 46:1176–1182.

Rasmussen, J.B. (1988). Habitat requirements of burrowing mayflies (Ephemeridae-Hexagenia) in lakes, with special reference to the effects of eutrophication. *J. North Am. Benthol. Soc.* 7:51–64.

Rigler, F.H. (1982). Recognition of the possible: an advantage of empiricism in ecology. *Can. J. Fish. Aquat. Sci.* 39:1323–1331.

Rørslett, B. (1987). A generalized spatial niche model for aquatic macrophytes. *Aquat. Bot.* 29:63–81.

Rose, G.A. and W.C. Leggett. (1990). The importance of scale to predator-prey spatial correlations: An example of Atlantic fishes. *Ecology* 71:33–43.

Seip, K.L. and H. Ibrekk. (1988). Regression equations for lake management — how far do they go?. *Verh. Int. Verein. Limnol.* 23:778–785.

Sokal, R.R. and F.J. Rohlf. (1969). *Biometry.* Freeman, San Francisco.

Thomas, W.H. (1979). Anomalous nutrient-chlorophyll interrelationships in the offshore eastern tropical Pacific Ocean. *J. Mar. Res.* 37:327–335.

Turing, A. (1952). The chemical basis of morphogenesis. *Philos. Trans. R. Soc. London B* 237:37–72.

Vaqué, D, C. Marrasé, V. Iñiguez, and M. Alcaraz. (1989). Zooplankton influence on the phytoplankton bacterioplankton coupling. *J. Plankton Res.* 11:625–632.

Vollenweider, R.A. (1987). Scientific concepts and methodologies pertinent to lake research and lake restoration. *Schweiz. Z. Hydrol.* 49:129–147.

16
Comparing Tropical and Temperate Forests

ARIEL E. LUGO AND SANDRA BROWN

Abstract

To avoid the pitfalls of traditional comparisons of tropical and temperate forests, we suggest an hierarchical approach that takes into consideration the great diversity of tropical forests and constrains comparisons by climatic (life zone), edaphic and topographic (plant associations), and temporal (successional) factors. The hierarchical approach is illustrated with comparative information on soil carbon, forest biomass, litterfall, forest complexity, and species richness. The analysis uncovers several misconceptions that have been adopted after incomplete comparisons between temperate and tropical forests.

"Las hayas y los arces de Europa están aquí reemplazados por las mas imponentes formas de la Ceiba y de las palmeras Praga e Irase" — Humboldt 1799

Like Humboldt, temperate zone visitors to tropical forests tend to evaluate them on the basis of familiar forests. These latitudinal comparisons have led to numerous generalizations about patterns of ecosystem structure and function, including greater richness of species, higher biomasses, and greater speed of ecological processes (e.g., succession, productivity, decomposition) in tropical forests than in temperate or boreal forests (Farnworth and Golley 1974). However, the comparisons of tropical with temperate forests have not always led to generalizations that withstand scrutiny. For example, the generalization that tropical soils are highly leached, infertile, and lower in organic matter content than temperate soils is a misconception (Sánchez 1976). Also, litter decomposition in some tropical forests can be as slow as that in temperate forest (compare data in Cole and Rapp 1981; Brown and Lugo 1982; Lugo and Murphy 1986).

Further, the belief that tropical forests accumulate higher biomasses than temperate ones is incorrect (Brown and Lugo 1982; Brown et al. 1989). In fact, it is becoming increasingly obvious that it is no longer possible to make simple comparisons between "tropical" and "temperate" forests. Such comparisons are bound to confuse rather than clarify the many interesting and at times counterintuitive similarities and differences between these two bioregions of the globe.

Many of the misconceptions that result from poor comparisons between tropical and temperate forests arise because little attention is given to the diversity of tropical environments (Holdridge 1967; Sánchez 1976; Lugo and Brown 1981). Instead, the tropics are often viewed as a fairly homogeneous region (generally hot and humid) with environmental diversity similar to that found in the temperate zone. Many of the generalizations about the tropics are based solely on assumptions about the humid tropics, a zone that is usually poorly defined and that can extend into superhumid zones of the biome with a rainfall range of more than 6000 mm, while ignoring the dry tropics, which comprise more than 40% of the tropical land mass (Brown and Lugo 1982).

To avoid the pitfalls of traditional comparisons of tropical and temperate forests, we suggest a hierarchical approach that takes into consideration the great diversity of tropical forests and constrains comparisons by climatic (life zone), edaphic and topographic, and temporal factors. The advantage of this approach is that the number of variables that influence the comparisons is reduced, and limitation is the paucity of quantitative data for tropical forests. Information is not available to treat each level of the hierarchy with equal comprehensiveness. For this reason, this analysis contains more information on the first level of the hierarchy (life zone or climatic), and less for the other two levels (edaphic and topographic, and temporal).

The System for the Classification of World Life Zones by L.R. Holdridge (1967) provides the basis for comparing tropical and temperate forests. This is an objective (empirical) system for classifying world climates (life zones), plant associations (usually edaphically or topographically controlled), and successional (temporal) stages of vegetation such that given plant association occurs in one life zone only. The determination of life zone conditions for any given land area provides an objective first step in attempting comparisons of that land area with other locations in the world. Life zones are delimited by two of three environmental factors: mean annual biotemperature, mean annual precipitation, and potential evapotranspiration. Biotemperature is obtained by assigning values of 0 to all mean daily temperatures above 30°C or below 0°C. Holdridge (1967) assumed that plant net photosynthesis is positive between 0° and 30°C. The scales of the life zone chart are geometric (Fig. 16.1A and 16.1B). For montane locations, biotemperature values are adjusted to sea level using the lapse rate of 6°C/1000 m.

A summary of the geographic distribution of life zones (Table 16.1) illustrates how life zone conditions can be useful for comparing ecosystems. The tropical region contains almost twice as many life zones as the temperate zone and about seven times the number in the boreal zone. In fact, more than half of the world's life zones are tropical. This richness of tropical

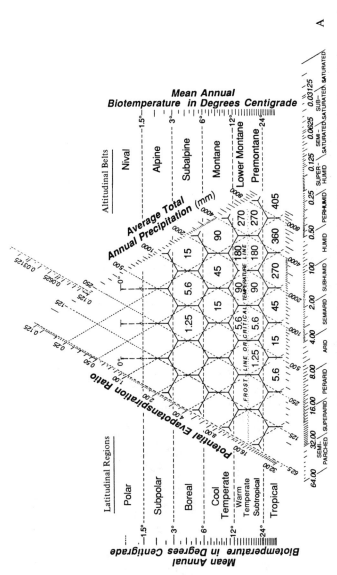

FIGURE 16.1. Complexity index (A) and tree species richness (B) in tropical and subtropical life zones of Costa Rica. Values are based on 0.1-ha plots, trees with diameter at breast height >10 cm, and plant associations without any excessively favorable or restrictive growth factors. Complexity index is product of basal area, tree height, tree density, and the number of tree species multiplied by 10^{-3}. Data are from Holdridge (1967) and Holdridge et al. (1971).

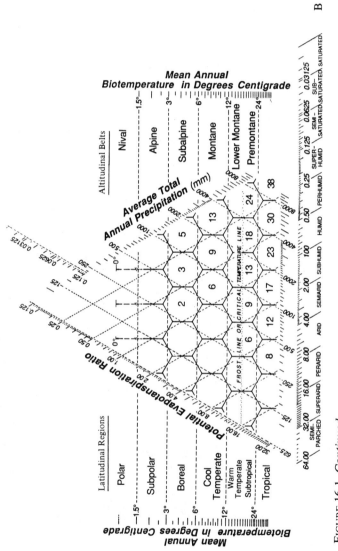

FIGURE 16.1. *Continued*

TABLE 16.1. Number of life zones by major geographic regions and climatic zones.

	Number of life zones			
Category	Tropics	Temperate	Boreal	Total
Dry[a]	33	16	3	52
Moist[b]	11	7	2	20
Wet[c]	22	14	4	40
Total	66	37	9	112
Forested	33	16	3	52

[a]Potential evapotranspiration-to-precipitation ratio (PET/P) >1.
[b]PET/P $= 0.5$–1.
[c]PET/P <0.5.

life zones, including forested life zones, extends over broad gradients of temperature and rainfall.

Comparisons of temperate and boreal forests are likely to require fewer precautions in terms of controlling intrinsic environmental variability because the number of life zones involved is small. The low number of life zones in these two regions increases the likelihood that a given forest type is close to being representative of the region as a whole. This is not true of the tropical region, where the number of life zones exceeds the total of all other regions of the world.

Because critical structural and functional characteristics of forests vary with life zone conditions, it is possible when comparing forests to reach almost any conclusion by ignoring life zone constraints. For example, tropical closed forests may appear to be less productive than temperate closed forests if the tropical forest studied was seasonally dry and the temperate one a moist maple forest. The difference in productivity would be a function of rainfall, not latitude. But comparing tropical moist and temperate moist forests could more closely reveal a latitudinal difference in productivity (assuming control for edaphic and age characteristics).

The importance of life zone conditions for ecosystem comparisons can be further illustrated with the variation of ecosystem complexity and tree species richness among life zones (see Fig. 16.1). The complexity index was devised by Holdridge (1967) for comparing mature vegetation (trees >10 cm in breast height diameter, dbh) sampled in 0.1-ha plots; it is the product of height (m), tree density, basal area (m^2), and number of tree species multiplied by 10^{-3}.

Each life zone appears to have a characteristic complexity index and tree species richness that increase toward the lowland tropics and with increasing rainfall. These trends have been confirmed by several authors. For example, Gentry (1982) and Gentry and Dodson (1987) found a linear relation between total plant species richness and rainfall in tropical environments. Rice and Westoby (1983) demonstrated that at the scale of 0.1 ha, a given plant association bounded by the same mean annual rainfall and temperature regimes (a hexagon in the life zone chart, Fig. 16.1) would always have the same number of species, whether it be in a temperate or a tropical environment. Itow (1988) used a warmth index (sensu Kira 1977) to show latitudinal changes in Fisher's alpha diversity (sensu Williams 1947) on the mainland and on Pacific islands. This temperature gradient segregated tropical and temperate forests, as well as forests in small isolated islands from forests on the mainland.

Use of life zones helps underscore other differences and similarities between tropical and temperate forests. Figure 16.2 shows the distribution of soil carbon content in tropical and temperate soils. The highest soil carbon contents are in the wet life zones (potential evapotranspiration-to-precipitation ratio, PET/P, <0.5), regardless of latitudinal region. However, in the moist and dry life zones, sharp deviations in soil carbon content across latitudinal regions are exhibited at the frost line (Fig. 16.2), and the contents in the temperate zone are generally higher than in the tropics. In this example, the latitudinal effect for one ecosystem parameter (soil carbon) results in similar as well as contrasting patterns when two moisture extremes are considered.

The rate of aboveground biomass accumulation during the first 40 years of succession appears to be much faster in tropical moist broadleaf forests than in temperate moist broadleaf forests (Fig. 16.3A; the slope of the regression line for tropical forest is 4.39 versus 1.72 for temperate forests). However, the database for young temperate forests is poor compared to tropical forests, whereas data for older, known-age tropical forests are few compared to temperate forests. In spite of this paucity of data, both geographic biomes appear to achieve similar biomasses at maturity, but because the tropical forest approaches steady state first, the temperate zone forests continue to exhibit net biomass increase for a longer time period. The implication of these results to the global carbon cycle is that the aboveground biomass compartment in comparable moist tropical and moist temperate forests may have similar potential as carbon sinks, but the tropical forests act as sinks that maximize this removal of carbon early in the succession (first 40 years?) whereas the temperate forests function as slower carbon sinks that operate over longer time periods (for about 100 years). It would thus appear that many "mature" temperate forests are still functioning as carbon sinks at rates comparable to their younger stages of succession.

The notion that litter production is linearly related to latitude and controlled by temperature was first proposed by Bray and Gorham (1964) and later confirmed by O'Neill and DeAngelis (1981), using the woodlands data set from the IBP program. However, a close examination of the data for tropical latitudes shows that all were from forests in the tropical moist life zones that have the highest rates of litter production (Brown and Lugo 1982). When a larger data set of tropical forest litter production is included,

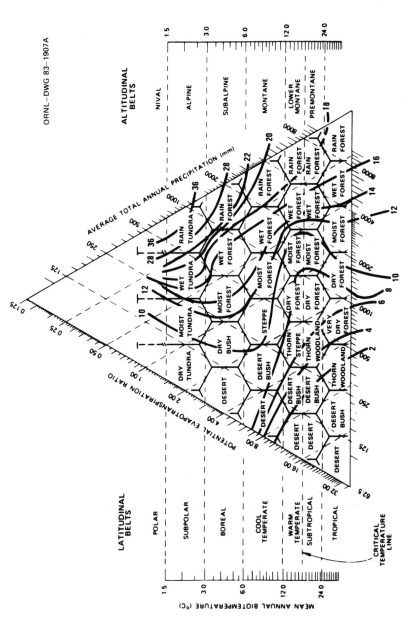

FIGURE 16.2. Isopleths of soil carbon density plotted in the Holdridge life zone chart. Dashed lines indicate probable extrapolation of isopleth in climatically extreme life zones for which no data were available. From Post et al. (1985).

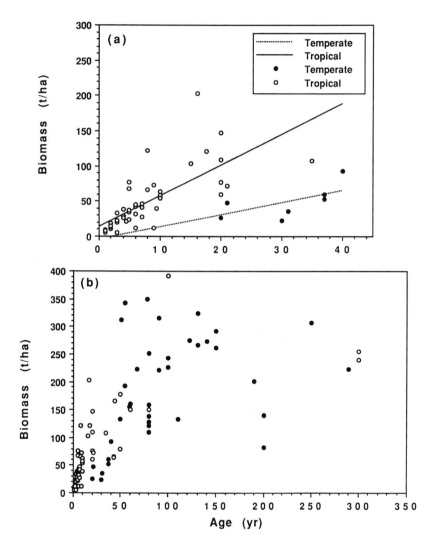

FIGURE 16.3. Relationship between stand age and aboveground biomass in tropical and temperate moist broadleaf forests. (A) Biomass versus age for forest < 40 years old. Equations for both regression lines (significant at $p = .05$): tropical biomass (in t/ha) = 13.31 + 4.39* age (in years) ($r^2 = .55$) and temperate biomass = −4.16 + 1.72* age (in years) ($r^2 = .65$). (B) Biomass versus age for complete data set. Data for tropical forests are summarized in Brown and Lugo (1982, 1990) and for temperate forests in DeAngelis et al. (1981).

the Bray and Gorham (1964) relationship between litterfall and latitude is considerably weakened, and, as suggested by O'Neill and DeAngelis (1981), the slope is less pronounced (Fig. 16.4A). The reason is that low litterfall production values from tropical dry and tropical rain forest life zones greatly increase the scatter of data at low latitudes. These life zones have litterfall rates that overlap with those of high-latitude forests.

An alternative hypothesis is that litterfall is responding nonlinearly to moisture availability. The pattern of response of litter production to moisture availability (as measured by the ratio of temperature to precipitation, T/P, [Brown and Lugo 1982] is similar in tropical and temperate zone forests (Fig. 16.4B). However, for any given life zone, tropical forests have higher rates of production with almost no overlap of values as in Fig. 16.4A. The response of this ecosystem parameter to latitude is an excellent example of the effect of life zone richness in the tropics (see Table 16.1) and the consequent wide range of litterfall values at low latitudes.

Because the biological complexity (see Fig. 16.1), carbon accumulation (Figs. 16.2 and 16.4), and rate of biomass turnover (Figs. 16.3 and 16.4) of forests are related to life zone, ecologists must consider the importance of climatic factors when comparing structure and processes in forests. To isolate the importance of latitude in determining species richness or any other forest parameter, it is necessary to compare life zones along temperature gradients with the same water balance, that is, along a constant PET/ P line in the life zone chart. Alternatively, one could explore the same temperature gradient, but along a constant precipitation line, that is, decreasing water balance resulting from increasing temperature in the tropical latitudes or from high elevation toward the lowland forests.

The second level in the hierarchical comparison of ecosystems was demonstrated by Holdridge et al. (1971) in a comparison of mature forests in Costa Rica from the same life zones but growing under different edaphic and topographic conditions. Holdridge et al. found trends in forest complexity and structure along gradients of soil moisture, depth of water table, and salinity. In dry forests, high water tables increased forest complexity, while in wet and rain forest life zones, the opposite occurred. Drainage improved wet and rain forest structure but stressed dry forest structure.

More rigorous analysis is required at the third level in the hierarchical approach when comparisons of tropical and temperate forests involve smaller scale or specific ecological phenomena. An example would be comparisons of plantations and natural forests or freshwater and saltwater forested wetlands. Small-scale comparisons are vulnerable to greater site factor interference caused by intersite variability. These comparisons thus require consideration of all three levels in the proposed hierarchical approach: (1) life zone (climatic), (2) edaphic and topographic, and (3) temporal or hydrological factors for terrestrial and wetland ecosystems, respectively (see Lugo et al. 1988 for comparisons of forest wetlands). For example, failure to control for soil texture or slope may invalidate compari-

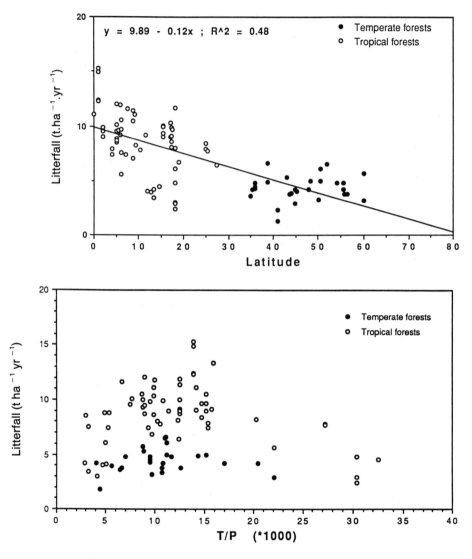

FIGURE 16.4. Relationship between (A) total litterfall and latitude and (B) ratio of temperature to precipitation (T/P) for tropical and temperate forests. Data for tropics are from Brown and Lugo (1982), with additions of Proctor et al. (1983) and Rai and Proctor (1986); data from temperate forests are from DeAngelis et al. (1981). Same data are used in both plots.

sons of soil fertility and soil organic matter content in tropical and temperate forests (Anderson and Coleman 1985). Similarly, it is questionable to compare young managed forests with mature natural forests unless the objective is to document the trivial differences between forests of different age. These kinds of comparisons must control for age of forests, life zone conditions, and edaphic and topographic factors so that the effects of management can be uncovered.

In summary, we have demonstrated that early comparisons of tropical and temperate forests were casual observations based on temperate experience and suffered from little if any rigorous control of the variables that influence the ecosystem properties being compared. In contrast, the use of a hierarchical approach based on Holdridge (1967) and consisting of the life zone (climatic factors), plant association (edaphic and topographic factors), and temporal stages of succession provides a powerful scheme for comparisons of forests.

REFERENCES

Anderson, D.W. and D.C. Coleman. (1985). The dynamics of organic matter in grassland soils. *J. Soil Water Conserv.* 40:211–216.

Bray, J.R. and E. Gorham. (1964). Litter production in forests of the world. *Adv. Ecol. Res.* 2:101–157.

Brown, S. and A.E. Lugo. (1982). The storage and production of organic matter in tropical forests and their role in the global carbon cycle. *Biotropica* 14:161–187.

Brown, S. and A.E. Lugo. (1990). Tropical secondary forests. *J. Trop. Ecol.* 6:1–32.

Brown, S., A.J.R. Gillespie, and A.E. Lugo. (1989). Biomass estimation methods for tropical forests with applications to forest inventory data. *For. Sci.* 35:881–902.

Cole, D.W. and M. Rapp. (1981). Elemental cycling in forest ecosystems. In: *Dynamic Properties of Forest Ecosystems*, D.E. Reichle, ed. Cambridge University Press, Cambridge, pp. 341–409.

DeAngelis, D.L., R.H. Gardner, and H.H. Shugart, Jr. (1981). Productivity of forest ecosystems studied during IBP: the woodlands data set. In: *Dynamic Properties of Forest Ecosystems*, D.E. Reichle, ed. Cambridge University Press, Cambridge, pp. 567–672.

Farnworth, E.A. and F.B. Golley. (1974). *Fragile Ecosystems.* Springer-Verlag, New York.

Gentry, A.H. (1982). Patterns of neotropical plant species diversity. *Evol. Biol.* 15: 1–83.

Gentry, A.H. and C. Dodson. (1987). Contribution of nontrees to species richness of a tropical rain forest. *Biotropica* 19:149–156.

Holdridge, L.R. (1967). *Life Zone Ecology.* Tropical Science Center, San Jose, Costa Rica.

Holdridge, L.R., W.C. Grenke, W.H. Hatheway, T. Liang, and J.A. Tosi, Jr. (1971). *Forest Environments in Tropical Life Zones, a Pilot Study.* Pergamon Press, Oxford.

Humboldt, A. (1799). *Viaje a Las Regiones Equinocciales del Nuevo Mundo.* Vol.

2. (Translated in 1941 by L. Alvarado.) Monte Avila Editores, Caracas, Venezuela, p. 85.

Itow, S. (1988). Species diversity of mainland and island forests in the Pacific area. *Vegetatio* 77:193-200.

Kira, T. (1977). A climatological interpretation of Japanese vegetation zones. In: *Vegetation Science and Environmental Protection*, A. Miyawaki and R. Tuxen, eds. Maruzen, Tokyo.

Lugo, A.E. and S. Brown. (1981). Tropical lands: popular misconceptions. *Mazingira* 5:10-19.

Lugo, A.E. and P.G. Murphy. (1986). Nutrient dynamics of a Puerto Rican subtropical dry forest. *J. Trop. Ecol.* 2:55-72.

Lugo, A.E., S. Brown, and M.M. Brinson. (1988). Forested wetlands in freshwater and saltwater environments. *Limnol. Oceanogr.* 33:894-909.

O'Neill, R.V. and D.L. DeAngelis. (1981). Comparative productivity and biomass relations of forest ecosystems. In: *Dynamic Properties of Forest Ecosystems*, D.E. Reichle, ed. Cambridge University Press, Cambridge, pp. 411-450.

Post, W.M., J. Pastor, P.J. Zinke, and A. Stangenberger. (1985). Global patterns of soil nitrogen storage. *Nature* (London) 317:613-616.

Proctor, J., J.M. Anderson, S.C.L. Fogden, and H.W. Vallack. (1983). Ecological studies in four contrasting lowland rain forests in Gunung Mulu National Park, Sarawak. II. Litterfall, litter standing crop and preliminary observation on herbivory. *J. Ecol.* 71:261-283.

Rai, S.N. and J. Proctor. (1986). Ecological studies on four rainforests in Karnataka, India. II. Litterfall. *J. Ecol.* 74:455-463.

Rice, B. and M. Westoby. (1983). Plant species richness at the 0.1 hectare scale in Australian vegetation compared to other continents. *Vegetatio* 52:129-140.

Sànchez, P.A. (1976). *Properties and Management of Soils in the Tropics*. Wiley, New York.

Williams, C.B. (1947). The logarithmic series and its application to biological problems. *J. Ecol.* 34:253-272.

17
Density-Dependent Positive Feedbacks between Consumers and Their Resources

THOMAS S. BIANCHI AND CLIVE G. JONES

Abstract

Increases in consumer population density are traditionally considered to result in decreases in the per capita supply of resources and consumer fitness. Here we review studies that demonstrate a density-dependent positive feedback between asocial species of consumers and their resources. In all cases in which consumer activity results in increased primary production or resource supply and the fitness of the consumer has been measured, fitness is found to be enhanced at some higher densities, in marked contrast to the traditional view. Such interactions occur with divergent consumer taxa living plants, detritus, and abiotic resources in very different types of ecosystems. Our cross-system analysis revealed that consumer-induced increases in resource productivity and supply is common, but effects on consumer fitness have rarely been examined. Based on the assumption that positive feedbacks on consumer fitness cannot occur unless resource productivity or supply increases, we examine known examples of consumer-induced increases in resource production and supply to identify common mechanisms.

Consumer-induced increases in nutrient cycling appear to be the most common mechanism. We therefore focused on this mechanism, examining the likely constraints on increased production and supply and the probability that this will result in increased consumer fitness. General system characteristics, intrinsic characteristics of resource and consumer, and spatial and temporal characteristics of consumer–resource interactions are postulated to be the major controlling factors. These characteristics are used to predict when and where increased resource production and supply will occur and when this will lead to positive feedbacks on consumer fitness.

Introduction

The fundamental objective of most cross-system analyses is to determine the similarity of system or component-level processes between different ecosystems. Much of the discussion at the Third Cary Conference provided interesting cross-system comparisons within a particular type of ecosystem, for example, within aquatic or terrestrial systems. However, comparisons across very different systems, such as deserts and lakes, received little attention. The generality of a particular process in nature is strengthened by evidence for its occurrence in a wide range of ecosystems and taxa. We recently reviewed studies demonstrating density-dependent positive feedbacks between consumers and their resources in several different systems. Positive feedbacks are defined as an increase in consumer growth and fitness as population density increases, across certain density ranges (Bianchi et al. 1989). This is in marked contrast to the traditional view of negative density-dependent feedbacks such as density-dependent intraspecific competition. Here we take a broader, cross-system perspective of the potential for positive interactions between consumers and their resources at the population level. We present some of the general patterns of consumer-induced increases in resource production and supply, examining the general mechanisms, and discuss when and where these are most likely to lead to positive feedbacks on consumer fitness.

Consumer–Resource Interactions and Positive Feedback on Consumer Fitness

The idea that the feeding activity of asocial animal species can increase the productivity of their resources has received considerable attention in past years (Chew 1974; Flint and Goldman 1975; Cargill and Jeffries 1984; Sterner 1986; Carpenter 1986; Carpenter and Kitchell 1987). For example, herbivory has been shown to increase overall primary production at high compared to low levels of consumption in both aquatic and terrestrial environments (Bianchi et al. 1989). Although most of the attention has focused on the effects of consumers on the fitness of living resources (reviewed in Belsky 1986), few studies have taken the extra step to examine whether or not a positive feedback on the fitness of the consumer occurs.

A number of studies have demonstrated that consumer fitness can increase with increasing consumer population density. Bianchi et al. (1989) argued that positive feedbacks in natural populations are not detectable from simple correlations between resource abundance and animal population density in space or time. This may explain why positive feedbacks are not commonly reported and have not received much attention. However, experimental manipulation of resource supply or animal density can reveal the presence of positive feedbacks. We have postulated that density-dependent positive feedbacks may result in higher equilibrium densities of animal populations and alter the density range over which negative feedbacks from intraspecific competition occur; such interactions can be described by a simple model (Bianchi et al. 1989).

Consumer-Induced Increases in Resource Production and Supply across Divergent Ecosystems and Taxa

Here we expand our review of the literature on interactions between resources and consumers to determine the extent to which positive feedbacks have been shown to occur in different systems. We focus on literature that shows increases in resource productivity or supply, based on the assumption that only in these cases was there potential for positive feedbacks on consumer growth or fitness. Table 17.1 illustrates that the phenomenon of increased resource production or supply resulting from consumer activity appears widespread in asocial organisms. Many different consumer taxa, resources, and widely divergent ecosystems are represented.

Researchers in different subdisciplines have demonstrated that consumer activity can increase resource production or supply, but the generality of this phenomenon across systems and taxa has not been previously recognized. Most importantly, Table 17.1 shows that the potential consequences of increased production or supply to consumer fitness have rarely been tested. Nevertheless, in the six cases that have been examined, consumer fitness was shown to increase. It would therefore seem reasonable that a concerted effort should be made to determine whether or not positive feedbacks on the consumer occur in the large number of examples we have found that demonstrate consumer-induced increases in resource production or supply. Further, most studies have focused on interactions between herbivorous consumers and living plant resources. Detrital and abiotic resources such as water have largely been ignored, despite the evidence that consumer activity can increase resource supply and consumer fitness (Table 17.1).

Given the dearth of studies examining positive feedbacks on consumer fitness, it is not possible to say exactly how widespread this phenomenon may be. Nevertheless, we can examine the patterns and mechanisms involved in consumer-induced increases in resource production and supply, with the recognition that these interactions have potential to increase consumer fitness. A number of mechanisms appear to generate density-dependent increases in resource production and supply. These include rapid recycling of nutrients that would otherwise be supplied at slower rates (polychaetes/sandflat and geese/marsh); stimulation of production by maintenance of highly productive stages of the resource (ungulates/grassland and sawflies/willows); diversion of resources to the consumer (aphids/agricultural crops); changes in resource suitability that increase supply (beetles/conifers); and modifications of the physical environment that increase the probability of finding the resource (isopods/desert) or its rate of supply (polychaetes/mudflat).

Our survey indicated that there may be relatively few dominant mechanisms. For example, 11 of the 15 examples presented involving consumers

TABLE 17.1. Examples of positive density-dependent interactions between divergent taxa of asocial species of consumers and their resources in different ecosystems. These interactions affect consumer growth and fitness or resource productivity and supply.

Ecosystem	Consumer (class)	Resource	Resource feedback	Feedback mechanism	Consumer fitness[a]	References
Open ocean	Copepods (Crustacea)	Phytoplankton	Increased production	Increased nutrient cycling	NT	Glibert et al. 1982
Coral reefs	Sea urchins (Echinoidea)	Benthic algae	Increased production	Increased nutrient cycling	NT	Carpenter 1986
Coral reefs	Damselfish (Osteichthyes)	Benthic algae	Increased production	Juvenilization[b] and nutrient cycling	NT	Montgomery 1980
Mudflats	Worms (Polychaeta)	Plant detritus	Increased flow	Increased sediment mixing	Increased	Rice 1986; Rice et al. 1986
Sandflats	Worms (Polychaeta)	Benthic diatoms	Increased production	Increased nutrient cycling	Increased	Bianchi and Rice 1988
Streams	Crayfish (Crustacea)	Benthic algae	Increased production	Increased nutrient cycling	NT	Flint and Goldman 1975
Lakes	Water fleas (Crustacea)	Phytoplankton	Increased production	Increased nutrient cycling	NT	Sterner 1986
Lakes	Cladocerans	Phytoplankton	Increased production	Increased nutrient cycling	NT	Bergquist and Carpenter 1986
Marshes	Snow geese (Aves)	Grass	Increased production	Increased nutrient cycling	NT	Cargill and Jeffries 1984
Grasslands	Sheep (Mammalia)	Grass	Increased production	Increased nutrient cycling	NT	Vickery 1972
Grasslands	Zebra and wildebeests (Mammalia)	Grass	Increased production	Increased nutrient cycling	NT	McNaughton 1985

						Reference
Grasslands	Gophers (Mammalia)	Grass	Increased production	Increased light and nutrients	NT	Huntly and Inouye 1988
Grasslands	Grasshoppers (Insects)	Grass	Increased production	Unknown	NT	Dyer and Bokhari 1976
Riparian zones	Sawflies (Insects)	Willow shrubs	Increased production	Juvenilization of resource	Increased	Craig et al. 1986
Agricultural crops	Aphids (Insects)	Vegetable crops	Increased flow	Increased sink pressure by aphids	Increased	Way and Cammell 1970
Coniferous forests	Pine beetles (Insects)	Trees	Defenses weakened	Increased resource suitability	Increased	Berryman 1976
Tundra	Voles (Mammalia)	Herbs	Increased production	Juvenilization of resources	NT	Smirnov and Tokmakova 1972
Desert	Isopods (Isopoda)	Water	Increased availability	Collective burrowing to water table	Increased	Shacak and Yair 1984
Soils	Nematodes (Nematoda)	Bacteria	Increased production	Increased nutrient cycling	NT	Ingham et al. 1985

[a]Fitness measured as increase in growth or fecundity. NT, not tested.

[b]Juvenilization: herbivores, via feeding activity, result in higher proportion of young, highly productive primary producers, compared to older, less productive standing crop, in absence of herbivores.

and primary producers (see Table 17.1) show increased production of the resource as a consequence of increased nutrient cycling. However, mechanisms such as juvenilization (i.e., younger, more productive resources) may also turn out to be important. There are too few examples of consumer interactions with detritus and abiotic resources to make generalizations about the mechanisms at this time.

Predicting When and Where Positive Feedbacks on Consumer Fitness Will Occur

Even though there have been very few studies that examine positive feedbacks on consumer fitness, we believe that it is possible to predict the circumstances in which positive feedbacks on consumer fitness are likely to occur. Our model of positive feedbacks (Bianchi et al. 1989) and the examples in Table 17.1 suggest that this can be accomplished by considering system constraints on both the consumer and the resource; intrinsic characteristics of consumer and resource; and temporal and spatial constraints on consumer–resource interactions. We illustrate these ideas by using the most commonly observed pattern of increased primary production that results from consumer-induced increases in nutrient cycling. We distinguish situations in which increased production can occur, as well as situations in which consumer growth and fitness will increase as a consequence of increased production.

SYSTEM CONSTRAINTS

If nutrient inputs to the system are very high, or rates of nutrient cycling via routes other than the consumer are high (i.e., not nutrient limited), and there are no other constraints on primary production (e.g., water, light), then rates of primary production will be close to the intrinsic maximum for the plant species in the system. Consumer-induced nutrient cycling is unlikely to affect primary production in these circumstances, and thus density-dependent positive feedbacks on consumer fitness will not occur. On the other hand, if nutrient inputs to the system are low, or rates of nutrient cycling via nonconsumer routes are low (i.e., nutrient limited, such as the open ocean), and there are no other constraints on primary production, then consumer activity is likely to increase primary production in a density-dependent manner over some density range. This may or may not result in increased consumer growth and fitness because intrinsic constraints, and temporal or spatial scales of consumer–resource interactions, will then be critical.

In addition, if other system constraints on the consumer affect growth rate or reproductive output to a greater extent than resource supply (e.g., climate, predators, parasites, diseases), positive feedback effects on consumer growth and fitness are unlikely to occur, even if there is increased resource production or supply. Further, intraspecific competition for the same resource from other consumers in the system may also prevent positive feedbacks, even if production or supply is increased. Last, systems in which the flow of abiotic resources, such as the rapid flow of water, carry regenerated nutrients away as fast as they are produced are also likely to prevent positive feedbacks.

INTRINSIC CONSTRAINTS

Inherently slow-growing producers, that is, plants which have low genetically determined rates of production, are unlikely to show marked increases in production even if nutrient cycling increases. On the other hand, fast-growing plants are likely to show increased production when their rates of nutrient supply increase. Similarly, consumers adapted to living on low-productivity resources are unlikely to show marked increases in growth or fitness, while consumers adapted to inherently fast-growing producers will show such increases.

TEMPORAL AND SPATIAL CONSTRAINTS ON CONSUMER-RESOURCE INTERACTIONS

If the turnover rate of the consumer population exceeds the rate with which production of the resource can increase, following enhanced nutrient recycling, then feedbacks will not occur. Thus consumers with rapid growth rates or reproductive outputs, such as many opportunistic species, are less likely to show positive feedback effects on growth or fitness if they feed on resources that turn over at a relatively slower rate. It is certainly possible that intergenerational effects on the consumer could occur (e.g., sawflies; see Table 17.1), provided that increased reproductive output from one generation is synchronized with subsequent increases in primary production. Further, consumers that have very high growth rates and reproductive outputs, such as many insect herbivores, are likely to overshoot increases in resource supply, experiencing intraspecific competition at high densities. These consumers will tend to show positive feedbacks over relatively short periods of time unless other factors, such as emigration, reduce the local population growth rate. This raises the interesting possibility that positive feedbacks in one system or part of a system could make important contributions to the establishment and dynamics of consumer populations in adjacent areas. Relatively slow-growing consumers (e.g., ungulates) consuming fast-growing plants (e.g., grasses) are likely to show positive feedback effects on growth within generations and reproduction across generations.

Spatial scales at which consumer and resource interact are also likely to be important. Consumer populations that move from one patch of resources to another may increase nutrient cycling and primary production in one patch, but may not benefit in terms of increased growth and fitness because they are now feeding elsewhere. When patch-to-patch movement is bounded, such that increased nutrient input at one time results in increased production that is then consumed by resident consumers, we would expect positive feedbacks. Alternatively, if the consumer is migratory (e.g., snow geese or ungulates in the Serengeti), positive feedbacks will occur if the herbivore returns to the same location, year after year. The benefit here will, certainly, depend on synchrony of consumer-induced nutrient cycling, subsequent increases in the production, and timing of return of the herbivore. When resources move continuously (e.g., phytoplankton), positive feedbacks will require either very rapid recycling and/or continuous spatial tracking of the resource by the consumer.

Summary of Predictions

Overall, the extent to which both increases in resource production and supply and positive feedbacks on consumer growth and fitness will occur should be determined by constraints imposed by the system, intrinsic characteristics of the resource and consumer, and the spatial and temporal scales of consumer–resource interactions. We predict that for interactions involving nutrient cycling, both of the aforementioned phenomena are most likely to occur in systems and taxa in which the following are combined: (1) potential primary production is high, but nutrient limited; (2) consumer growth and fitness is primarily determined by the resource; (3) producers are relatively fast growing; (4) consumers respond to increased resource availability, but increase their growth or fitness at rates lower than or equal to the rates of increased production. This would be particularly true for individual consumers. Alternatively, increases in resource availability may affect subsequent generations of consumers: (5) consumer reproductive rates are not so high as to rapidly result in negative density-dependent intraspecific competition; (6) the resource and consumer are spatially and temporally connected in terms of local nutrient recycling and increased primary production or supply. In principle, it should be possible to qualitatively estimate these characteristics for many of the examples shown in Table 17.1, and it should also be possible to model these interactions and constraints. Future studies in this area could therefore test whether or not these assumptions and the concomitant predictions are correct and evaluate the importance of positive feedbacks from the viewpoint of the consumer, the producer, and the system.

Acknowledgments. We thank George McManus and David Strayer for their constructive criticisms. We also thank Richard Ostfeld for insightful comments on the ramifications of positive feedbacks on local versus regional population dynamics. We thank the Mary Flagler Cary Charitable

Trust for financial support. This article is a contribution to the program of the Institute of Ecosystem Studies.

REFERENCES

Belsky, A.J. (1986). Does herbivory benefit plants? A review of the evidence. *Am. Nat.* 127:870-892.

Bergquist, A.M. and S.R. Carpenter. (1986). Limnetic herbivory: effects on phytoplankton populations and primary production. *Ecology* 67:1351-1360.

Berryman, A.A. (1976). Theoretical explanation of mountain pine beetle dynamics in lodgepole pine forests. *Environ. Entomol.* 5:1225-1233.

Bianchi, T.S. and D.L. Rice. (1988). Feeding ecology of *Leitoscoloplos fragilis* II. Effects of worm density on benthic diatom production. *Mar. Biol.* 99:123-131.

Bianchi, T.S., C.G. Jones, and M. Shachak. (1989). The positive feedback of consumer population density on resource supply. *Trends Ecol. Evol.* 4:234-238.

Cargill, S.M. and R.L. Jeffries. (1984). The effects of grazing by lesser snow geese on the vegetation of a sub-arctic salt marsh. *J. Appl. Ecol.* 21:669-686.

Carpenter, R.C. (1986). Partitioning herbivory and its effects on coral reef algal communities. *Ecol. Monogr.* 56:345-363.

Carpenter, S.R. and J.F. Kitchell. (1987). The temporal scale of variance in limnetic primary production. *Am. Nat.* 129:417-433.

Chew, R.M. (1974). Consumers as regulators of ecosystems: an alternative to energetics. *Ohio J. Sci.* 74:359-370.

Craig, T.P., P.W. Price, and J.K. Itami. (1986). Resource regulation by a stem-galling sawfly on the arroyo willow. *Ecology* 67:419-425.

Dyer, M.I. and U.G. Bokhari. (1976). Plant-animal interactions: studies of the effects of grasshopper grazing on blue gramma grass. *Ecology* 57:762-772.

Flint, R.W. and C.R. Goldman. (1975). The effects of a benthic grazer on the primary productivity of the littoral zone of Lake Tahoe. *Limnol. Oceanogr.* 20: 935-944.

Glibert, P.M., D.C. Biggs, and J.J. McCarthy. (1982). Utilization of ammonium and nitrate during austral summer in the Scotia Sea. *Deep-Sea Res.* 29:837-850.

Huntly, N. and R. Inouye. (1988). Pocket gophers in ecosystems: patterns and mechanisms. *BioScience* 38:786-793.

Ingham, R.E., J.A. Trofymow, J.A. Ingham, and D.C. Coleman. (1985). Interactions of bacteria, fungi, and their nematode grazers: effects of nutrient cycling and plant growth. *Ecol. Monogr.* 55:119-140.

McNaughton, S.J. (1985). Ecology of a grazing ecosystem: the Serengeti. *Ecol. Monogr.* 55:259-294.

Montgomery, W.L. (1980). The impact of non-selective grazing by the giant blue damselfish, *Microspathodon dorsalis*, on algal communities in the Gulf of California, Mexico. *Bull. Mar. Sci.* 30:290-303.

Rice, D.L. (1986). Early diagenesis in bioadvective sediments: relationships between the diagenesis of beryllium-7, sediment reworking rates, and the abundance of conveyor-belt deposit-feeders. *J. Mar. Res.* 44:149-184.

Rice, D.L., T.S. Bianchi, and E. Roper. (1986). Experimental studies of sediment reworking and growth of *Scoloplos* spp. (Orbiniidae: Polychaeta). *Mar. Ecol. Prog. Ser.* 30:9-19.

Shachak, M. and A. Yair. (1984). Population dynamics and the role of *Hemilepistus reaumuri* in a desert ecosystem. In: *The Biology of Terrestrial Isopods*, S.L. Sutton and D.M. Holdrich, eds. Oxford, London, pp. 295–314.

Smirnov, U.S. and S.G. Tokmakova. (1972). Influence of consumers on natural phytocoenoses production variation. In: *Proceedings of the IVth International Meeting on Biological Production in the Tundra, Leningrad*, F.E. Wielgolaski and T. Rosswall, eds., pp. 122–127.

Sterner, R.W. (1986). Herbivores' direct and indirect effects on algal populations. *Science* 231:605–607.

Vickery, P.J. (1972). Grazing and net primary production of a temperate grassland. *J. Appl. Ecol.* 9:307–314.

Way, M.J. and M. Cammell. (1970). Aggregation behaviour in relation to food utilization by aphids. In: *Animal Populations in Relation to Their Food Resources*, A. Watson, ed. Blackwell, Oxford, pp. 229–247.

Part VI   Reports from Discussion Groups

18
Institutional Structures*

JAMES T. CALLAHAN

The emergent question for this discussion group came to be: "Are there innovations to be made in institutional and organizational structures that would advance comparative research on ecosystems more rapidly?" The discussants devoted several hours to the airing of the kinds of problems, both general and specific, that appear to play roles in constraining comparative ecosystem research. Questions such as these were explored:

1. Should most of the research on intersystem comparison be finely focused short-term experiments or observations, or should it be oriented more toward the long term?
2. Should the research focus more on specific target organisms or on whole-ecosystem characteristics?
3. Should the research be focused on purely local phenomena or should it have a more global perspective?
4. Are single-discipline or multidiscipline approaches more likely to be productive?
5. Is there a reason to distinguish between basic and applied research?
6. How can institutions best mesh the roles of research performance and education?

A number of overall objectives for comparative ecosystem research were also identified. These objectives tended to take on a somewhat "applied" tone, as they were finally phrased. Mainly, they had to do with the related topics of appraising the health of the environment, predicting the consequences of ecological change, and identifying possible ameliorative measures that would lead to ecological recovery.

The convergence of participant opinion seemed to be that comparative research should consider mainly longer-term, broader-scale approaches to the comparison of whole ecosystems by interdisciplinary teams of scientists.

---

*Discussion Group 1: J.T. Callahan (leader), O.W. Heal, B. Huntley, G.M. Lovett (reporter), W.H. Schlesinger, R.H. Waring, J.S. Warner.

Further, it was held that there is little sense in trying too hard to distinguish between basic and applied research, especially because the best and most meaningful research at the most appropriate scales of space and time would likely find near-term application to real-world problems. Finally, opinions were virtually unanimous that, although different institutions may vary in proportion of effort, one cannot realistically separate their complementary roles in both research and education.

The next major stage in the evolution of this discussion concerned the question, "Can we design an optimum institution to best foster comparative ecosystem research?" The unifying answer seemed to be "No," although the specific reasons voiced by participants were quite variable. "Too rigid" and "too costly" were most prominent among these reasons. A better articulated version of the original "No" answer was that "there is no single institutional type or structure that would or could do all the necessary things."

Subsequently, several significant impediments to comparative ecosystem research were identified. These included:

1. The difficulty of responsibly archiving and providing easy and reliable access to data, documentation, and models deriving from previous research efforts.
2. The lack of continuous, effective communication among disciplines.
3. The dearth of truly interdisciplinary education curricula.
4. The reluctance of employing institutions to recognize and allocate professional credit properly to those who participate in genuinely interdisciplinary research and publication efforts.
5. The apparent lack of reviewer enthusiasm, and therefore financial support, for multisite, long-term research projects.

All discussants suggested most strongly that these impediments should be addressed by the most direct means possible. Such means would almost invariably require the allocation of money to efforts aimed at countering the problem as a first level of approach. However, several suggestions had to do with the better utilization of nonfiscal resources which, of course, translates mostly into personal effort by individuals and groups. For example:

1. In addition to more frequent conferences and workshops on cross-system topics, there should be more and better goal-directed communication between meetings.
2. More cross-system research proposals should be produced to sensitize the reviewing and funding system and to educate the community of peer scientists.
3. Ecosystem-focused education should be increased at all levels, and interdepartmental, even interinstitutional, curricula should be developed to underwrite the necessary interdisciplinary research approach.
4. A roster of awards and, perhaps, fellowships should be instituted to openly and specifically recognize those who have produced significant cross-system accomplishments.

AUTHOR'S OPINION

It is the opinion of this discussion leader that the suggestions related to nonfiscal means are all useful ideas. However, because they are so highly dependent on generating and sustaining personal and institutional effort, primarily on an altruistic plane, they are also subject to the normal lapses in directed effort that so commonly occur. Some sort of institutionalization and routinization of such efforts must take place. Perhaps the most appropriate auspices would be those of a professional society or organization like the Ecological Society of America or the Association of Ecosystem Research Centers.

Many of the earlier stated impediments (e.g., data, documentation, models), when taken together with the distinct inclination toward large scales of time and space, interdisciplinarity, and the necessity for continuity, strike this reporter as demanding an institutional type of solution. This would seem to be especially so when one considers that the comparative syntheses that are needed so badly can best be fostered in an environment where collaborators have the information resources immediately available, where they have access to the latest in data reduction and display hardware and software, and where they are unencumbered by ordinary interruptions and day-to-day problems so common at their institutional homes.

Perhaps it would be appropriate to end this report by asking the original question again in a somewhat modified manner:

ISN'T IT TIME WE BECAME WILLING TO TRY SOME INNOVATION IN OUR INSTITUTIONAL STRUCTURES, WITH THE OBJECTIVE OF IMPROVING THE WAYS WE DO COMPARATIVE RESEARCH ON ECOSYSTEMS?

19
Cross-System Comparisons of Detritus Food Webs*

LARS O. HEDIN

Detritus is a quantitatively important source of energy and carbon in most aquatic and terrestrial ecosystems. Despite the general importance of detritus-based food webs, much of the information underlying our current models of decomposition processes and detrital food webs has been derived from a fairly narrow range of ecosystems. For example, aquatic models are strongly influenced by the study of leaf litter decomposition in headwater streams or *Spartina* marshes. In addition, there have been few efforts to compare detritus food webs among different ecosystems.

This group considered comparisons of detritus food webs and detrital dynamics between different types of aquatic and terrestrial ecosystems, including desert stream, headwater stream, deciduous forest, coniferous forest, freshwater lake (mesotrophic), ocean (upper 100 m), salt marsh, and grassland. Input, storage, and fate (decomposition, transport losses or accumulation) of detritus in each of these ecosystems were considered.

This discussion suggested patterns that appeared to be common to most ecosystems and raised some questions for further comparative studies:

1. A dominant fraction of primary production enters the detrital food web, even in the case of "grazer-dominated" systems (e.g., lake epilimnion). The generally low assimilation efficiency of grazers indicates that even if a large proportion of primary production is consumed, the majority is egested and thus enters the detritus pool. Further, there are suggestions that, in some cases, detrital pathways may be more important for grazers than the direct feeding on living biomass (e.g., zooplankton feeding in lakes, or macroinvertebrate grazing in streams). However, the relative importance of detritus versus living biomass in supporting secondary production in different ecosystems is generally unknown, and comparative studies are needed.
2. A major fraction of the annual input of organic matter is generally

*Discussion Group 2: T. Bianchi, N. Caraco, S. Findlay (leader), S. Fisher, D. Fontvieille, L. Hedin (reporter), D. Johnson, J. Melillo, J. Meyer, L. Pomeroy.

mineralized, with only small amounts remaining for permanent or long-term storage (e.g., in soil or sediments) or transport. However, for high-gradient streams that lack structures which retain inputs of organic matter, such as organic debris dams, losses of organic matter by downstream transport may be significant. Permanent loss of organic matter by burial may be important in cases where rates of sedimentation (or litter input) are too rapid to permit significant mineralization in surface sediments (or soils).

3. The importance of metazoan consumption of detritus is poorly known for many ecosystems. The group thought that, in general, the utilization of detritus by metazoans is likely to be very low in most ecosystems. The questions of why (or why not) this may be the case, and whether there are significant differences between different types of ecosystems, were identified as important to address. Answers to these questions may help to determine why (and if) certain ecosystems show higher efficiency of metazoan production (e.g., fish production) than others.

In addition, the group discussed some general problems that are associated with comparisons of detrital dynamics between different ecosystems. The group noted the lack of a common framework and terminology for comparing detrital dynamics between very different types of ecosystems (e.g., aquatic and terrestrial).

Furthermore, the group noted that certain types of ecosystem parameters are easier to apply to cross-ecosystem comparisons than others. Parameters that are strongly dependent on the boundaries of the ecosystem (e.g., standing stock of organic matter) are less appropriate because it is not always obvious where to draw the boundaries of an ecosystem. For example, comparing the relative amount of carbon mineralized to CO_2 between different ecosystems is more easily accomplished by a parameter that is relatively insensitive to the system boundaries (e.g., CO_2 output/carbon input, both expressed per square meter), compared to a parameter which is more sensitive to these boundaries (e.g., CO_2 output/carbon standing stock, both expressed per square meter). Further, nonhomogeneity within ecosystems may make it hard to define ecosystem boundaries in an objective way. For example, the organic matter stored in soils (or sediments) often decreases in nutritional quality with depth in the soil (or sediment) and, depending on the question being asked, it may not be suitable to consider only a single "pool" of organic matter within the soil.

Finally, the group noted that many of the potential problems that were identified result from the general lack of questions or hypotheses broad enough to compare and contrast the dynamics of detrital food webs between different types of ecosystems. There is much room for both conceptual and empirical improvement.

20
Improving Use of Existing Data*

CATHY M. TATE AND CLIVE G. JONES

One of the barriers to comparative analysis is the difficulty in obtaining existing data. Our discussion group goal was to suggest ways that exchange of data could be facilitated. Data in the published literature are public domain and easily accessible if you have a library card, although finding enough published studies with exactly the data you need is ofttimes harder than doing the research yourself! Qualitative data (e.g., site descriptions, species lists) and public domain qualitative data (hydrologic and soil surveys, weather data) do not present much of a problem. However, obtaining unpublished quantitative data from individuals or research sites can be difficult and requires time, energy, and diplomacy.

We urge researchers to recognize that data sets have value. Much time, energy, and many resources have been expended to collect the data, and investigators therefore have strong vested interests. The easiest way to obtain the data is to carry out truly collaborative research. Failing that, you have to find other ways to get the data you need. It is obviously worthwhile spending a fair amount of time before you request data, thinking about the specific questions you will ask, what data you need (and what you do not need), and how the data will be analyzed. A telegram to the investigators demanding all their raw data for vague purposes is not likely to elicit the sort of response you would like, nor is requesting previously published data in the format you need! At minimum, you should send a courteous letter outlining your project, why the data are critical to you, exactly how you will use the data, and assuring the investigators that proper acknowledgment will be given (crossing your fingers that your request will not suddenly stimulate them to ask the same question!). This may be sufficient, particularly if you need limited amounts of data. If you need large data sets, or data that require prior manipulation, you may want to offer payment for the time and effort involved. If your project is large in scope, that is, you

*Discussion Group 3: J.A. Downing, C.G. Jones (reporter), J.J. Magnuson (co-leader), R. Myster, K. Morse, K. Reckhow, D.L. Strayer, C.M. Tate (leader), K.C. Weathers.

need masses of data from many investigators, you may do well to visit the sites. This allows the investigators to explain their exciting research, and allows you to see if the data are appropriate, how they were collected, and how best to analyze them, as well as giving you the opportunity of assuring the investigators of your wholesome intentions.

There are several ways investigators and research sites can facilitate this process, thereby enhancing the quid pro quo exchange of data. A protocol for requesting data will prevent misunderstandings. A time limit for exclusive use of the data by the investigators will also help. Perhaps unpublished data or raw data associated with publications could become public domain after a set number of years – say, 5 years – thereby encouraging investigators to publish. Data sets should, of course, be well documented, error free, and in a flexible format for multiple use. We also encourage investigators to deposit their raw data in a recognized repository (e.g., EDEX; see Siccama 1987), and recording this information in the Methods sections of their publications.

The discussion group also identified other issues but did not address them in any detail. These included philosophical and legal issues (ownership, intellectual property, copyrights, patents, and royalties) and pragmatic issues (data management, storage, and standardization) (see Michener 1986).

Given that everyone now has all the data needed, what is the best way to maximize use of the data to advance ecological understanding? There is a tendency for researchers to make comparisons among similar types of ecosystems (e.g., lakes) rather than divergent systems (e.g., deserts and lakes). There are plenty of excuses for why this is so – it is too complex, there are no common units, etc., etc. We think ecologists should not be so timorous and should not be afraid to go boldly where no ecologist has been before. Seek new patterns among divergent ecosystems. Naturally, there will be problems and uncertainties, but finding a new, broad pattern that begs explanation is exciting and important (e.g., McNaughton, Chapter 7, this volume).

Our final advice is to heed the words of the "Ten Commandments of Comparative Analysis":

The Ten Commandments of Comparative Analysis

1. Thou shalt honor the advancement of ecology.
2. Thou shalt seek great patterns among divergent ecosystems.
3. Thou shalt not kill creativity.
4. Thou shalt not commit adulteration of data.
5. Thou shalt not steal.
6. Thou shalt make thy data available even unto they enemies.
7. Thou shalt not assume all ecosystems are different until you do the residual analysis.
8. Thou shalt honor probability theory whenever possible.

9. Thou shalt release they data from bondage.
10. Thou shalt not convet they neighbors' data until they've had a crack at them.

Acknowledgments. We thank the organizers of the Third Cary Conference for providing a stimulating environment and thought-provoking presentations and discussions; the discussion group members for their insights, feedback, and biblical knowledge; and NSF (BSR-8900400) and the Mary Flagler Cary Charitable Trust for financial support. Contribution to the program of the Institute of Ecosystem Studies, The New York Botanical Garden.

REFERENCES

Michener, W.K. (1986). Research Data Management in the Ecological Sciences. The Belle W. Baruch Library in Marine Science, Number 16, University of South Carolina Press, Columbia.
Siccama, T.G. (1987). EDEX: Ecological data exchange – A proposed resource. *Bull. Ecol. Soc. Am.* 68:10–11.

21
Comparative Analysis of Ecosystems along Gradients of Urbanization: Opportunities and Limitations*

Mark J. McDonnell and Steward T.A. Pickett

Introduction

Humans have altered ecosystems to varying degrees, and the resulting array of natural, seminatural, and humanmade ecosystems within a landscape can be conceived as constituting both a readily measurable gradient of land use and a more complex gradient of anthropogenic effects. Urbanization is a massive, unplanned experiment that already affects large acreages and is spreading in many areas of the United States. Environmental consequences of urbanization are becoming important public issues, as evidenced by recent media coverage. The role of ecologists in any future discussion or study of the effects of urbanization on ecological systems in North America is unclear because ecologists have historically avoided urban systems. Yet many of the nonurban research sites we cherish are becoming engulfed by urban spread.

The study of ecosystems along urban to rural land-use gradients provides an opportunity not only to address practical problems related to anthropogenic impacts on ecological systems, but it also provides an opportunity to help answer fundamental ecological questions concerning the structure, function, and organization of ecosystems in general.

The group discussed the opportunities and limitations of comparative studies along urban to rural land-use gradients. Such studies can both document complex environmental gradients that result from urbanization and determine ecologically important changes in ecosystem structure and function along those gradients. The following questions were addressed: (1) Why have ecologists not worked on urban versus nonurban ecosystem contrasts? (2) Why should ecologists work on urban versus nonurban ecosystem contrasts? (3) What questions or hypotheses are best addressed using

*Discussion Group 4: E. Ames, B. Boeken, G. Likens, M. McDonnell (leader), S. Pickett (reporter), R. Waring, and C. White.

anthropogenic gradients? (4) What is the best strategy for developing research programs to address these questions?

Studying Ecosystems along Gradients of Urbanization

The structure of metropolitan areas (e.g., a 50-km area around New York City) and their fringes consists of a variety of components, ranging from totally built environments to "natural" or seminatural areas. Natural areas in an urban context are those not intentionally managed by people (e.g., wooded natural areas in city parks, lakes, ponds, streams, etc.). Urban to rural land-use gradients exist in metropolitan areas and are characterized by a dense, highly developed core surrounded by irregular rings of diminishing development or impact. Established and successful gradient analysis techniques can compare and contrast the structure and function of natural areas along urban to rural land-use gradients. The familiar concepts and theories appropriate to natural continua provide a powerful organizing tool for research on gradients of anthropogenic influences on ecosystems.

Disadvantages and Advantages of Working on Urban versus Nonurban Ecosystem Constrasts

North American ecologists have been reluctant to work on natural ecosystems (e.g., lakes, streams, meadows, and forests) within metropolitan areas for a variety of philosophical and practical reasons (Table 21.1). Philosophically, some of the most loudly voiced concerns center on the assumption that these ecosystems are too heavily impacted by humans and thus are not "natural." On the practical side, there are many problems associated with

TABLE 21.1. Perceptions that may discourage ecologists from working on urban versus nonurban ecosystem contrasts.

1. Not "natural."
2. Research doesn't contribute to ecological theory, only to management questions.
3. Security of personnel and instruments.
4. Unknown histories.
5. Unknown sources of stress.
6. Fragmentation may limit questions.
7. Fragments not representative.
8. Cannot publish in mainstream ecology journals.
9. Not an established community of "peers."
10. Urban ecology is perceived as second-rate.
11. Aesthetic preference of ecologists.
12. Systems perceived to be actively managed.

working in densely populated areas, including security of sites, unknown histories and disturbance regimes, and limited size of study areas. But, when the positive aspects of examining urban versus nonurban ecosystem contrasts are compiled, the list is surprisingly long (Table 22.2). Some of the most compelling reasons for working in these areas include the fact that people live there; an infrastructure of roads and power is available; research results are relevant to current environmental and societal problems; stress gradients are long and provide new end points for existing studies; and, because people live in close proximity to research sites, a unique opportunity exists to educate the public about important ecological issues.

In the final analysis it was concluded that virtually all the negative aspects of working in urban areas could be overcome. Many of the problems listed were in fact common to all ecological studies. There appears to be no compelling reason for excluding ecosystems within urban and suburban environments from comparative ecological studies.

Questions and Hypotheses

Comparative studies of ecosystems along urban to rural land-use gradients can appropriately address important ecological questions relating to whole system attributes (e.g., structure, function, response to disturbance, and succession), as well as system components (e.g., demography, physiology).

TABLE 22.2. Why should ecologists work on urban versus nonurban ecosystem contrasts?

1. People live in urban and suburban areas.
2. Underused systems for research.
3. Stress gradients are long and provide new endpoints for existing gradients.
4. Stress gradients relatively obvious and can be readily quantified.
5. Power (e.g., electricity) and infrastructure available.
6. Source of novel stresses and organisms.
7. Good historical records (e.g., air photos, etc.).
8. Primeval remnants provide historical reference.
9. Understand system response to change in stress and disturbance regimes. Many replicates and treatments available.
10. Evolution versus acclimation of systems.
11. Equivalence of certain variables (concrete is concrete).
12. Information relevant to societal concerns.
13. Education of the public.
14. Accessibility to a large public and to researchers.
15. Possible expansion of funding base.
16. Urbanization is spreading, providing an opportunity for before-and-after studies.
17. Test predictions for system restoration.
18. Urban and suburbanization ubiquitous. All biomes are affected.
19. Research addresses important societal and ecological issues.

Some specific questions that appeared especially compelling to the group include:

1. Does productivity of ecosystems change along anthropogenic gradients?
2. How does landscape fragmentation affect ecosystem structure and function?
3. Does species diversity change along anthropogenic gradients? Are limits to diversity predominately generated by biotic interactions (e.g., introduction of exotic species) or by physical stresses?
4. How do fluxes in CO_2, heat, water vapor, and anthropogenic sources of carbon vary along anthropogenic gradients? Do thresholds in various stress factors coincide?
5. How do trophic web structures vary along anthropogenic gradients? Do heterotrophic versus autotrophic food chains shift importance on the gradient?
6. Are there biotic responses that precede stress-induced mortality?
7. How do the limits on community regeneration and succession change along the gradient?

The richness of this comparative approach is evident when these questions are explored further. For instance, the effects of fragmentation on ecosystem structure and function can be used to test the applicability of island biogeography theory along the gradient. Because the size of discernible ecosystems varies along the gradient, questions can be asked concerning the minimum size of a functional ecosystem. In addition, the ecological importance of corridors between similar ecosystems could be addressed.

Questions concerning biological diversity are particularly germane to the anthropogenic urban–rural gradient. Interesting ones include: How does the number of higher trophic level species vary along anthropogenic gradients? Do nonnative species replace or add to native species diversity? and, Do ecosystems respond first to changes in disturbance regimes or stress along the gradient?

Whole-system questions concerning the flux of organisms, nutrients, water, and energy can readily be addressed along the gradient because boundaries between adjacent ecosystems are relatively sharp in urban and suburban regions. These areas are especially good for establishing mixed pixel coefficients for fluxes of CO_2, heat, water vapor, albedo, advection, and convection.

Finally, because an anthropogenic gradient commonly parallels the land-use gradient, this approach would be especially effective in testing for stress indicators. Waring (Chapter 11, this volume) suggests that physiological changes should precede any observed change in ecosystem or community structure. Thus, physiological assessment of plants along an anthropogenic gradient could identify physiological warning signs that indicate the systems are under stress.

Predictions of Ecosystem Response

Based on current ecological knowledge, it is appropriate to make predictions concerning ecosystem response to anthropogenic gradients. The discussion group focused on forest ecosystems and made predictions concerning the effect of a new stress (i.e., evolutionarily novel; anthropogenically derived) on forest ecosystems:

1. Variance in photosynthesis, growth, respiration as a function of the steepness of the gradient would increase.
2. Selection response would be in the direction of maintaining physiological balance while reducing growth.
3. Stemwood production per unit leaf area would be less efficient.
4. Ecosystem ability to absorb radiation on a yearly basis would decrease.
5. Markers in stable isotopes would show change in photosynthesis.
6. Selection would result in loss of sensitive species.

The foregoing sequence of events should occur in that order, as urbanization spreads through a region. In addition, if the structure of the landscape (e.g., amount of buildings versus natural ecosystems) changed albedo and emissivity along the land-use gradient, a change in CO_2 and H_2O flux/area would result.

Research Strategy

The group believed strongly that the comparative analysis of ecosystems along urban to rural land-use gradients was a workable and productive direction for ecologists. Such studies should be geographically extensive to cover the range of conditions along the gradients, but they should also be anchored on intensive studies of a few sites. Studies should be based on mechanistic understanding of the systems and should not be just "fishing trips." All significant correlations should be tested for causal significance, with follow-up experiments. Finally, to get the most out of the approach, researchers should initially concentrate on the simplest gradients available.

In conclusion, the comparative analysis of ecosystems along gradients of urbanization provides exciting new opportunities to not only address practical problems but also to address basic ecological questions concerning how ecosystems are organized.

22
Comparison between Tropical and Temperate Ecosystems*

ROBIN WELCOMME AND ALAN R. BERKOWITZ

One of the central problems besetting tropical ecology has been the way in which it has been conditioned by ideas formulated during the study of temperate ecosystems. This is to some degree natural, in that the body of information available on the temperate zone is vastly greater than that on the tropics, the number of researchers is considerably larger in the temperate zone, and most hypotheses on the functioning of ecosystems have been formulated for temperate ecology. Furthermore, until recently, much of the research carried out in the tropics was done by temperate scientists visiting the lower latitudes for comparatively short stays and carrying with them much of the intellectual baggage acquired in the higher latitudes.

As a result of this orientation, studies on the tropics have been mostly discontinuous and of short duration. Sample sizes have been comparatively small, and the range of systems studied has been restricted. Comparative studies between ecosystems in the tropics have been relatively few, and the full range of variation within any one type of ecosystem rarely has been explored. As a result, a coherent ecology of the tropics has failed to emerge, despite a disquieting feeling that many of the hypotheses that have been formulated in the temperate zone do not explain the observed functioning of tropical systems. Instead, much of the synthesis has appeared as a series of paradigms comparing tropical with temperate ecosystems.

A group of 13 scientists met during the Cary Conference to discuss the applicability of the commoner paradigms to their experience. Ecosystem types covered by the participants included forests, savanna grasslands, mountain ecosystems, rivers, lakes, salt marshes, coral reefs, and deserts. The following paradigms were identified as being commonly encountered in generalized statements about the tropics.

1. Tropical floras and faunas are more diverse than temperate ones in any given ecosystem type.

---

*Discussion Group 5: A.R. Berkowitz (reporter), B. Huntley, J. Kalff, A.E. Lugo, P. Matson, J.M. Melack, J.M. Mellilo, S.J. McNaughton, W.J. Parton, S.P. Seitzinger, P. Vitousek, R. Welcomme (leader), W.D. Williams.

2. Tropical ecosystems are more benign because of (i) a greater predictability and (ii) more favourable ambient environmental conditions.
3. Tropical ecosystems are more productive.
4. Tropical ecosystems are nutrient limited, as opposed to temperate ecosystems, which are thermally limited.
5. In the tropics altitude is equivalent to latitude, that is, changes in community structure and processes that occur with altitude in the tropics are homologous with those that occur as one moves from the equator to the poles.

Examination of the range of tropical systems showed that more parameters lay within the values expected for a similar range of temperate systems of the same general type. The only exceptions to this were that of diversity, which is nearly always greater at low than at high latitudes for equivalent systems, and productivity, which is sometimes higher in the tropics. The group considered therefore that many of the commonly held views of tropical ecology (particularly regarding ways in which tropical systems are expected to *differ* from temperate ones) are not borne out by firsthand experience. It also concluded that the lack of careful comparative studies has limited our knowledge of any possible differences and functions within the tropical zone as well as between tropical and temperate zones. To resolve this situation, it was believed that more emphasis should be placed on funding such studies in the short term as well as setting up longer-term comparative studies to detect the range of variation over time.

23
Legitimizing Cross-System Comparison in Ecology*

JONATHAN J. COLE

This group gathered to argue about the need to prove or state the legitimacy of the comparative approach in ecology. Within the group were both enthusiasts of the approach in general and those who found the approach either often abused or not often useful, but it struck this reporter that no one in the group appeared to consider the approach per se to be illegitimate. On the other hand, enthusiasts and detractors alike realized that comparative ecology, while perhaps not illegitimate in anyone's book, has had a tainted image. That is, comparative ecology is sometimes perceived to be less scientific than direct experimentation. Some even felt that comparative ecology is discriminated against in terms of unfair journal and proposal reviews. Others countered here that perhaps all ecology is "undervalued."

We set out to define what is effective use of the comparative approach, seriously discuss the limits and short-comings of the approach, and pinpoint what it is that may have given this approach a bad reputation. It was thought that putting this discussion into writing would be worthwhile. To this end, much of our discussion has been distilled by R.H. Peters into an essay (Chapter 4, this volume) that endeavors to put the use of the comparative approach into its proper perspective both within ecology as a field and within science at large.

---

*Discussion Group 6: J. Armesto, B. Boecken, S.R. Carpenter, J.J. Cole (reporter), C.T. Driscoll, C.M. Duarte, T.M. Frost, J.P. Grime, S.H. Hurlbert, J. Kolasa, C. Ochs, M.L. Pace, R.H. Peters (leader), E. Prepas, D. Tilman, W.G. Sprules.

Part VII Concluding Remarks

24
Concluding Remarks

MICHAEL L. PACE

"The essence of knowledge is generalization"　　　— Hans Reichenbach

The goal of comparisons among ecosystems is to arrive at generalizations. These generalizations can take the form of qualitative or quantitative patterns. These patterns, in turn, become the grist for ideas about mechanisms and theories describing how ecosystems work. The purpose of the Cary Conference was to explore the relationship among these three components of our science (patterns, mechanisms, and theories) and to ask how comparative analysis might enhance our efforts.

Goals — A Reprise

At the initiation of the conference, three goals were presented:

1. to encourage the application of comparative methods to ecosystem studies;
2. to address methodological and philosophical issues related to the comparative approach;
3. to provide an opportunity for cross-fertilization among ecologists who study a diversity of environments, including terrestrial, freshwater, and marine ecosystems.

We can review each of these goals in turn and ask how successful the conference was at achieving these aims.

Applications of the Comparative Method

The talks presented many comparisons of a variety of processes. For example, quantitative studies by Carpenter et al. (Part II, Chapter 5) and McNaughton et al. (Part II, Chapter 7) were presented, of aquatic primary production and terrestrial herbivory, respectively. In these studies, among-systems comparisons were made of processes. Carpenter et al. demon-

361

strated that physical-chemical factors and herbivory explain about equal portions of the variation in phytoplankton biomass and production observed at the scale of a summer season in a lake. McNaughton et al. documented general relationships between primary production and herbivory in a variety of terrestrial systems. This study demonstrated the feasibility of comparing processes in systems as different as African grasslands and temperate forests.

Qualitative generalizations emerged from Howarth's (Part III, Chapter 9) survey of oil spill studies. Functional properties of ecosystems such as production, respiration, and nutrient cycling do not show a consistent response to these perturbations. Structural attributes of the system, on the other hand, underwent consistent change, as manifested by the reduction or elimination of "sensitive" benthic species. Systems vary in the number of "sensitive" species they contain, leading Howarth to postulate that responses to pollution depend on how open or closed systems are with regard to flux across system boundaries of limiting nutrients. Relatively open systems (marshes) are more stress tolerant than relatively closed systems (central ocean gyres). These ideas require further investigation, but the important point is that qualitative comparisons of the existing studies of oil spills provide interesting hypotheses as a future starting point for examining stresses on ecosystems.

Methodological and Philosophical Issues

The second goal of the conference was to examine methodological and philosophical problems with comparative approaches. This goal could be rephrased as a question: Why are ecologists not more aggressively using the comparative approach? Before the meeting the participants were surveyed and asked what they thought the critical issues were in comparative studies. A remarkable number of the participants identified three general problems, which I think answers the question concerning limited use of the comparative approach (Table 24.1). First and foremost, there is the problem of how to make appropriate comparisons. For example, the process of evapotranspiration occurs in all terrestrial ecosystems. The context, meaning, and consequence of evapotranspiration, however, are very different in a dry grassland than in a wet tropical forest. Nonsensical comparisons similar to

TABLE 24.1. A survey of common responses to the question, "What are the critical issues in comparative ecological studies?" Question was posed to the Cary Conference participants before the meeting, $n = 55$.

1. Make appropriate comparisons (17)
2. Data limitations (15)
3. Standardize methods (10)

those so familar in advertising are easy to imagine. Nevertheless, Downing's fruit metaphor (Part I, Chapter 3) points out that careful formulation of our questions is the key to avoiding inappropriate comparisons. Two other related concerns limiting the comparative approach are insufficient or inappropriate data for carrying out comparisons and the lack of standardized methods (Table 24.1). Undoubtedly, data and methods will be always be limiting. It is often the case that data are simply not available for comparative analyses of more complex processes or issues. It seems, however, that an ecological literature of 40,000 English pages per year will rapidly wipe out this limitation (see Strayer, Part I, Chapter 1). Case studies of all kinds are accumulating in the literature. The limitations cited in Table 24.1 on comparative studies may be less serious than we perceive.

The conference was remarkably free of the philosophical rancor that lurks about much ecological discussion. We had envisioned that some arguments would emerge, given the diversity of our participants. Instead, fire fights were few. Quite significantly, there was general agreement in the utility of the comparative approach and the need for more of it. What becomes clear in reading the chapters of this book is the need for rigor and interaction among comparative, observational, experimental and theoretical studies (Caraco et al., Part IV, Chapter 12; Carpenter et al., Part II, Chapter 5; Duarte, Part V, Chapter 15; Downing, Part I, Chapter 3; Pomeroy, Part II, Chapter 6).

Cross-Fertilization

The third general goal of the conference was a hope for cross-fertilization among participants who study a variety of habitats and employ diverse approaches and methods. Success here is difficult to judge. For the most part, there was a failure in the presentations to make comparisons across systems of different types. When the speakers were given their charge, we assigned them very general topics (e.g., the relationship between primary and secondary production), and did not suggest that they restrict their analysis to a particular environment. Most, naturally, chose to draw comparisons for systems they know best. There was a consensus that we need to go forward and begin to compare across ecosystem types. A simple example: How do respiration rates vary among terrestrial, freshwater, and marine ecosystems? Few studies have been posed at this scale. One result of our meeting was a recognition of the desirability of such studies and the need for ecologists from diverse areas of the discipline to work together on these problems.

The Title

An additional way to assess the conference is to reflect on the title, which was "Comparative Analysis of Ecosystems: Patterns, Mechanisms, and Theories." I stated at the beginning of this essay that the comparative approach helps us discover patterns and arrive at generalizations. In that sense, comparative ecology can be viewed as a method, or better, a strategy for studying ecological phenomena. Does it also help us to identify mechanisms, to develop and test theory?

The conference provided several excellent examples of how the elements of pattern, mechanism, experiment, and theory interact. Caraco et al. (Part IV, Chapter 12) began with the observation of patterns of phosphorus regeneration across a variety of marine and freshwater ecosystems. They then postulated several mechanisms that could produce the observed patterns. These mechanisms lead to the development of a theoretical model based on biogeochemical interactions, which potentially explains the observed cross-system patterns. The sequence in their work was pattern, mechanism, and theory. Smith (Part IV, Chapter 13) began with the general theory of the stoichiometry of organic matter formation, degradation, and elemental cycling. The theory generated a series of observations to identify patterns of behavior of autotrophic versus heterotrophic ecosystems. These considerations lead Smith to a novel conclusion that the carbon cycle drives the nitrogen cycle. In mechanistic terms, nitrogen fixation and recycling will provide nitrogen sufficient to supply the demands of carbon production. The sequence in Smith's work was theory, pattern, mechanism. Finally, Carpenter et al. (Part II, Chapter 5) evaluated the relative importance of nutrient loading and herbivory in determining primary production. This research was motivated by a model of cascading trophic interactions (theory) that led to a series of experiments on specific lakes to test the model (Carpenter and Kitchell 1984; Carpenter et al. 1987). The comparative study presented in this volume represents an expansion of the tests to comparative data from many systems. In this work the sequence of theory, experiments, and pattern was followed.

Interestingly, we see that the sequencing of these elements occurs differently in the work of Caraco et al., Smith, and Carpenter et al., suggesting that the comparative approach can be usefully exploited at a number of points in the scientific process. In summary, comparative analyses are useful for both developing new ideas or testing the generality of existing ideas.

Taxonomy of Comparative Studies

An important point of discussion throughout the conference concerned clear identification of types of comparisons. Heal and Grime (Part I, Chapter 2) distinguished between comparisons of components of ecosystems (e.g., biomass of herbivores) and system properties (e.g., ratio of autotrophic to heterotrophic biomass). Components are more easily compared, but a science of ecosystems requires comparison of system properties to fully distinguish similarities and differences.

Fisher and Grimm (Part III, Chapter 10) described a hierarchy of system types that must be carefully identified when performing comparative studies. Comparisons may be of the same type of system within a given disturbance or seasonal regime (e.g., desert streams), or comparisons may be of similar systems in very different disturbance or seasonal regimes (e.g., desert streams versus temperate woodland streams). Finally, comparisons may be of very different systems (e.g., lakes versus streams). Although these distinctions are obvious, what may not be obvious is that results of comparative analyses will often be quite different, given the scale and type of comparison (Fisher and Grimm, Part III, Chapter 10). It will be important to be specific in describing comparisons and to scale the comparisons appropriately to the question.

Ontogeny of a Comparative Study

The steps in a typical comparative study are outlined in Fig. 24.1. After developing an initial idea, there is often an exciting expansion phase during which new ideas, tests, and corollaries are generated. The balloon, however, is often burst when one confronts the available data or faces the prospect of collecting the necessary data to test the idea. Some retrenchment or contraction follows, and then the drudgery of collecting the data sets in. The phase of analysis, pattern generation, and new ideas is a final loop in the sequence.

I facetiously note that another endpoint of comparative studies is often statistical criticism. This raises the serious question: are the statistical procedures we employ adequate? Unfortunately, the answer is often no. Many comparative studies are of necessity based on imperfect samplings, nonran-

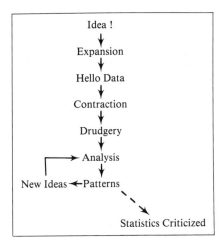

FIGURE 24.1. Ontogeny of a typical comparative study.

dom observation sets, and unsatisfactory statistical models. Application of many standard statistical procedures such as regression is seriously compromised by these types of problems (see Carpenter et al., Part II, Chapter 5). Downing (Part I, Chapter 3) pointed out a number of shortcomings in a survey of contemporary papers that are largely related to study design. These problems include bias in the variables we choose to measure, too few ecosystems sampled, too many independent variables sampled, and informal analyses. Equally daunting are problems associated with building quantitative relationships when independent and dependent variables are measured with large error. This can lead to the "errors in variables" problem discussed by Walters (1986) wherein underlying relationships are entirely obscured by error. The statistical methodology applied to comparative studies needs to be carefully considered and priority given to the development of better methods and to providing guidelines for use of existing methods.

At the conference, McNaughton (Part II, Chapter 7) remarked that 9–12 person-months were required to exhume the data for their analysis. This comment suggests a serious limitation for comparative studies. These studies require substantial effort, even when the data are drawn from the literature. Comparative studies often do not fit well into existing grant programs because they may not involve the collection of any new data. There is an important need to support comparative efforts and to recognize both the demands and potential rewards of these types of studies.

Challenges for Comparative Ecology

A number of themes which emerged at the conference represent challenges for comparative ecology in particular and the field of ecology in general. The first of these concerned the problem of integrating species characteristics and population dynamics with ecosystem processes. We are well aware that there are key species in any system whose dynamics are major determinants in driving material cycles and structuring the system. There was a sense among many of the conference participants that we need to search for appropriate ways to study populations as they relate to questions pertaining to ecosystems. Such an initiative will require the development of new theory or at least the hybridization of existing theories and collaboration among population and ecosystem ecologists (Heal and Grime, Part I, Chapter 2; Pomeroy, Part II, Chapter 6).

Many times during the conference questions were raised about conceptual issues and their operationalization. There is a need to pay careful attention to how we use concepts like scale, disturbance, and stress. We need work on structuring and translating these ideas into operational questions that can be tested. An example of this kind of effort can be found in some of the recent work on disturbance (e.g., Pickett et al. 1989).

Comparative studies arise from theories, concepts, or perhaps just hunches about how systems work. A concern voiced frequently at the conference was what theories were motivating the comparisons. How explicit are these theories? Can these theories be tested effectively in the light of comparative studies? There needs to be more interaction between comparative studies, theory development, and theory testing. If useful theories are available, these theories need to be more explicitly analyzed and presented as we carry out complementary comparative studies.

We typically think of ourselves as forest ecologists, stream ecologists, oceanographers, etc. This point of view makes us willing and capable of performing comparative studies within our environments of interest but reluctant to consider problems across environments. There was a strong sentiment at the conference that there are interesting problems to be addressed across ecosystem types and that it is time for this work to begin. The need for collaboration is obvious, but equally important is the need to recognize the value and potential significance of comparison of different types of ecosystems.

A final point concerns the difficulty of detecting and interpreting change in ecological systems. This issue is not specifically a problem for comparative ecology but affects all aspects of the discipline. The problem was well illustrated by Howarth (Part III, Chapter 9), where it was demonstrated in marine systems that certain benthic species were the most reliable indicators of response to stress and therefore reliable indicators of change. To correctly interpret such changes, however, it is necessary to place species declines as a response to an impact in the context of the "background" invasions and extinctions that regularly occur in any particular community. Comparative studies may be valuable in this effort in providing alternate systems to evaluate change against. There has been some excellent recent thinking on how time series data from systems with and without impacts can be compared (Stewart-Oaten et al. 1986; Carpenter et al. 1989). The detection and interpretation of change in ecological systems remains a major challenge, and it appears that comparative studies will play an increasing role in addressing this problem.

Epilogue

I have been wondering lately what we will say in the future to our children and grandchildren as they attempt to cope with a warmer world of untold environmental disruptions. As old ecologists, we will be forced to admit, if only to ourselves, that we were in a position to know something of what was ahead. Clearly, it is time to shake loose from some of the narrow concerns that have dominated the discipline and to attack the big problems.

Many of the concerns we must address are occurring at the management scale of specific ecosystems and at regional and global scales. I foresee that comparative ecology will be valuable in the effort to study problems on these scales. Comparative studies provide evidence for the strongest variables driving ecosystems and these, in turn, are useful in management. The study of phosphorus loading to lakes provides an excellent example of how both whole-system experiments (Schindler 1974) and cross-system analysis (Vollenweider 1968; Dillon and Rigler 1974) provided the information and predictive models necessary for management. Suggested remedies have been applied and the predictive models tested against management actions (Smith and Shapiro 1981). At regional and global scales, quantitative summaries of patterns from many existing studies will provide the information necessary to formulate models describing larger-scale phenomena. Qualitative summaries of patterns from existing studies will provide insights necessary to test ideas at these larger scales.

It is important that we recognize the value of comparative studies now so that the effort can accelerate. The merits of the comparative approach have been amply demonstrated. There is a need to bring comparative studies into the front line of ecological methods as an equal partner with observational, experimental, and theoretical approaches.

REFERENCES

Carpenter, J.F. and J.F. Kitchell. (1984). Plankton community structure and limnetic primary production. *Am. Nat.* 124:159–172.

Carpenter, S.R., T.M. Frost, D. Heisey, and T.K. Kratz. (1989). Randomized intervention analysis and the interpretation of whole ecosystem experiments. *Ecology* 70:1142–1152.

Carpenter, S.R., J.F. Kitchell, J.R. Hodgson, P.A. Cochran, J.J. Elser, M.M. Elser, S.M. Lodge, S. Kretchmer, X. He, and C.N. von Ende. (1987). Regulation of lake primary productivity by food web structure. *Ecology* 68:1863–1876.

Dillon, P.J. and F.H. Rigler. (1974). The phosphorus–chlorophyll relationship in lakes. *Limnol. Oceanogr.* 19:767–773.

Pickett, S.T.A., J. Kolosa, J.J. Armesto, and S.L. Collins (1989). The ecological concept of disturbance and its expression at various hierarchical levels. *Oikos* 54: 129–136.

Schindler, D.W. (1974). Eutrophication and recovery in experimental lakes: implications for lake management. *Science* 184:897–899.

Smith, V.H. and J. Shapiro. (1981). Chlorophyll–phosphorus relations in individual lakes. Their importance to lake restoration strategies. *Env. Sci. Technol.* 15:444–451.

Stewart-Oaten, A., W. Murdoch, and K. Parker. (1986). Environmental impact assessment: "pseudoreplication" in time? *Ecology* 67:929–940.

Vollenweider, R.A. (1968). Scientific fundamentals of the eutrophication of lakes and flowing waters with particular reference to nitrogen and phosphorus as factors in eutrophication. Organization for Economic Cooperation and Development, Paris.

Walters, C. (1986). *Adaptive Management of Renewable Resources*. Macmillan, New York.

Index